CMOS Time-Mode Circuits and Systems

Fundamentals and Applications

Devices, Circuits, and Systems

Series Editor
Krzysztof Iniewski
Emerging Technologies CMOS Inc.
Vancouver, British Columbia, Canada

PUBLISHED TITLES:

Atomic Nanoscale Technology in the Nuclear Industry
Taeho Woo

Biological and Medical Sensor Technologies
Krzysztof Iniewski

Building Sensor Networks: From Design to Applications
Ioanis Nikolaidis and Krzysztof Iniewski

**Cell and Material Interface: Advances in Tissue Engineering,
Biosensor, Implant, and Imaging Technologies**
Nihal Engin Vrana

Circuits at the Nanoscale: Communications, Imaging, and Sensing
Krzysztof Iniewski

CMOS: Front-End Electronics for Radiation Sensors
Angelo Rivetti

**CMOS Time-Mode Circuits and Systems: Fundamentals
and Applications**
Fei Yuan

Design of 3D Integrated Circuits and Systems
Rohit Sharma

Electrical Solitons: Theory, Design, and Applications
David Ricketts and Donhee Ham

Electronics for Radiation Detection
Krzysztof Iniewski

Electrostatic Discharge Protection: Advances and Applications
Juin J. Liou

**Embedded and Networking Systems:
Design, Software, and Implementation**
Gul N. Khan and Krzysztof Iniewski

Energy Harvesting with Functional Materials and Microsystems
Madhu Bhaskaran, Sharath Sriram, and Krzysztof Iniewski

Gallium Nitride (GaN): Physics, Devices, and Technology
Farid Medjdoub

FORTHCOMING TITLES:

Radio Frequency Integrated Circuit Design
Sebastian Magierowski

Silicon on Insulator System Design
Bastien Giraud

Semiconductor Devices in Harsh Conditions
Kirsten Weide-Zaage and Malgorzata Chrzanowska-Jeske

Smart eHealth and eCare Technologies Handbook
Sari Merilampi, Lars T. Berger, and Andrew Sirkka

Structural Health Monitoring of Composite Structures Using Fiber Optic Methods
Ginu Rajan and Gangadhara Prusty

Tunable RF Components and Circuits: Applications in Mobile Handsets
Jeffrey L. Hilbert

Wireless Medical Systems and Algorithms: Design and Applications
Pietro Salvo and Miguel Hernandez-Silveira

CMOS
Time-Mode
Circuits
and Systems

Fundamentals and Applications

EDITED BY **FEI YUAN**
Ryerson University, Toronto, Canada

KRZYSZTOF INIEWSKI
MANAGING EDITOR
CMOS Emerging Technologies Research Inc.
Vancouver, British Columbia, Canada

CRC Press
Taylor & Francis Group
Boca Raton London New York

CRC Press is an imprint of the
Taylor & Francis Group, an **informa** business

CRC Press
Taylor & Francis Group
6000 Broken Sound Parkway NW, Suite 300
Boca Raton, FL 33487-2742

First issued in paperback 2020

© 2016 by Taylor & Francis Group, LLC
CRC Press is an imprint of Taylor & Francis Group, an Informa business

No claim to original U.S. Government works

ISBN-13: 978-1-4822-9873-4 (hbk)
ISBN-13: 978-0-367-73760-3 (pbk)

Contents

Preface

The rapid scaling of complementary metal oxide semiconductor (CMOS) technology has resulted in the sharp increase of time resolution and the continuous decrease of voltage headroom. As a result, time-mode circuits where information is represented by the time difference between the occurrence of digital events, rather than the nodal voltages or branch currents of electric networks, offer a viable and technology-friendly means to combat scaling-induced difficulties encountered in the design of mixed-mode systems. Time-mode approaches have found a broad spectrum of applications since their inception in time-of-flight measurement several decades ago. These applications include digital storage oscillators, laser-based vehicle navigation systems, analog-to-digital data converters, signal processing, medical imaging, instrumentation, infinite and finite impulse response filters, all digital phase-locked loops, giga-bit-per-second (Gbps) serial links, and channel select filters for software-defined radio, to name a few. Various architectures and design techniques of time-mode circuits have emerged recently; a comprehensive examination of the principles of time-based signal processing and the design techniques of time-mode circuits, however, is not available. This book provides the fundamentals of time-based signal processing with an emphasis on the design techniques and applications of CMOS time-mode circuits.

Chapter 1 examines the fundamentals of time-mode circuits. The definition of time-based signal processing is provided. The characteristics of time-mode circuits are examined and compared with those of their voltage-mode and current-mode counterparts. Challenges encountered in time-based signal processing are investigated. The key building blocks of time-mode circuits are briefly examined with a detailed study of these building blocks in later chapters. The applications of time-based approaches in mixed-mode signal processing are discussed briefly.

Chapter 2 deals with voltage-to-time converters. An emphasis is given to the techniques that improve the linearity of voltage-to-time converters. Voltage-to-time converters using voltage-controlled delay units are studied, and their pros and cons are examined in detail. Voltage-to-time converters using voltage-controlled delay units and source degeneration are also investigated. Relaxation voltage-to-time converters that provide a better linearity are studied. Reference voltage-to-time converters that also exhibit a good linearity are examined. The applications of voltage-to-time converters in time-mode comparators are investigated.

Chapter 3 provides a comprehensive treatment of the principles, architectures, and design techniques of time-to-digital converters (TDCs) with an emphasis on the critical assessment of the advantages and limitations of each class of TDCs. It first provides the classification of TDCs. The key performance indicators of TDCs are then depicted. Sampling TDCs where time variables are digitized directly such as counter TDCs, delay line TDCs, TDCs with interpolation, vernier delay line TDCs, pulse-shrinking TDCs, pulse-stretching TDCs, successive approximation TDCs, flash TDCs, and pipelined TDCs are investigated in detail. Noise-shaping TDCs that suppress in-band quantization noise such as gated ring oscillator TDCs, switched ring oscillator TDCs, gated relaxation oscillator TDCs, MASH TDCs, and $\Delta\Sigma$ TDCs are studied.

Chapter 4 starts with a brief review of TDC architectures. A three-step TDC with phase interpolation is introduced to improve the resolution and reduce the power consumption and die area. The resolution of the three-step TDC with phase interpolation is improved by using a phase interpolator and a time amplifier for the improvement of the in-band phase noise when used in all-digital phase-locked loops.

Chapter 5 introduces some important performance parameters of time interval measurements. It is followed by the presentation of basic counting methods for time measurement. Interpolation methods for performance improvement are presented and analyzed in a greater detail. We show that the combination of the counter method with the interpolation of timing pulse positions within the clock period provides a very efficient method for realizing a high-precision, accurate TDC with a wide operation range. This approach combines the inherently good single-shot resolution of a short-range interpolator based on digital delay line techniques with the excellent accuracy and wide linear range of the counting method. It is shown that in a general measurement situation where the timing pulses are asynchronous with respect to the system clock, the effect of interpolator nonlinearities on the final averaged output is strongly suppressed due to the inherent averaging effect of the interpolation method. On the other hand, these nonlinearities widen the distribution of the measured single-shot results and in many cases limit the single-shot precision of the TDC. It is also pointed out that the careful synchronization of the timing signals is needed in order to get unambiguous measurement results that are free from systematic errors. Finally, two case studies show that, with the aforementioned approaches, a TDC realized in standard nonaggressive CMOS technologies can achieve a ps-level resolution and a single-shot precision better than 10 ps (sigma value) over a wide operation range of hundreds of microseconds.

Chapter 6 explores the time-mode techniques that overcome the difficulties encountered in the realization of multibit voltage-mode analog-to-digital converters. The chapter starts with the close examination of the key parameters and figure of merits that quantify the performance of analog-to-digital converters. It is followed by the detailed study of the principles and properties of multibit quantizers realized using voltage-controlled ring oscillators. Both voltage-controlled oscillator (VCO)-based phase and frequency quantizers are studied. The chapter continues with the investigation of open-loop analog-to-digital converters utilizing VCO phase and frequency quantizers. The chapter first reviews the fundamentals of $\Delta\Sigma$ modulators in closed-loop time-mode analog-to-digital converters. Time-mode $\Delta\Sigma$ modulators are then introduced, and their characteristics are investigated in detail. Time-mode $\Delta\Sigma$ modulators with VCO phase and frequency quantizers are explored. $\Delta\Sigma$ modulators with phase feedback are also examined. $\Delta\Sigma$ modulators with pulse-width modulation for linearity improvement are explored. Multistage, also known as MASH time-mode $\Delta\Sigma$ modulators, both single-rate and multirate are examined. Dynamic element matching, an effective technique to minimize the effect of the mismatch of digital-to-analog converters, is briefly studied with the inclusion of an exhaustive list of published studies on dynamic element matching. The chapter ends with the comparison of the performance of some recently reported time-mode $\Delta\Sigma$ modulators.

Chapter 7 describes $\Delta\Sigma$ converters that adopt time-mode signal processing techniques. A key advantage of time-mode $\Delta\Sigma$ converters is that they are realized using digital circuits and process information in the form of time-difference intervals. As a consequence of using digital circuits, this technique benefits from low-voltage operation without concern for reduced signal swings, sensitivities to thermal noise effects, or switching noise sensitivity. Recently, several studies on time-mode $\Delta\Sigma$ converters are conducted showing that such methodology has high potential in low-voltage design. The noise-shaping behavior demonstrated by this technique can be implemented and extended in various ways, including voltage-controlled delay unit or gated-ring-oscillator-based implementations of TM $\Delta\Sigma$ converters. In this chapter, after a brief review of $\Delta\Sigma$ ADC specifications, the different architectures of TM $\Delta\Sigma$ converters that have been recently proposed are examined.

Chapter 8 covers the fundamentals of all-digital phase-locked loops. The chapter starts with a close examination of the drawbacks of charge-pump phase-locked loops. It is followed by a detailed examination of the phase noise of phase-locked loops. The basic configuration of all-digital phase-locked loops is then studied. An investigation of digitally controlled oscillators is followed. The phase noise of all-digital phase-locked loops is studied. The chapter ends with a brief examination of all-digital frequency synthesizers.

Chapter 9 outlines the general concepts related to time-mode signal processing and some of its state-of-the-art applications. These provide a very good alternative to conventional techniques, which suffer from problems such as linearity and accuracy limitations among others. The ultimate goal of this chapter is to arrive at all-digital time-domain circuits that can be synthesized using existing digital computer-aided design tools and to make the design process fully automated in contrast to its conventional analog counterpart.

Chapter 10 studies time-mode-integrated temperature sensors. Both relaxation oscillator temperature sensors and ring oscillator temperature sensors are investigated. Temperature sensors that utilize TDCs are also studied. It further investigates digital set point temperature sensors. The chapter ends with a comparison of the performance of some recently reported time-mode temperature sensors.

The book provides a comprehensive treatment of the principles and design techniques of CMOS time-mode circuits. Readers are assumed to have a fundamental knowledge of electrical networks, semiconductor devices, CMOS analog and digital integrated circuits, feedback systems, signals and systems, and communication systems. As time-mode circuits and systems are still a domain of active research, new architectures and implementations continue to emerge. The book by no means attempts to provide a complete collection of time-mode circuits and systems; it rather provides the fundamentals of time-based signal processing and the design techniques of CMOS time-mode circuits and systems and, therefore, is intended to serve as a source for those who are interested in time-based signal processing to explore further in this exciting field of research. A rich collection of recently published work on time-mode circuits and systems is provided at the end of each chapter so that readers can seek further information on the subjects covered in the book.

Although an immense amount of effort was made in the preparation of the manuscript, flaws and errors might exist due to erring human nature and time constraints. Suggestions and corrections from readers are gratefully appreciated by the editors and authors.

Fei Yuan
Krzysztof (Kris) Iniewski
Toronto, Ontario, Canada

MATLAB® is a registered trademark of The MathWorks, Inc. For product information, please contact:

The MathWorks, Inc.
3 Apple Hill Drive
Natick, MA 01760-2098 USA
Tel: 508-647-7000
Fax: 508-647-7001
E-mail: info@mathworks.com
Web: www.mathworks.com

Acknowledgments

The editors are deeply grateful to all the authors for their contributions to this book. A special thank-you goes to Professor Juha Kostamovaara of the University of Oulu, Finland, and Professor Gordon Roberts of McGill University, Canada, who are great pioneers in the fields of time-to-digital conversion and time-based signal processing, for their contributions to this book.

The editorial staff of Taylor & Francis Group/CRC Press, especially Nora Konopka, the publisher of engineering and environmental sciences; Jessica Vakili, senior project coordinator, editorial project development; and Michele Smith, senior editorial assistant (engineering), have been warmly supportive from the initial approval of the book proposal to the publishing of the book. It has been a wonderful experience working with Taylor & Francis Group/CRC Press.

Finally, and most importantly, this book could not have been possible without the unconditional support of our families.

Editors

Dr. Fei Yuan earned his BEng in electrical engineering from Shandong University, Jinan, Shandong, China, in 1985, and MASc in chemical engineering and PhD in electrical engineering from the University of Waterloo, Canada, in 1995 and 1999, respectively. He was a lecturer in the Department of Electrical Engineering, Changzhou Institute of Technology, Jiangsu, China, during 1985–1989. In 1989, he was a visiting professor at Humber College of Applied Arts and Technology, Toronto, Ontario, Canada, and Lambton College of Applied Arts and Technology, Sarnia, Ontario, Canada. He worked with Paton Controls Limited, Sarnia, Ontario, Canada, as a controls engineer during 1989–1994. Since 1999, he has been with the Department of Electrical and Computer Engineering, Ryerson University, Ontario, Canada, where he is currently a professor and the chair. Dr. Yuan is the author of *CMOS Current-Mode Circuits for Data Communications* (Springer, 2007), *CMOS Active Inductors and Transformers: Principle, Implementation, and Applications* (Springer, 2008), and *CMOS Circuits for Passive Wireless Microsystems* (Springer, 2010) and the principal coauthor of *Computer Methods for Analysis of Mixed-Mode Switching Circuits* (Kluwer Academic, 2004). In addition, he has authored/coauthored approximately 200 research papers in refereed journals and conference proceedings. Dr. Yuan was awarded a postgraduate scholarship by Natural Science and Engineering Research Council of Canada during 1997–1998, the Teaching Excellence Award by Changzhou Institute of Technology in 1988, and the Dean's Research Excellence Award and the Ryerson Research Chair Award in 2004 and 2005, respectively, by Ryerson University. Dr. Yuan is a registered professional engineer in the province of Ontario, Canada. He can be reached at fyuan@ryerson.ca.

Dr. Krzysztof (Kris) Iniewski manages R&D at Redlen Technologies Inc., a start-up company in Vancouver, Canada. Redlen's revolutionary production process of advanced semiconductor materials enables a new generation of more accurate, all-digital, radiation-based imaging solutions. Dr. Iniewski is also president of CMOS Emerging Technologies Research Inc. (www.cmosetr.com), an organization covering high-tech events on communications, microsystems, optoelectronics, and sensors. In his career, Dr. Iniewski has held numerous faculty and management positions at the University of Toronto, the University of Alberta, Simon Fraser University, and PMC-Sierra Inc. He has published more than 100 research papers in international journals and conferences. He holds 18 international patents granted in the United States, Canada, France, Germany, and Japan. He is a frequent invited speaker and has consulted for multiple organizations worldwide. He has written and edited several books for publishers such as CRC Press, Cambridge University Press, IEEE Press, Wiley, McGraw Hill, Artech House, and Springer. His personal goal is to contribute to healthy living and sustainability through innovative engineering solutions. In his leisure time, he can be reached at kris.iniewski@gmail.com.

Contributors

Moataz Abdelfattah
Department of Electrical and Computer
  Engineering
McGill University
Montréal, Québec, Canada

Ghyslain Gagnon
École de Technologie Supèrieure
Universitè du Quèbec
Montrèal, Quèbec, Canada

Jussi-Pekka Jansson
Electronics Laboratory
Department of Electrical Engineering
University of Oulu
Linnanmaa, Finland

Pekka Keränen
Electronics Laboratory
Department of Electrical Engineering
University of Oulu
Linnanmaa, Finland

Juha Kostamovaara
Electronics Laboratory
Department of Electrical Engineering
University of Oulu
Linnanmaa, Finland

Kang-Yoon Lee
College of Information and
  Communication Engineering
Sungkyunkwan University
Seoul, Republic of Korea

Antti Mäntyniemi
Electronics Laboratory
Department of Electrical Engineering
University of Oulu
Linnanmaa, Finland

Gordon W. Roberts
Department of Electrical and Computer
  Engineering
McGill University
Montrèal, Quèbec, Canada

Fei Yuan
Department of Electrical and Computer
  Engineering
Ryerson University
Toronto, Ontario, Canada

Soheyl Ziabakhsh
École de Technologie Supèrieure
Universitè du Quèbec
Montrèal, Quèbec, Canada

Symbols and Abbreviations

Symbol	Description
C_{ox}	Gate-oxide capacitance per unit area
g_m	Transconductance
g_o	Output conductance
I_{DS}	Drain-source channel current (DC)
i_{DS}	Drain-source channel current (DC + AC)
i_{ds}	Drain-source channel current (AC)
k	Boltzmann constant (1.38066×10^{-23} J/K)
K_{LF}	Gain of loop filters
K_{PD}	Gain of phase detectors
K_{VCO}	Phase-voltage gain of voltage-controlled oscillators
L	Channel length of MOS transistors
n_i	Concentration of intrinsic charge carriers in silicon (1.5×10^{10} at 300 K)
P_e	Power of quantization error
p_e	Probability density function of quantization error
q	Charge of electron ($1.60217657 \times 10^{-19}$ C)
Q	Quality factor
R_{on}	Channel resistance of MOS transistors in triode
T_o	Reference temperature, typically 300 K or 27°C (room temperature)
t_{ox}	Thickness of gate oxide
V_{DD}	Supply voltage
V_{GS}	Gate-source voltage (DC)
v_{GS}	Gate-source voltage (DC + AC)
v_{gs}	Gate-source voltage (AC)
V_T	Threshold voltage of MOS transistors
V_{Tn}	Threshold voltage of NMOS transistors
V_{Tp}	Threshold voltage of PMOS transistors
W	Channel width of MOS transistors
α_{V_T}	Temperature coefficient of threshold voltage of MOS transistors
γ	Body effect coefficient of MOSFETs
$\Gamma(\omega_o \tau)$	Impulse sensitivity function
Δ	Quantization error
ϵ_{ox}	Dielectric constant of oxide (3.5×10^{13} F/cm)
ϵ_{si}	Dielectric constant of silicon (1.05×10^{12} F/cm)
λ	Channel length modulation coefficient
μ_n	Surface mobility of electrons
μ_p	Surface mobility of poles
ξ	Damping factor of phase-locked loops
τ_{PHL}	High-to-low propagation delay

τ_{PLH}	Low-to-high propagation delay
ϕ_F	Fermi potential
ω_n	Loop bandwidth of phase-locked loops

Abbreviations

$\Delta\Sigma$	Delta-sigma
AAF	Antialiasing filter
AC	Alternating current
ADC	Analog-to-digital converter
ADPLL	All-digital phase-locked loop
ASP	Analog signal processing
BiDWA	Bidirectional data-weighted averaging
BP	Band pass
CAD	Computer-aided design
CLA	Conventional clocked average
CMOS	Complementary metal oxide semiconductor
CRO	Coupled ring oscillator
CP	Charge pumps
CT	Continuous time
DAC	Digital-to-analog converter
dB	Decibel
DC	Direct current
DCDL	Digital-controlled delay line
DCO	Digitally controlled oscillator
DD	Double delay
DEM	Dynamic element matching
DFE	Decision feedback equalization
DFF	D-type flip-flip
DLL	Delay-locked loop
DNL	Differential nonlinearity
DR	Dynamic range
DT	Discrete time
DTC	Digital-to-time converter
DTF	Distortion transfer function
DWA	Data-weighted averaging
ED	Edge detector
ENOB	Effective number of bits
FCW	Frequency control word
FE	Forward Euler
FET	Field effect transistor
FFT	Fast Fourier transform
FGRO	Fast gated-ring oscillator
FIR	Finite impulse filter
FOM	Figure of merit
FSR	Full-scale range

Gbps	Giga bit per second
GHz	Gigahertz
GRO	Gated-ring oscillator
GSRO	Gated switched ring oscillator
HP	High pass
Hz	Hertz
IC	Integrated circuit
IIR	Infinite impulse filter
ILA	Individual level averaging
INL	Integral nonlinearity
ISF	Impulse sensitivity function
KHz	Kilohertz
LDI	Lossless discrete integrator
LP	Low pass
LSB	Least significant bit
LTI	Linear time-invariant
LUT	Look-up table
MASH	Multistage noise-shaping
MDLL	Multiplying delay-locked loop
MHz	Megahertz
MIM	Metal-insulator-metal
MOS	Metal oxide semiconductor
MSB	Most significant bit
MUX	Multiplexer
NMOS	n-type metal oxide semiconductor
NTF	Noise transfer function
OPAMP	Operational amplifier
OSR	Oversampling ratio
OTA	Operational transconductance amplifier
PD	Phase detector
PFD	Phase frequency detector
PHL	Propagation delay high-to-low
PI	Phase interpolator
PLH	Propagation delay low-to-high
PLL	Phase-locked loop
PMOS	p-type metal oxide semiconductor
PSD	Power spectral density
PTAT	Proportional to absolute temperature
PVT	Process, voltage, and temperature
PWM	Pulse-width modulation
QPSK	Quadrature phase shift keying
RC	Resistor–capacitor
RDA	Random averaging
RDWA	Rotated data-weighted averaging
RF	Radio frequency
RFID	Radio frequency identification

RJ	Random jitter
RLC	Resistor, inductor (L), and capacitor
RMS	Root mean square
RnDWA	Randomized data-weighted averaging
RS	Reset-set
RTA	Resistor tuning array
SA	Successive approximation
SAR	Successive approximation register
SC	Switched capacitor
SDU	Switched delay unit
SFDR	Spurious-free dynamic range
SGRO	Slow gated-ring oscillator
S/H	Sample and hold
SNDR	Signal-to-noise-plus-distortion ratio
SNR	Signal-to-noise ratio
SPICE	Simulation program with integrated circuit emphasis
SR	Set-reset
SRO	Switched ring oscillator
SSB	Single side band
STF	Signal transfer function
T2B	Thermometer to binary
TA	Time amplifier
TAC	Time-to-amplitude converters
TCC	Temperature coefficient of current
TDC	Time-to-digital converter
TM	Time mode
TMSP	Time-mode signal processing
T-Reg	Time register
TSPC	True single phase logic
TVC	Time-to-voltage converter
VCDU	Voltage-controlled delay unit
VC-GRO	Voltage-controlled gated ring oscillator
VCO	Voltage-controlled oscillator
ZTC	Zero temperature coefficient

1 Introduction to Time-Mode Signal Processing

Fei Yuan

CONTENTS

The rapid scaling of CMOS technology has resulted in the sharp increase of time resolution and the continuous reduction of voltage resolution. As a result, time-mode circuits where information is represented by the time difference between the occurrences of digital events rather than the nodal voltages or branch currents of electric networks offer a viable and technology-friendly way to reduce scaling-induced performance degradation of mixed-mode systems. This chapter examines the fundamentals of time-mode circuits. The definition of time-based signal processing is provided in Section 1.1. Section 1.2 examines the characteristics of time-mode circuits and compares them with those of their voltage-mode and current-mode counterparts. Challenges encountered in time-based signal processing are investigated in Section 1.3. Section 1.4 browses through the key building blocks of time-mode circuits. These building blocks include time-to-digital converters (TDCs), digital-to-time converters (DTCs), time amplifiers, time quantizers, and time-mode arithmetic units. Section 1.5 briefly explores the applications of time-based approaches in mixed-mode signal processing. These applications include analog-to-digital converters, phase-locked loops, frequency synthesizers, and temperature measurement. The detailed analysis of these building blocks and applications will be provided in later chapters. The chapter is summarized in Section 1.6.

1.1   WHAT IS TIME-MODE?

Time-mode circuits depict an analog signal using the difference between the time instants at which two digital events take place. The amount of the time difference is linearly proportional to the amplitude of the analog signal ideally. A time variable is a pulse-width-modulated signal with its pulse width directly proportional to the amplitude of the signal that it represents. A time variable possesses a unique duality characteristic; specifically, it is an analog signal as the continuous amplitude of the analog signal is represented by the duration of the pulse, but, it is also a digital signal as it only has two largely distinct values. The duality of time variables enables them to conduct analog signal processing in a digital environment. This unique characteristic is clearly not possessed by neither analog nor digital variables.

Time-mode signal processing deals with addition, multiplication, amplification, integration, quantization, etc., of time variables. It is interesting to note that time-based signal processing is quite similar to information transmission through the neuron systems of human brains [1]. Since information to be processed by time-mode circuits is represented by the time difference of digital signals, these circuits are essentially digital systems that perform analog and mixed analog–digital signal processing without using power-greedy and speed-impaired digital signal processors.

1.2   WHY TIME-MODE?

The advancement of CMOS technology is always geared toward optimizing the performance of digital systems. As a result, CMOS analog circuits are continuously losing the benefits of specialized and process-controlled components critical to the performance of these circuits. In addition, they must also cope with a rapidly decreasing voltage headroom, that is, the difference between the given supply voltage of a circuit and the minimum supply voltage of the circuit required for MOS transistors to operate in saturation, caused by the slow decline of the device threshold voltage and the aggressive reduction of the supply voltage while meeting ever-stringent performance specifications [2–4]. The shrinking voltage headroom not only limits the maximum achievable signal-to-noise ratio (SNR), it also signifies that the effect of the nonlinear characteristics of MOS devices subsequently reduces the dynamic range of voltage-mode circuits. Further, technology scaling raises the thermal noise floor quantified by kT/C where k is Boltzmann's constant, T is temperature in Kelvin, and C is the minimum capacitance. As a result, the accuracy of voltage-mode circuits, loosely defined as the ratio of the minimum detectable voltage, typically set by the noise floor, to the maximum available voltage headroom, scales poorly with technology.

Current-mode approaches where analog information to be processed is represented by the branch current of electric networks offer an alternative means to cope with the challenges induced by the dropping voltage headroom. These circuits achieve a low nodal voltage swing by lowering the impedance of nodes. The existence of low-impedance nodes throughout current-mode circuits, however, gives rise to large branch currents. As a result, current-mode circuits typically consume more power as compared with their voltage-mode counterparts. Lowering the power consumption of current-mode circuits while meeting other design constraints at the same time is rather difficult [5]. The characteristics of the low-impedance nodes of current-mode circuits, on the other hand, offer an intrinsic advantage of a low time constant at every node of the circuits. As a result, current-mode circuits are suitable for applications where speed rather than power consumption is most critical. These applications include high-speed serial links, current-steering logic, and current-mode arithmetic units such as current-mode adders in decision feedback equalization (DFE), to name a few. Since voltages and currents are inherently related to each other via impedance or conductance, the characteristics of voltage-mode circuits and current-mode circuits do not differ fundamentally. As a result, the performance of both circuits does not scale well with technology.

Although the detrimental effect of technology scaling on the performance of analog circuits, regardless of whether they are voltage-mode or current-mode, can be compensated to some extent using digitally assisted means, such as digitally tuned resistor arrays for the calibration of impedance matching of serial links, digitally tuned capacitor arrays or current arrays for the cancellation of the offset voltage of comparators, and digitally tuned capacitor arrays for the coarse frequency tuning of oscillators; these approaches are costly both in terms of silicon and power consumption as the switching transistors in these digital networks must be sufficiently large in order to minimize the effect of the channel resistance of these switches. The addition of digitally assistance blocks also has a negative impact on the performance of analog circuits such as increasing the capacitance of the critical nodes through which high-frequency signals propagate. On top of that, the performance of voltage-mode analog or mixed analog-digital circuits continues to decline with technology scaling, further demanding for more digitally assisted compensation.

The intrinsic gate delay of digital circuits, on the other hand, has been the primary beneficiary of technology scaling. The improved switching characteristics of MOS transistors offer an excellent timing accuracy such that the time resolution of digital circuits has well surpassed the voltage resolution of analog circuits implemented in nanoscale CMOS technologies. "In a deep-submicron CMOS process, the time-domain resolution of a digital signal edge transition is superior to the voltage resolution of an analog signal" as stated by Dr. R. Staszewski [6]. Time-mode approaches where information is represented by the difference between the time instants at which digital events take place rather than the nodal voltages or branch currents of electric networks offer a new means to neutralize the scaling-induced challenges that once seemed unconquerable. Since time-mode circuits perform analog signal processing in the digital domain, not only the performance of these circuits scales well with technology, time-mode circuits also offer a number of attractive characteristics including full programmability, the ease of portability, low-power consumption, and high-speed operation.

As information to be processed by time-mode circuits is represented by the time difference between the occurrence of digital events, time-mode circuits are essentially digital circuits. The detrimental effect of technology scaling on the performance of voltage-mode or current-mode analog signal processing disappears in time-mode circuits. Time-mode circuits are less sensitive to interferences such as cross talk, switching noise, and substrate coupling, which have a severe impact on the performance of voltage-mode or current-mode circuits.

The full programmability of time-mode circuits, attributable to their digital realization, allows them to be deployed in a broad spectrum of applications where tunable characteristics are mandatory. If voltage-mode or current-mode circuits are used in these applications, complex digitally assisted tuning mechanisms will be required.

In addition to programmability, portability is of a critical importance in order to minimize design turn-around time. The digital nature of time-mode circuits allows them to be migrated from one generation of technology to another with the minimum design time, subsequently lowering the cost.

As the intrinsic gate delay of digital circuits benefits the most from technology scaling, time-mode circuits are capable of carrying out rapid signal processing. For example, the oscillation frequency of ring oscillators implemented in state-of-the-art CMOS technologies has reached tens of GHz. Voltage-controlled oscillator (VCO)-based multibit quantizers can provide a large oversampling ratio while consuming a small amount of power.

It is evident from the preceding experiment that time-based signal processing has many desirable characteristics such as excellent scalability with technology, good immunity from interferences and imperfections, full programmability, the ease of portability, low-power consumption, and high-speed operation. All are vital to mixed analog-to-digital signal processing and all are not possessed by either voltage-mode or current-mode circuits.

1.3   CHALLENGES IN TIME-MODE SIGNAL PROCESSING

Although time-based signal processing possesses a number of critical advantages as compared with its voltage-mode or current-mode counterparts, a number of stiff challenges are yet to be overcome in order for time-mode circuits to be deployed in a broad range of applications. In this section, we briefly examine these challenges.

Although the intrinsic gate delay of digital circuits benefits the most from technology scaling, device mismatch arising mainly from process spread deteriorates with technology scaling. In order to minimize the effect of device mismatch, minimum-sized unit delay cells should be avoided. This inevitably has a detrimental impact on the speed and subsequently on the resolution of time-mode circuits.

As most time-mode circuits are built upon basic delay cells, such as static CMOS inverters and current-starved CMOS inverters, the propagation delay of these delay cells is a strong function of supply voltage fluctuation. In design of analog circuits, cascode is a convenient, economical, and effective means to minimize the effect of a fluctuating supply voltage. For time-mode circuits, delay-locked loops (DLLs) are widely used to minimize the effect of process, supply, and temperature (PVT) variations on the delay of delay lines. Although DLLs can be conveniently used to stabilize the delay of the delay lines, it is difficult to use them to minimize the effect of PVT on the delay of the logic gates that are often also part of time-mode circuits and control the operation of time-mode circuits. For example, in cyclic vernier TDCs, the delay of the control logic gates has to be made negligibly small as compared with that of the delay cells in order to minimize their effect.

Perhaps one of the stiffest challenges in time-based signal processing is the design of time-mode arithmetic units, especially time integrators. The accumulation of a variable in the voltage domain can be conveniently realized by representing the variable as a current and integrating the current onto a capacitor. The voltage of the capacitor gives the result of the integration

$$v_c(t) = \frac{1}{C} \int_0^t i_c(\tau)d\tau$$

Withholding or storing a time variable is rather difficult due to the irretrievable nature of time. In contrast, a voltage-mode analog variable can be conveniently stored indefinitely using a capacitor if the leakage of the capacitor is negligible. A voltage-mode discrete variable can be stored even more conveniently using a latch. Time registers that store time variables and read out the stored time variables upon the arrival of a readout command become critical.

1.4 BUILDING BLOCKS OF TIME-MODE CIRCUITS

A complex analog circuit is typically constructed from a set of building blocks such as common-source amplifiers, common-gate amplifiers, common-drain amplifiers (source followers), cascode amplifiers, and differentially configured amplifiers, to name a few. Similarly, a time-based signal-processing system is made of a set of building blocks that perform tasks such as interfacing with voltage-mode and current-mode circuits, time amplification, time arithmetic operations, time quantization, time-to-digital, and digital-to-time conversion. We briefly browse through them in this section. A detailed examination of these building blocks will be provided in later chapters.

1.4.1 VOLTAGE-TO-TIME CONVERTERS

One of the key building blocks of time-mode circuits is voltage-to-time converters (VTCs) that map a voltage to a time variable. A VTC serves as a gateway bridging voltage-mode and time-mode domains. The most important performance indicators of VTCs are linearity and conversion gain. As time-mode circuits are digital circuits, the linearity of a time-mode system is largely dominated by that of its VTC. The dynamic range of a delay-line TDC is lower-bound by the per-stage time delay and upper-bound by the total delay of the delay line of the TDC. Since the former scales well with technology, the resolution of a time-mode system is largely determined by the conversion gain of VTCs. The higher the conversion gain, the better is the resolution.

1.4.2 TIME-TO-DIGITAL CONVERTERS

TDCs map a time variable to a digital code. The deployment of TDCs in nuclear science research dates back to 1970s [7,8]. The application of TDCs has extended well beyond nuclear science to digital storage oscillators [9,10], laser range finders [11], and digital frequency synthesizers [12], to name a few. Similar to analog-to-digital converters, the performance of TDCs is quantified by a number of parameters such as SNR and signal-to-noise-plus-distortion ratio (SNDR) for noise-shaping TDCs, and differential nonlinearity (DNL) and integral nonlinearity (INL) for sampling TDCs. To compare the performance of TDCs of different architectures, the amount of the power consumption per conversion step of TDCs is the most widely used figure-of-merit (FOM).

1.4.3 DIGITAL-TO-TIME CONVERTERS

The opposite of time-to-digital conversion is digital-to-time conversion. DTCs map a digital code to a time variable. A digital-to-time operation is needed in applications such as time-mode successive approximation TDCs. A DTC assumes a similar role as that of a digital-to-analog converter in voltage-mode successive approximation analog-to-digital converters (ADCs) or multibit $\Delta\Sigma$ modulators to establish negative feedback.

1.4.4 TIME AMPLIFIERS

Similar to a voltage amplifier that amplifies a voltage, a time amplifier amplifies a time variable. Time amplification is often needed in applications such as the precision measurement of the width of a narrow pulse where the pulse width is often too small to be quantized accurately. Time amplification can also be employed for accurately quantizing the output of TDC-based phase detectors in the vicinity of the lock state so as to establish a precision phase lock. Time amplifiers play a pivotal role in improving the resolution of time-mode circuits. Similar to voltage amplifiers, the gain and linearity of time amplifiers are the most important design specifications. For high-speed time digitization, the bandwidth of time amplifiers is also of a great importance.

1.4.5 TIME QUANTIZERS

A single-bit voltage quantizer maps a voltage to a Boolean variable by comparing it with a reference voltage. Similarly, a single-bit time quantizer maps a time variable to a Boolean output by comparing it with a time reference. Time quantizers can be realized using a time comparator with the time reference with which the input compares coming from a voltage-controlled oscillator (VCO) of a constant frequency. To conduct the multibit quantization of a time variable, the time variable can be used as a gating signal to activate or deactivate a multistage ring oscillator of a constant oscillation frequency, known as gated ring oscillator (GRO). Since the number of the oscillation cycles of the oscillator and the output of each stage of the oscillator uniquely correspond to the duration of the gating signal, a multibit time quantizer can be constructed. GRO-based multibit time quantizers offer the intrinsic advantage of first-order noise-shaping. In addition, as compared with voltage-mode multibit quantization, which requires a total of 2^N voltage comparators where N is the number of quantization bits, VCO-based multibit time quantization offers the advantage of low-power consumption, fast quantization resulting in a large oversampling ratio (OSR), good linearity, first-order noise-shaping, and all-digital realization.

1.4.6 TIME-MODE ARITHMETIC UNITS

Time-mode arithmetic units such as time adders and time integrators are critically needed in time-based signal processing. The accumulation of a variable in the voltage domain can be realized by representing the variable as a current and integrating the current onto a capacitor, that is,

$$v_c(t) = \frac{1}{C}\int_0^t i_c(\tau)d\tau$$

The resultant voltage of the capacitor is the integration of the variable. Similarly, the accumulation of a variable in the current domain can be implemented by representing the variable as a voltage and integrating the resultant voltage onto an inductor, that is,

$$i_L(t) = \frac{1}{L}\int_0^t v_L(\tau)d\tau$$

The resultant current of the inductor is the integration of the variable. Withholding a time variable, on the other hand, is rather difficult due to the irretrievable nature of time. Recent work by Hong et al. [13] and Kim et al. [14] has opened the door for time registers and later time integrators, a key component of all-digital $\Delta\Sigma$ modulators.

1.5 APPLICATIONS OF TIME-MODE SIGNAL PROCESSING

In this section, we briefly browse through some of the key applications of time-based signal processing. These applications include analog-to-digital converters, phase-locked loops, frequency synthesizers, and temperature measurement. An in-depth investigation of them will be given in later chapters.

1.5.1 ANALOG-TO-DIGITAL CONVERTERS

Driven by the benefits of time-mode signal processing, analog-to-digital conversion using time-mode approaches has received a special attention from both academia and industry recently. A key difference between conventional voltage-mode ADCs and time-mode ADCs is the replacement of voltage comparator-based quantizers with VCO-based quantizers as the former suffer from a number of drawbacks including high power consumption especially for multibit quantization and the need for digitally assisted mismatch compensation. Unlike voltage comparator-based quantizers, in VCO-based quantization, the voltage to be quantized is the control voltage of the VCO. Since for each control voltage, there exists a corresponding oscillation frequency of the oscillator, the number of the cycles of the oscillator within the duration in which the input voltage is held and the output of the delay stages of the oscillator at the end of the sample-and-hold interval of the control voltage provide the digital representation of the control voltage. VCO-based quantizers offer a number of attractive intrinsic advantages including built-in first-order noise-shaping, inherent multibit quantization with a good linearity, fast quantization subsequently a large over-sampling ratio, low-power consumption, and full scalability with technology.

Architecturally, time-mode ADCs can be loosely categorized into open-loop ADCs and closed-loop ADCs. The former perform analog-to-digital conversion in an open-loop fashion, while the latter utilize negative feedback to improve performance, in particular, to suppress the effect of nonlinearities in the forward path of the feedback loop and in-band noise through the noise-shaping of the quantization noise. Open-loop time-mode ADCs offer the advantages of rapid conversion. They, however, suffer from a small dynamic range largely due to in-band harmonics and a high level of in-band quantization noise. Although closed-loop time-mode ADCs provide a better SNDR as compared with their open-loop counterparts, a high-order operational amplifier (op-amp)-based voltage-mode integrator is still needed in the forward path in most recent designs in order to have an adequate loop gain to suppress the effect of the nonlinearities and quantization noise. These ADCs are therefore not all-digital. As a result, their performance does not scale naturally with technology. As the performance of voltage-mode integrators scales poorly with technology, time integrators with a large in-band gain are critically needed for all-digital time-mode ADCs.

1.5.2   ALL-DIGITAL PHASE-LOCKED LOOPS

Similar to analog-to-digital converters, phase-lock loops are another key building block of mixed analog-digital systems. In a phase-locked loop, a low-pass loop filter with large capacitors is typically needed to filter out high-frequency components of the control voltage of the VCO so as to minimize spurs and improve phase noise performance. The programmability constraint also demands that the loop bandwidth of the phase-locked loop be tunable. It is therefore highly desirable to have the loop filter realized digitally in order to reduce silicon cost and provide programmability. In a conventional phase-locked loop with a linear phase detector, the phase difference between the input reference and the output of the VCO is represented by a pulse obtained using a phase detector with pulse width proportional to the phase difference. The resultant pulse is then converted to an analog signal, specifically the control voltage of the oscillator of the phase-locked loop, using both a charge pump and a low-pass loop filter. The power consumption of the charge pump typically constitutes a large portion of the overall power consumption of the phase-locked loop. This is because in order for the phase-locked loop to provide an adequate and timely correction, the current of the charge pump must be sufficiently large.

One might argue that the same can be achieved by lowering the capacitance of the loop filter. Lowering the capacitance of the loop filter, however, will degrade the ability of the loop filter to filter out high-frequency components present on the control voltage line of the oscillator. It is therefore highly desirable from a low-power consumption point of view to have the charge pump removed. The removal of the charge pump also eliminates the source of reference spurs. As the output of the phase detector is a pulse with its pulse width directly proportional to the phase difference between the input of the phase-locked loop and the output of the VCO, this time variable can be digitized by a TDC. The digital output of the TDC can then be fed to a digital loop filter. As a result, not only power-greedy charge pump and silicon-consuming loop filter are removed, the loop bandwidth can also be made fully programmable. The resultant phase-locked loop is now all-digital and enjoys the full merits of technology scaling. If noise-shaping TDCs are used, the low quantization noise of the TDCs will also improve the overall phase noise of the phase-locked loop.

1.5.3 ALL-DIGITAL FREQUENCY SYNTHESIZERS

Frequency synthesizers are one of the core subsystems of wireless systems. An all-digital phase-locked loop can be migrated to an all-digital frequency synthesizer by including a frequency divider in the feedback path. Integer-N frequency synthesizers suffer from a large frequency adjustment step dictated by the reference frequency, while fractional-N frequency synthesizers yield a fine frequency resolution but suffer from fractional spurs [15,16]. Randomizing the control bits of the frequency divider widens the output of the frequency divider so that fractional spurs are reduced. This, however, is at the expense of increased in-band noise [17]. To reduce the in-band noise, a noise-shaping $\Delta\Sigma$ modulator with a DC input can be employed to generate the random bits for selecting the modulus of the frequency divider [18]. The noise-shaping characteristics of the $\Delta\Sigma$ modulator ensure a low level of in-band quantization noise is achieved by moving the excessive quantization noise to high frequencies outside the signal band, which can be removed effectively by the loop dynamics [19,20]. The recent architecture of all-digital frequency synthesizers places a TDC in the feedback path [6,21]. The TDC performs frequency division. The removal of an explicit frequency divider in the feedback path and its associated performance enhancement blocks such as $\Delta\Sigma$ modulators not only greatly simplifies synthesizers, it also eliminates the source of fractional spurs [20,22,23].

1.5.4 TIME-BASED TEMPERATURE SENSORS

Integrated temperature sensors are very important in applications such as medical implants, smart sensors for environment monitoring, and on-chip temperature monitoring of VLSI (very-large-scale integration) systems, to name a few. Traditionally, integrated temperature sensors are manufactured using a temperature-dependent voltage-mode circuit whose output voltage is a linear function of temperature, a temperature reference circuit whose output voltage is independent of temperature and whose temperature set point can be adjusted, and a voltage comparator that compares the output voltage of the temperature-dependent circuit and that of the temperature reference circuit. The continuous dip in supply voltage greatly reduces the resolution of voltage-mode circuits, making it very difficult to improve the measurement accuracy of these temperature sensors.

Temperature can be measured by comparing the frequency of a PTAT (proportional to absolute temperature) oscillator whose frequency is proportional to temperature with that of a reference oscillator whose frequency is independent of temperature [24–26]. Alternatively, one can first convert temperature to a pulse with pulse width proportional to temperature and then digitize the width of the resultant pulse using a low-power TDC such as a pulse-shrinking TDC. The removal of oscillators greatly lowers the power consumption, making TDC-based temperature sensors particularly attractive in low-power-requiring applications [12,27]. A time-mode integrated temperature sensor can also be realized using a temperature-dependent delay line whose delay is a linear function of temperature, a temperature reference line whose delay is independent of temperature and can be adjusted by changing its temperature set-point, and a time comparator that discriminates the delay of the two delay lines. For a given temperature, the delay of the temperature reference can be made identical to that of the temperature-dependent delay line by digitally adjusting the delay of the temperature reference, that is, the temperature set point of the temperature reference line. Once this occurs, the digital code for adjusting the temperature set point of the temperature reference line gives the digital representation of the temperature [28,29].

1.6   SUMMARY

In this chapter, we briefly examined technology scaling-induced challenges encountered in design of voltage-mode or current-mode circuits for mixed-mode signal processing. We showed that although technology scaling results in a reduced voltage accuracy, which leads to the deteriorating performance of both voltage-mode and current-mode circuits, it sharply improves the switching accuracy of digital circuits at the same time. As a result, the performance of time-based signal processing not only scales well with technology, it also surpasses that of voltage-mode or current-mode circuits. The challenges encountered in design of time-mode circuits such as device mismatches, PVT effect, and the storage of time-mode variables were explored. The key building blocks of time-mode circuits including VTCs, TDCs, DTCs, time amplifiers, time quantizers, and time-mode arithmetic units were examined briefly. The applications of time-mode circuits in analog-to-digital converters, phase-locked loops, frequency synthesizers, and temperature measurement were briefly explored.

REFERENCES

1. V. Ravinuthula, V. Garg, J. Harris, and J. Fortes. Time-mode circuits for analog computation. *International Journal of Circuit Theory and Applications*, 37:631–659, 2009.
2. G. Shahidi. Challenges of CMOS scaling at below 0.1 μm. In *Proceedings of International Microelectronics Conference*, 2000, Tehran, pp. 5–8.
3. R. Gera and D. Hoe. An evaluation of CMOS adders in deep sub-micron processes. In *Proceedings of Southeastern Symposium on System Theory*, 2012, Jacksonville, FL, pp. 126–129.

4. B. Jonsson. On CMOS scaling and A/D converter performance. In *Proceedings of NORCHIP*, 2010, Tampere, Finland, pp. 1–4.
5. F. Yuan. *CMOS Current-Mode Circuits for Data Communications*. Springer, New York, 2006.
6. R. Staszewski, K. Muhammad, D. Leipold, C. Hung, Y. Ho, J. Wallberg, C. Fernando et al. All-digital TX frequency synthesizer and discrete-time receiver for Bluetooth radio in 130-nm CMOS. *IEEE Journal of Solid-State Circuits*, 39(12):2278–2291, December 2004.
7. T. Yoshiaki and A. Takeshi. Simple voltage-to-time converter with high linearity. *IEEE Transactions on Instrumentation and Measurement*, 20(2):120–122, May 1971.
8. D. Porat. Review of sub-nanosecond time-interval measurements. *IEEE Transactions on Nuclear Science*, NS-20:36–51, September 1973.
9. K. Park and J. Park. 20 ps resolution time-to-digital converter for digital storage oscillator. *Proceedings of IEEE Nuclear Science Symposium*, 2:876–881, Toronto, ON, 1998.
10. P. Chen, C. Chen, and Y. Shen. A low-cost low-power CMOS time-to-digital converter based on pulse stretching. *IEEE Transactions on Nuclear Science*, 53(4):2215–2220, August 2006.
11. C. Chen, P. Chen, C. Hwang, and W. Chang. A precise cyclic CMOS time-to-digital converter with low thermal sensitivity. *IEEE Transactions on Nuclear Science*, 52(4):834–838, August 2005.
12. P. Chen, C. Chen, C. Tsai, and W. Lu. A time-to-digital-converter-based CMOS smart temperature sensor. *IEEE Journal of Solid-State Circuits*, 40(8):1642–1648, August 2005.
13. J. Hong, S. Kim, J. Liu, N. Xing, T. Jang, J. Park, J. Kim, T. Kim, and H. Park. A 0.004 mm² 250 W $\Delta\Sigma$ TDC with time-difference accumulator and a 0.012 mm² 2.5 mW bang-bang digital PLL using PRNG for low-power SoC applications. In *IEEE International Conference on Solid-State Circuits Digest of Technical Papers*, 2012, pp. 240–242.
14. K. Kim, W. Yu, and S. Cho. A 9 bit, 1.12 ps resolution 2.5 b/stage pipelined time-to-digital converter in 65 nm CMOS using time-register. *IEEE Journal of Solid-State Circuits*, 49(4):1007–1016, April 2014.
15. G. Gillette. The digiphase synthesizer. In *Proceedings of 23rd Annual Frequency Control Symposium*, 1969, Fort Monmouth, NJ, pp. 25–29.
16. J. Gibbs and R. Temple. Frequency domain yields its data to phase-locked synthesizer. *Electronics*, 27:107–113, April 1978.
17. V. Reinhardt. Spur reduction technique in direct digital synthesizer. In *Proceedings of 47th Frequency Control Symposium*, October 1993, Salt Lake City, UT, pp. 230–241.
18. T. Riley, M. Copeland, and T. Kwanieski. Delta-sigma modulation in fractional-N frequency synthesis. *IEEE Journal of Solid-State Circuits*, 28(5):553–559, 1991.
19. C. Hsu, M. Straayer, and M. Perrott. A low-noise wide-BW 3.6-GHz digital $\Delta\Sigma$ fractional-N frequency synthesizer with a noise-shaping time-to-digital converter and quantization noise cancellation. *IEEE Journal of Solid-State Circuits*, 43(12):2776–2786, December 2008.
20. E. Temporiti, C. Wu, D. Baldi, R. Tonietto, and F. Svelto. A 3 GHz fractional all-digital PLL with a 1.8 MHz bandwidth implementing spur reduction techniques. *IEEE Journal of Solid-State Circuits*, 44(3):824–834, 2009.
21. R. Staszewski, J. Wallberg, S. Rezeq, C. Hung, O. Eliezer, S. Vemulapalli, C. Fernando et al. All-digital PLL and transmitter for mobile phones. *IEEE Transactions on Nuclear Science*, 40(12):2469–2482, December 2005.
22. M. Lee, M. Heidari, and A. Abidi. A low-noise wideband digital phase-locked loop based on a coarse-fine time-to-digital converter with subpicosecond resolution. *IEEE Journal of Solid-State Circuits*, 44(10):2808–2816, October 2009.

23. E. Temporiti, C. Wu, D. Baldi, R. M. Cusmai, and F. Svelto. A 3.5 GHz wideband ADPLL with fractional spur suppression through TDC dithering and feed-forward compensation. *IEEE Journal of Solid-State Circuits*, 45(12):2723–2736, 2010.
24. S. Zhou and N. Wu. A novel ultra low power temperature sensor for UHF RFID tag chip. In *Proceedings of IEEE Asian Solid-State Circuits Conference*, November 2007, Jeju, Korea, pp. 464–467.
25. C. Kim, B. Kong, C. Lee, and Y. Jun. CMOS temperature sensor with ring oscillator for mobile DRAM self-refresh control. In *Proceedings of IEEE International Symposium on Circuits and Systems*, May 2008, Seattle, WA, pp. 3094–3097.
26. S. Park, C. Min, and S. Cho. A 95 nW ring oscillator-based temperature sensor for RFID tags in 0.13 μ m CMOS. In *Proceedings of IEEE International Symposium on Circuits and Systems*, May 2009, pp. 1153–1156.
27. P. Chen, C. Hwang, and W. Chang. A precise cyclic CMOS time-to-digital converter with low thermal sensitivity. *IEEE Transactions on Nuclear Science*, 52(4):834–838, August 2005.
28. P. Chen, T. Chen, Y. Wang, and C. Chen. A time-domain sub-micro watt temperature sensor with digital set-point programming. *IEEE Sensors Journal*, 9(12):1639–1646, 2009.
29. P. Chen, C. Chen, Y. Peng, K. Wang, and Y. Wang. A time-domain SAR smart temperature sensor with curvature compensation and a 3 σ inaccuracy of 0.4°C–0.6°C over a 0°C to 90°C range. *IEEE Journal of Solid-State Circuits*, 45(3):600–609, March 2010.

2 Voltage-to-Time Converters

Fei Yuan

CONTENTS

One of the key building blocks of time-mode systems for processing analog signals is voltage-to-time converters (VTCs) that map a voltage to a time variable with its value directly proportional to the amplitude of the voltage. Although VTCs can be realized using complex circuits to achieve good performance such as a large conversion range and a good linearity, operational amplifiers whose performance scales poorly with technology are often required [1–3]. As time-mode circuits are digital circuits, the linearity of a time-mode system that processes analog quantities is largely dominated by that of its VTC. VTCs whose performance scales well with technology are critical to time-based signal processing.

This chapter deals with VTCs with an emphasis on the techniques that improve the linearity of the converters. The chapter is organized as follows: Section 2.1 investigates VTCs implemented using the voltage-controlled delay units (VCDUs). VTCs using VCDUs and source degeneration are discussed in Section 2.2. Section 2.3 looks into relaxation voltage-to-time converters. Reference VTCs are examined in Section 2.4. The applications of VTCs in time-mode comparators are investigated in Section 2.5. The chapter is summarized in Section 2.6.

2.1 VCDU VOLTAGE-TO-TIME CONVERTERS

A voltage can be mapped to a time variable using the VCDU shown in Figure 2.1 [4–11]. The VCDU consists of a current-starved inverter, that is, a static inverter with its charging or discharging current controlled by a current source, a load capacitor, and a static inverter. The load capacitor should be linear and its capacitance should be much larger as compared with the capacitances of the transistors such that the effect of the nonlinearity of the device capacitances is negligible as

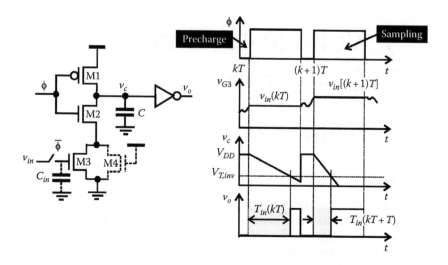

FIGURE 2.1   Voltage-to-time converter using voltage-controlled delay cells.

compared with that of the load capacitor. The current-starved inverter is clocked by a periodic signal ϕ that presets the VCDU prior to a conversion operation. In the precharge phase where ϕ is low, the load capacitor is precharged to V_{DD} and the output of the static inverter is set to logic-0. The input voltage v_{in} is sampled by C_{in}. In the following discharge phase where $\phi = 1$, the load capacitor is discharged by the current of M3 whose value is set by the input voltage $v_{in}(kT)$ at the end of the precharge phase.

The voltage of the gate of M3 is kept unchanged during the discharge phase and so is the discharge current of the load capacitor. As a result, the voltage of the capacitor C decreases linearly with time. When v_c drops below the threshold voltage $V_{T,inv}$ of the following static inverter, v_o will be set to logic-1. The value of the resultant time variable T_{in}, which is measured from the time instant at which $\phi = 1$ to the time instant at which v_c is submerged by the threshold voltage of the inverter, is given by

$$T_{in}(kT) = \frac{C}{i_{DS3}(kT)}\left(V_{DD} - V_{T,inv}\right), \tag{2.1}$$

where

V_{DD} is the supply voltage

$i_{DS3}(kT)$ is the channel current of M3 in kth period and is given by

$$i_{DS3}(kT) = k_{n3}\left[v_{in}(kT) - V_T\right]^2\left(1 + \lambda v_{DS3}\right), \tag{2.2}$$

where

$$k_{n3} = \frac{1}{2}\mu_n C_{ox}\left(\frac{W}{L}\right)_3, \qquad (2.3)$$

where
    μ_n is the surface mobility of free electrons
    C_{ox} is gate-oxide capacitance per unit area
    W_3 and L_3 are the width and length of M3, respectively
    V_T is the threshold voltage of MOSFETs
    λ is the channel-length modulator constant
    v_{DS3} is the drain-source voltage of M3

For simplicity, we assume that NMOS and PMOS transistors have the same threshold voltage, that is, $V_{Tn} = |V_{Tp}| = V_T$. Substituting Equation 2.2 into Equation 2.1 yields

$$T_{in}(kT) = \frac{C(V_{DD} - V_{T,inv})}{k_{n3}\left[v_{in}(kT) - V_T\right]^2 \left(1 + \lambda v_{DS3}\right)}. \qquad (2.4)$$

It is observed from Equation 2.4 that for each $v_{in}(kT)$, there is a corresponding $T_{in}(kT)$. The need for the sample-and-hold operation of v_{in} becomes apparent. Also it is observed from Equation 2.4 that the relation between T_{in} and v_{in} is inherently nonlinear. Further, not only is T_{in} set by v_{in}, T_{in} also varies with v_c as $v_c = v_{DS2} + v_{DS3} \approx v_{DS3}$, further signifying the effect of the nonlinearity of the transconductance. Figure 2.2a shows the dependence of T_{in} on the input voltage of Figure 2.1. The nonlinear characteristic observed in the figure is consistent with our findings in the investigation of the nonlinear relation between T_{in} and v_{in}. As can be seen, an excessive amount of delay exists when the input voltage is low. Also, linearity improves when the input voltage is large.

To improve the performance, an additional transistor M4 can be added in parallel with M3, as shown in Figure 2.1 [12–14]. To understand the effect of M4, let us examine its mode of operation. To ensure that M4 is in triode, $v_{DS4} < v_{GS4} - V_T$ is required. This translates to $v_c < V_{DD} - V_T$. In the precharge phase where $\phi = 0$, we have $v_c = V_{DD}$. In the discharge phase where $\phi = 1$, the load capacitor is discharged through M3 with part of the discharge current controlled by the input and the other part set by M4. If $V_{DD} - V_T < v_c$, M4 will be in saturation. It will be in triode if $v_c < V_{DD} - V_T$. We use the typical parameters of a 130 nm CMOS technology as an example to exemplify the region of the operation of M4. Since for a typical 130 nm CMOS technology, $V_{DD} = 1.2$ V and $V_T \approx 0.4$ V, M4 will be in saturation if $v_c < 0.8$ V and in triode if $v_c < 0.8$ V. It is evident from Equation 2.1 that since the discharge current of the load capacitor consists of both the channel current of M3, which is set by v_{in}, and the channel current of M4, which is independent of v_{in}, we have

$$T_{in}(kT) = \frac{C}{i_{DS3}(kT) + i_{DS4}(kT)}(V_{DD} - V_{T,inv}). \qquad (2.5)$$

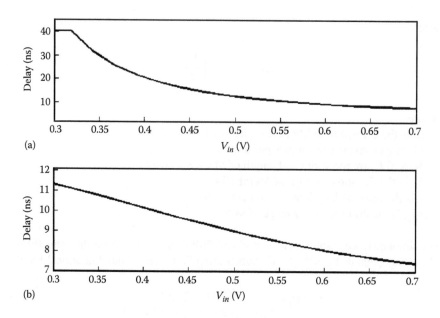

(a)

(b)

FIGURE 2.2 Dependence of the delay of VCDU of Figure 2.1 implemented in an 130 nm CMOS technology. (a) Without M4. (b) With M4.

It is seen from Equation 2.5 that when $i_{DS3}(kT) \ll i_{DS4}(kT)$,

$$T_{in}(kT) \approx \frac{C}{i_{DS4(kT)}}\left(V_{DD} - V_{T,inv}\right). \tag{2.6}$$

Also, when $i_{DS3}(kT) \gg i_{DS4}(kT)$,

$$T_{in}(kT) \approx \frac{C}{i_{DS3}(kT)}\left(V_{DD} - V_{T,inv}\right). \tag{2.7}$$

Since hyperbolic function $f(x) = 1/x$ exhibits a good linearity if $x \ll 1$ or $x \gg 1$ and a poor linearity in the neighborhood of $x = 1$, the addition of M4 improves the linearity of T_{in} when i_{DS3} is small. When i_{DS3} is large, a better linearity is also obtained. This is evident in Figure 2.2b where the dependence of the time delay on the input voltage of Figure 2.1 with M4 added is shown. When i_{DS3} is small, T_{in} is dominated by i_{DS4}, the dependence of T_{in} on i_{DS3} is reduced, so is the conversion gain. The larger i_{DS4}, the worse is the loss of the conversion gain.

To further improve the linearity, a differentially configured VTC consisting of two identical single-ended VTCs of Figure 2.1 can be used, as shown in Figure 2.3 [15]. To demonstrate the improvement of linearity, we neglect the effect of channel length modulation and assume a perfect match between the two transconductors

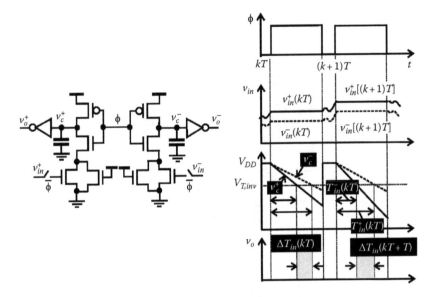

FIGURE 2.3 Voltage-to-time converter using differential voltage-controlled delay cells. (From Macpherson, A. et al., A 5 GS/s 4-bit time-based single-channel CMOS ADC for radio astronomy, *Proceedings of IEEE Custom Integrated Circuit Conference*, 2013, pp. 1–4.)

forming the differential configuration. Let $v_{in}^+ = V_{in} + \Delta v$ and $v_{in}^- = V_{in} - \Delta v$ where Δv denotes the variation of v_{in} from its nominal value V_{in}. From Equation 2.4, we have

$$\Delta T_{in}(kT) = T_{in}^+(kT) - T_{in}^-(kT)$$

$$= -\frac{C(V_{DD} - V_{T,inv})}{k_{n3}} \frac{4(\Delta v)\left[v_{in}(kT) - V_T\right]}{\left\{\left[v_{in}(kT) - V_T\right]^2 - (\Delta v)^2\right\}^2}. \qquad (2.8)$$

If $(\Delta v)^2$ is negligibly small as compared with $[V_{in}(kT) - V_T]^2$, we have

$$\Delta T_{in}(kT) \approx -\left\{\frac{4C(V_{DD} - V_{T,inv})}{k_{n3}\left[v_{in}(kT) - V_T\right]^3}\right\}\Delta v. \qquad (2.9)$$

It is seen from Equation 2.9 that as long as $\Delta v \ll V_{in} - V_T$, a linear relation between ΔT_{in} and Δv will exist. It should be emphasized that the relation between T_{in} and v_{in} given by Equation 2.4 is nonlinear, while that between ΔT_{in} and Δv given by Equation 2.9 is linear.

Another approach to improve the linearity is to use the symmetrical load proposed by Maneatis in [16], as shown in Figure 2.4a [17]. Although voltage-to-current conversion is solely carried out by M3, the addition of diode-connected M4 improves the linearity of $i_c \sim v_c$ relation, as illustrated graphically in Figure 2.4c. It should be noted that since the discharge current should be solely set by the input,

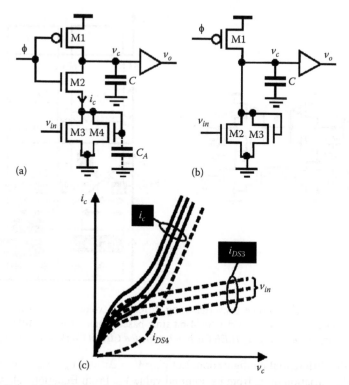

FIGURE 2.4 (a) Voltage-to-time converter. (From Taillefer, C. and Roberts, G., *IEEE Trans. Circuits Syst. I*, 56(9), 1908, 2009.) (b) Voltage-to-time converter. (From Konishi, T. et al., A 40-nm 640-μm² 45-dB opampless all-digital second-order MASH ΔΣ ADC, in *Proceedings of IEEE International Symposium on Circuits and Systems*, 2011, pp. 518–521.)

the dependence of i_c on v_c introduces unwanted distortion. Konishi et al. pointed out that the voltage of the drain of M3 and M4 in Figure 2.4a cannot be fully recovered to the supply voltage when $\phi = 0$, resulting in the variation of the propagation delay [18]. This is because when $\phi = 1$, although C is fully charged, the charge sharing between C and the parasitic capacitor at the drain of M3 and M4 will result in a dip of v_c immediately after ϕ becomes logic-1. To demonstrate this, let the capacitance at the drains of M3 and M4 be C_A. Further let the voltage of C and C_A be $v_C(0^-) = V_{DD}$ and $v_A(0^-)$, respectively, when $\phi = 0$. If we neglect the channel resistance of all transistors when they are in triode, then immediately after $\phi = 1$, C and C_A will have the same voltage, that is, $v_C(0^+) = v_A(0^+) = v(0^+)$. Using charge conservation, we have

$$(C + C_A)v(0^+) = CV_{DD} + C_A v_A(0^-),$$ (2.10)

from which we obtain

$$v(0^+) = \frac{CV_{DD} + C_A v_A(0^-)}{C + C_A}.$$ (2.11)

The instantaneous change of v_c immediately after $\phi = 1$ is evident in Equation 2.11. Equation 2.11 also reveals that if $C \gg C_A$, then the effect of charge sharing will be negligible. The modified VTC shown in Figure 2.4b eliminates this drawback.

2.2 VCDU VOLTAGE-TO-TIME CONVERTERS WITH SOURCE DEGENERATION

It is well understood that negative feedback improves linearity [19]. To improve the linearity of the preceding VCDU, negative feedback formed by the source degeneration shown in Figure 2.5a can be used [20]. Source degeneration is an effective and economical means to improve the linearity of MOS transconductors and is widely used in a broad range of applications such as radio-frequency down-conversion mixers for linearity improvement. The mechanism of source degeneration on improving the linearity of the MOSFET transconductor can be briefly depicted as follows: When v_{in} rises, v_{GS3} will rise accordingly. The increase of v_{GS3} will signify the nonlinear effect of $v_{GS3} \sim i_{DS3}$ relation as MOSFETs are inherently nonlinear transconductors and will exhibit a good linearity only when v_{GS} is small. To demonstrate this, we neglect the effect of channel length modulation and let $v_{GS} = V_{GS} + v_{gs}$ where V_{GS} and v_{gs} are the dc and ac components of v_{GS}, respectively. Notice that

$$i_{DS3} = k_{n3}\left(V_{GS3} + v_{gs3} - V_T\right)^2$$

$$= I_{DS3} + g_{m3}v_{gs3} + k_{n3}v_{gs3}^2, \qquad (2.12)$$

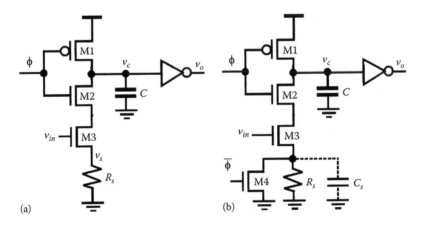

(a)

(b)

FIGURE 2.5 Voltage-to-time converter using voltage-controlled delay cells with source degeneration. (a) Without residual charge removal. (From Belloni, M. et al., A voltage-to-pulse converter for very high frequency DC-DC converters, *Proceedings of International Symposium on Power Electronics, Electrical Devices, Automation, and Motion*, 2008, pp. 789–791.) (b) With residual charge removal. (Agnes, A. et al., *Analog Integrated Circuits and Signal Process*, 54, 183, 2010.)

where

$$I_{DS3} = k_{n3}\left(V_{GS3} - V_T\right)^2 \tag{2.13}$$

is the dc current of M3, and

$$g_{m3} = 2k_{n3}\left(V_{GS3} - V_T\right) \tag{2.14}$$

is the transconductance of M3. It is seen from Equation 2.12 that the second-order term can be neglected if v_{GS3} is small. The increase of v_{GS3} will give rise to an increase in i_{DS3}, which will in turn boost the voltage of the source of M3. v_{GS3} will then drop due to the negative voltage feedback provided by the source degeneration resistor R_s. The reduced swing of v_{GS} will improve the linearity of $i_{DS3} \sim v_{GS3}$ characteristics. Like any negative feedback, source degeneration reduces the conversion gain. This is the price paid for linearity improvement.

In addition to linearity improvement, another added bonus of source degeneration is the improved bandwidth of the transconductor. In fact, source degeneration is a technique widely used to boost the bandwidth of continuous-time linear equalizers in multi-Gbps serial links [21].

To provide a quantitative comparison of the improvement of linearity obtained from source degeneration, we examine the current of the transconductor with and without source degeneration resistor R_s. To simplify analysis, we again neglect the effect of channel length modulation. When R_s is absent, we have

$$i_{DS3} = k_{n3}\left(v_{in} - V_T\right)^2. \tag{2.15}$$

When R_s is present, we have

$$i_{DS3} = k_{n3}\left(v_{in} - R_s i_{DS3} - V_T\right)^2, \tag{2.16}$$

Rearranging Equation 2.16 in the standard quadratic equation of i_{DS3}

$$k_{n3}R_s^2 i_{DS3}^2 - \left[1 + 2k_{n3}(v_{in} - V_T)\right]R_s i_{DS3} + k_{n3}(v_{in} - V_T)^2 = 0, \tag{2.17}$$

and solving Equation 2.17 for i_{DS3} yield

$$i_{DS3} = \frac{1}{2k_{n3}R_s^2}\left[1 + 2k_{n3}\left(v_{in} - V_T\right)\right]$$

$$\pm \sqrt{\left[1 + 2k_{n3}(v_{in} - V_T)\right]^2 - 4k_{n3}^2 R_s^2\left(v_{in} - V_T\right)^2}. \tag{2.18}$$

It becomes evident from Equation 2.18 that the second-order relation between i_{DS3} and v_{in} present in Equation 2.15 vanishes in Equation 2.18.

The residual charge of the parasitic capacitor C_s, which consists of the gate-source capacitance and source-substrate junction capacitance of M3, as well as the parasitic capacitance of R_s, exists at the end of the discharge phase. This is particularly true if the duration of the discharge phase, which is often constrained by sampling rate, is short and the time constant $R_s C_s$ is large. Note that a large R_s is desired in providing a strong negative feedback, and subsequently the better linearity of the transconductor. As a result, a large source voltage of M3 might exist at the onset of the next discharge phase. This initial voltage sets the lower bound of the gate voltage of M3 to

$$v_{in,\min} = v_{C_s}(0^-) + V_T \tag{2.19}$$

where $v_{C_s}(0^-)$ is the voltage of C_s at the beginning of the discharge phase. Also, the current of M3 is no longer solely determined by v_{in} but rather by both v_{in} and $v_{C_s}(0^-)$

$$i_{DS3} = k_{n3} \left[v_{in} - v_{CS}(0^-) - V_T \right]^2 (1 + \lambda v_{DS3}). \tag{2.20}$$

The residual charge of C_s can be removed by adding a reset transistor M4 in parallel with R_s, as shown in Figure 2.5b [22]. M4 is gated by ϕ. In the precharge phase where $\phi = 0$, C_s is fully reset and $v_s = 0$. In the discharge phase where $\phi = 1$, the reset operation is disabled so that M4 has no effect on the discharge of the load capacitor. Note that M4 needs to be sufficiently large in order to minimize its ON resistance so that the reset operation can be completed in the precharge phase, that is, $(R_s \parallel R_{on4}) C_s$ should be much smaller as compared with the duration of the precharge phase. A downside of large M4 is the increase of C_s.

2.3 RELAXATION VOLTAGE-TO-TIME CONVERTERS

The preceding VTCs precharge the load capacitor first and then drain the charge of the capacitor with the current controlled by the input voltage. Alternatively, voltage-to-time conversion can be performed by first charging the load capacitor with the input voltage and then discharging the capacitor with a constant current, as shown in Figure 2.6 [8,11,23,24]. We term these VTCs relaxation VTCs due to their resemblance to the operation of relaxation oscillators. Note that at the onset of the discharge phase, $v_c > V_{ref}$ must be satisfied. v_o is thus at logic-0 initially. The comparator will change its output when v_c drops below the reference voltage V_{ref}. The value of the resultant time variable T_{in}, which is defined from the time instant at which $\phi_1 = 0$ to the time instant at which v_c drops below V_{ref}, depends upon the input voltage v_{in}, the drain current I, and the reference voltage V_{ref}. For a given I in kth period, it can be shown that

$$T_{in}(kT) = \frac{C}{I} \left[v_{in}(kT) - V_{ref} \right]. \tag{2.21}$$

It is evident from Equation 2.21 that $T_{in}(kT)$ is directly proportional to the sampled value of the input voltage $v_{in}(kT)$.

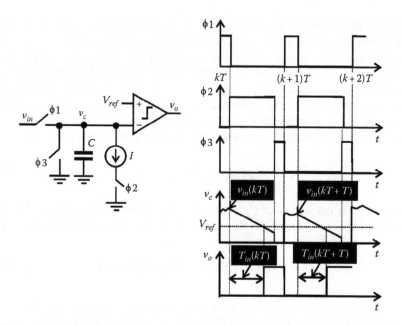

FIGURE 2.6 Relaxation voltage-to-time converter. (From Park, M. and Perrott, M., A single-slope 80 Ms/s ADC using two-step time-to-digital conversion, in *IEEE International Symposium on Circuits and Systems*, 2009, pp. 1125–1128; Min, Y. et al., A 5-bit 500-Ms/s time-domain flash ADC in 0.18 μm CMOS, in *Proceedings of IEEE International Symposium on Circuits and Systems*, 2011, pp. 336–339; Huang, H. and Sechen, C., A 22 mW 227 MSps 11b self-tuning ADC based on time-to-digital conversion, in *Proceedings of IEEE Circuits and Systems Workshop*, 2009, pp. 1–4; Mohamad, S. et al., A low power temperature sensor based on a voltage-to-time converter cell, in *Proceedings of IEEE International Conference of Microelectronics*, 2013, pp. 1–4.)

The residual charge of the load capacitor is removed in the reset phase where $\phi_3 = 1$. The removal of the residual charge of the sampling capacitor is important as it enables the establishment of an one-to-one mapping between v_{in} and v_c in each clock cycle. To demonstrate this, let the ON resistance of MOSFET switches be R_{on}. Assume that the residual charge of C at the assertion of the sampling phase where $\phi_1 = 1$ is $v_c(0^-)$. It is elementary to show that

$$V_c(s) = \frac{V_{in}(s)}{sR_{on}C+1} + \frac{R_{on}Cv_c(0^-)}{sR_{on}C+1}. \tag{2.22}$$

If we assume that the input is a unit step, that is, $v_{in} = u(t)$, it can be shown that

$$v_c(t) = v_c(0^-)e^{-t/(R_{on}C)} + \left[1 - e^{-t/(R_{on}C)}\right], \quad t \geq 0. \tag{2.23}$$

It is evident from Equation 2.23 that the effect of $v_c(0^-)$ will vanish only when $t \to \infty$. Since the period of the clock is finite, the residual charge of C in kth clock cycle will impact the voltage of the capacitor in $(k + 1)$th clock cycle. The removal of the residual charge of the sampling capacitor at the end of each clock cycle becomes vital to ensure the one-to-one mapping between v_{in} and v_c in each clock cycle.

The VCDU-based VTCs in Figure 2.1 perform voltage-to-time conversion in two steps: They first perform voltage-to-current conversion using a transconductor and then current-to-time conversion using a current-starved delay cell. Although the latter exhibits a perfect linearity ensured by both the constant discharge current and the linear load capacitor, the voltage-to-time conversion performed in the first step suffers from a poor linearity. The VTC in Figure 2.6 offers the advantage of a better linearity as the input voltage v_{in} is sampled by a linear capacitor C directly. No voltage-to-current conversion is performed on v_{in}. Another distinct characteristic of the relaxation VTC is that the gain of the VTC can be adjusted by varying I. For a given $v_{in}(kT)$, the larger the discharge current I, the smaller is the resultant time variable and the lower the conversion gain. On the other hand, the larger the discharge current I, the faster is the voltage-to-time conversion. The gain and conversion time of VCDU-based VTCs, however, cannot be adjusted.

2.4 REFERENCE VOLTAGE-TO-TIME CONVERTERS

The preceding VTC uses a fixed reference voltage with which input-dependent voltage v_c compares. Alternatively, one can perform voltage-to-time conversion by making the input voltage v_{in} the reference voltage and the voltage that compares with v_{in} comes from a ramping voltage generator with a constant slope, as shown in Figure 2.7 [25]. We term these VTCs reference VTCs, as the voltage to be converted is now the reference voltage with which the ramping voltage compares. Reference VTCs are also known as pulse-width-modulation generators as they map the amplitude of an input voltage to the width of a pulse linearly [26].

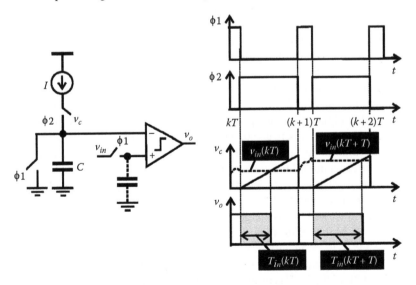

FIGURE 2.7 Voltage-to-time converter with input the reference voltage. (From Oh, T. and Hwang, I., *IEEE Trans. VLSI Syst.*, 2014.)

The input voltage v_{in} is fed to the noninverting terminal of the comparator. At the assertion of $\phi_1 = 1$, capacitor C is reset and v_{in} is sampled by the input capacitor of the comparator. The output voltage v_o is set to logic-1. When $\phi_2 = 1$, the capacitor is charged by a constant current I and v_c rises linearly with time. When v_c exceeds $v_{in}(kT)$, the output of the comparator will change from logic-1 to logic-0. The resultant time variable T_{in}, which is measured from the time instant at which $\phi_2 = 1$ is asserted to the time instant at which v_c exceeds $v_{in}(kT)$, is given by

$$T_{in}(kT) = \frac{C}{I} v_{in}(kT). \tag{2.24}$$

It is evident from Equation 2.24 that $T_{in}(kT)$ is directly proportional to the sampled input voltage $v_{in}(kT)$.

Reference VTCs possess the similar properties as those of relaxation VTCs in Figure 2.6, specifically, a good linearity as no voltage-to-current conversion is performed on v_{in} and a tunable conversion gain obtained by adjusting I. In addition, reference VTCs offer an infinite input impedance, an attractive property for many applications.

2.5   TIME-MODE COMPARATORS

Having investigated VTCs, in this section, we explore the applications of VTCs in time-mode comparators. It is well understood that voltage-mode comparators suffer from a number of drawbacks, in particular, the mismatch-induced offset voltage that deteriorates with technology scaling [27]. To improve the accuracy of voltage-mode comparators, digitally assisted calibration, which is typically power and area hungry, is needed. It was shown by Agnes et al. that voltage-mode comparators can also be implemented first by converting both the input and reference voltages to time variables using VTCs. The resultant time variables are then compared using a single-bit time quantizer, which accepts two time-mode inputs A and B and outputs a logic-1 if A leads B or a logic-0 otherwise. Figure 2.8 shows the schematic of the time-mode comparator proposed by Agnes et al. [22,28]. The waveforms of the critical nodes are sketched in Figure 2.9. The operation of the comparator is briefly depicted here: In the precharge phase where $\phi = 0$, both C_1 and C_2 are fully charged ($v_{c1}, v_{c2} = V_{DD}$). M6a and M6b, in the mean time, short the output nodes to the ground ($v_{o1}, v_{o2} = 0$). The parasitic capacitors at the nodes of R_{s1} and R_{s2} are reset by M4a and M4b, respectively. In the discharge phase where $\phi = 1$, C_1 and C_2 are discharged with their discharge currents controlled by v_{in} and v_{ref} respectively. Assume in kth discharge phase, $v_{in} < v_{ref}$, C_1 will be discharged slower than C_2 and $v_{c1} > v_{c2}$ will follow. When $v_{c2} < V_{DD} - V_T$, M5b will turn on. C_{out2} will be charged, and v_{o2} will arise. Similarly, when $v_{c1} < V_{DD} - V_T$, M5a will turn on. C_{out1} will be charged, and v_{o1} will arise. Since v_{o2} rises first, it will reach the threshold voltage $V_{T,inv}$ of the downstream inverter first. As a result, Clk of the DFF is reset first and $Q = 1$ is set when Clk = 0. Similarly one can show in $(k + 1)$th sampling period, Clk will be reset after $D = 0$ is set. The duration of Q is bordered by the time instants at which by Clk = 0.

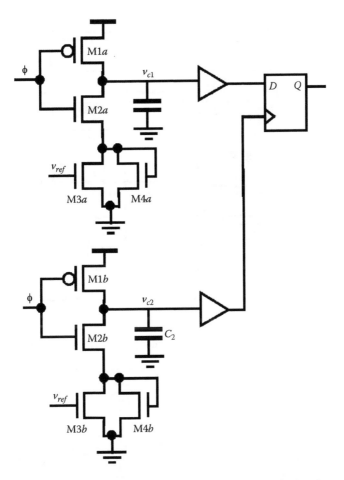

FIGURE 2.8 Time-mode comparator. (From Agnes, A. et al., *Analog Integr. Circuits Signal Process.*, 54, 183, 2010; Agnes, A. et al., A 9.4-ENOB 1V 3.8 μV 100 kS/s SAR ADC wit time domain comparator, in *IEEE International Solid-State Circuits Conference Digest of Technical Papers*, 2008, pp. 246–247.)

The performance of the preceding time-mode comparator is largely determined by the linearity of the VTCs. A large source degeneration resistor is often required in order to yield the desired linearity. As seen in Figure 2.9, when v_{in} and v_{ref} are close, the slope of v_{c1} will be close to that of v_{c2}. As a result, the time difference between the time instants at which Clk and D switch to logic-0 will be small. To avoid violating the metastate timing constraint of the DFF, the difference between the time instants at which Clk and D switch to logic-0 must not be overly small. This limits the accuracy of the comparator. Further, since mismatches between the two VCDUs exist, the accuracy of the comparator is inevitably affected by the mismatch-induced offset time.

The time-mode comparator proposed in [29] and shown in Figure 2.10 employs an automatic compensation mechanism to eliminate the mismatch-induced offset time of the comparator using transistors M4a and M4b, which are in parallel with

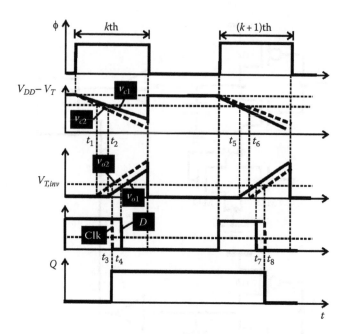

FIGURE 2.9   Waveforms of time-mode comparator proposed by Agnes et al. in Figure 2.8.

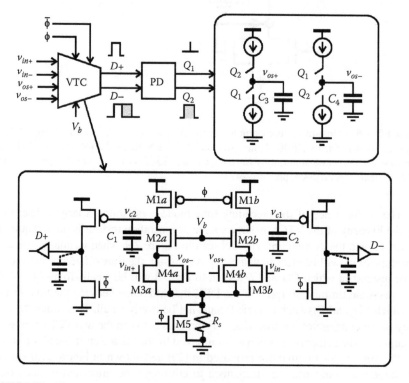

FIGURE 2.10   Time-mode comparator. (From Yang, S. et al., *J. Semicond.*, 32(3), 035002-1, 2011.)

the input transistors M3a and M3b. In addition, only one source degeneration resistor is used for both VCDUs. Further, to minimize power consumption, transistors M2a and M2b are employed to switch off the comparator once the output of the comparator assumes its logic state.

The elimination of mismatch-induced offset time is achieved by employing a phase detector and a charge pump that control the gate voltages of M4a and M4b, subsequently their currents. We briefly depict its operation here: Assume that due to mismatches, C_1 is discharged faster than C_2. The pulse width of $D+$ will be smaller as compared with that of $D-$. The phase detector will output a pulse Q_2 whose width is proportional to the difference between $D+$ and $D-$, while the pulse width of Q_1 will be zero. C_3 will then be charged and C_4 will be discharged. As a result, v_{os+} will rise, while v_{os-} will drop. The discharge current of C_2 will increase while that of C_1 will decrease, thereby eliminating the effect of the mismatches. The preceding offset time compensation technique bears a strong resemblance to the offset voltage compensation techniques for differential voltage-mode amplifiers.

Lee et al. proposed an elegant time-mode comparator shown in Figure 2.11 [30]. The comparator consists of two current-starved delay lines; each has 10 stages, 5 current-starved delay stages with a current source and 5 current-starved delay stages with a current sink. Upon the assertion of $\phi = 1$, $v_{1,3,5,7,9} = V_{DD}$ and $v_{2,4,6,8,10} = 0$. To simplify presentation, we assume that v_k, $k = 1, 2, \ldots, 10$, varies with time linearly. Further, we assume that all inverters have the same threshold voltage $V_{T,inv}$. Let the delay of the current-starved inverters with a current sink be τ_1 when the gating voltage of the current sink is V_{in} and that with a current source be τ_2 when the gating voltage of the current source is V_{in}, as shown in Figure 2.12.

When the gating voltage is changed from $v_{in} = V_{in}$ to $v_{in}^+ = V_{in} + \Delta v$, the delay of the current-starved inverters with a current sink will change from τ_1 to $\tau_1 - \Delta\tau_1$. Similarly, that with a current source will change from τ_2 to $\tau_2 + \Delta\tau_2$, as shown in Figures 2.13 and 2.14. Similarly, when the gating voltage is changed from $v_{in} = V_{in}$

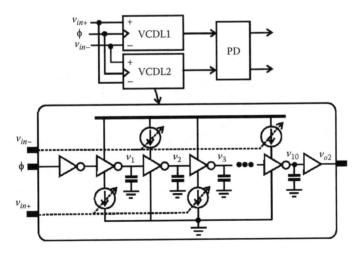

FIGURE 2.11 Time-mode comparator. (From Lee, S. et al., *IEEE J. Solid-State Circuits*, 46(3), 651, 2011.)

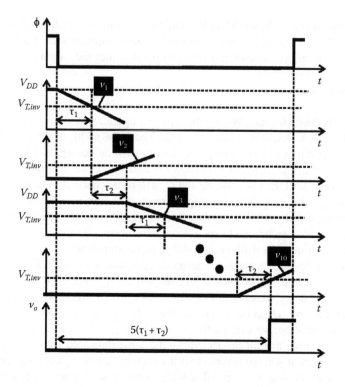

FIGURE 2.12 Waveforms of time-mode comparator proposed by Lee et al. shown in Figure 2.11.

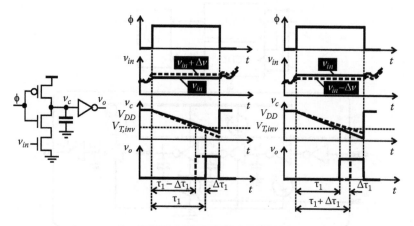

FIGURE 2.13 Voltage-controlled delay unit with a current sink.

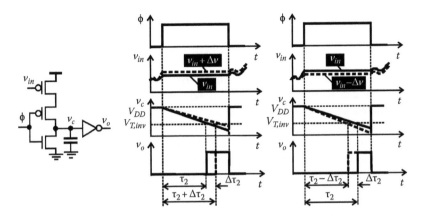

FIGURE 2.14   Voltage-controlled delay unit with a current source.

to $v_{in}^- = V_{in} - \Delta v$, the delay of the current-starved inverters with a current sink will change τ_1 to $\tau_1 + \Delta\tau_1$. Similarly, that with a current source will change from τ_2 to $\tau_2 - \Delta\tau_2$.

Consider VCDL1. Since v_{in}^+ is routed to the gate of current-sinking transistors and v_{in}^- is routed to the gate of current-sourcing transistors, we have the total delay

$$\tau_{VCDL1} = 5(\tau_1 + \tau_2) - 5(\Delta\tau_1 + \Delta\tau_2). \tag{2.25}$$

Similarly for VCDL2, as v_{in}^- is routed to the gate of current-sinking transistors and v_{in}^+ is routed to the gate of current-sourcing transistors, we have the total delay

$$\tau_{VCDL2} = 5(\tau_1 + \tau_2) + 5(\Delta\tau_1 + \Delta\tau_2). \tag{2.26}$$

The difference between τ_{VCDL1} and τ_{VCDL2} is given by

$$\Delta\tau_{VCDL} = \tau_{VCDL1} - \tau_{VCDL2} = -10(\Delta\tau_1 + \Delta\tau_2). \tag{2.27}$$

If we switch the polarity of the differential input, that is, v_{in}^+ is applied to the gate of the current-sourcing PMOS transistors and v_{in}^- is connected to the gate of the current-sinking NMOS transistors of VCDL1, we have

$$\tau_{VCDL1} = 5(\tau_1 + \tau_2) + 5(\Delta\tau_1 + \Delta\tau_2). \tag{2.28}$$

Similarly, when v_{in}^+ is applied to the gate of the current-sinking NMOS transistors and v_{in}^- is connected to the gate of the current-sourcing PMOS transistors of VCDL2, we have

$$\tau_{VCDL2} = 5(\tau_1 + \tau_2) - 5(\Delta\tau_1 + \Delta\tau_2). \tag{2.29}$$

We have

$$\Delta\tau_{VCDL} = 10(\Delta\tau_1 + \Delta\tau_2). \tag{2.30}$$

It becomes evident that the arrangement of the polarity of the inputs of the two VCDLs maps a differential voltage to a differential time variable. Also, cascading 10 voltage-controlled delay stages provides a time amplification of 10, allowing a better detection of whether $\Delta\tau_{VCDL} > 0$ or not or equivalently $v_{in} > 0$ or $v_{in} < 0$. The price paid for the improvement of time accuracy through time amplification is the loss of speed. It should also be emphasized that since v_{in}^+ and v_{in}^- are applied to both VCDLs, the effect of mismatches between the two VCDLs is minimized.

2.6 SUMMARY

Various design techniques for voltage-to-time conversion were investigated. We first showed that although voltage-to-time conversion can be realized using complex circuitry to achieve good performance such as a good linearity, operational amplifiers whose performance scales poorly with technology are often required. VTCs that utilize voltage-controlled current-starved delay units were investigated. We showed that these VTCs suffer from the drawback of a poor linearity, mainly due to the non-linear characteristics of $i_{DS} \sim v_{GS}$ of MOSFETs. Linearity can be improved to some extent using source degeneration feedback. This is at the expense of the reduced gain of VTCs. An added bonus of source degeneration is bandwidth improvement. Relaxation VTCs improve linearity by removing voltage-to-time conversion. These VTCs also offer the advantage of a tunable conversion gain, an attractive characteristic not possessed by VCDU-based VTCs. The residual charge at the end of discharge phase must be removed in order to ensure an one-to-one mapping between the input voltage and the voltage of the sampling capacitor. Reference VTCs possess the same properties as those of relaxation VTCs, specifically, a good linearity and a tunable conversion gain. In addition, they have an infinite input impedance, an attractive property for many applications. A direct application of VTCs is time-mode comparators implemented using two VTCs and a time-mode comparator. Mismatches between the VTCs give rise to the offset time that deteriorates the accuracy of time-mode comparators. Offset time compensation is therefore required. In addition to mismatches, the timing constraints of DFFs used for single-bit time quantizers also affect the accuracy of time-mode comparators. The use of time amplification in voltage-to-time conversion improves the resolution of time-mode comparators. This is at the expense of the speed of the comparators.

REFERENCES

1. J. Begueret, M. Benbrahim, Z. Li, F. Rodes, and J. Dom. Converters dedicated to long-term monitoring of strain gauge transducers. *IEEE Journal of Solid-State Circuits*, 32(3):349–356, March 1997.
2. X. Li and G. Meijer. A low-cost and accurate interface for four-electrode conductivity sensors. *IEEE Transactions on Instrumentation and Measurement*, 54(6):2433–2437, December 2005.

3. H. Yang and R. Sarpeshkar. A time-based energy-efficient analog-to-digital converter. *IEEE Journal of Solid-State Circuits*, 40(8):1590–1601, August 2005.
4. M. Maymandi-Nejad and M. Sachdev. A digitally programmable delay element: Design and analysis. *IEEE Transactions on VLSI Systems*, 11(5):871–878, October 2003.
5. M. Maymandi-Nejad and M. Sachdev. A digitally programmable delay element. *IEEE Journal of Solid-State Circuits*, 40(11):2212–2219, November 2005.
6. N. Mahapatra, S. Garimella, and A. Tareen. An empirical and analytical comparison of delay elements and a new delay element design. In *Proceedings of IEEE Workshop VLSI Circuits and Systems*, Orlando, FL, 2000, pp. 81–86.
7. H. Pekau, A. Yousif, and J. Haslett. A CMOS integrated linear voltage-to-pulse-delay-time converter for time-based analog-to-digital converters. In *Proceedings of IEEE International Symposium on Circuits and Systems*, Kos, Greece, 2006, pp. 2373–2376.
8. M. Park and M. Perrott. A single-slope 80 Ms/s ADC using two-step time-to-digital conversion. In *IEEE International Symposium on Circuits and Systems*, Taipai, Taiwan, 2009, pp. 1125–1128.
9. A. Macpherson, K. Townsend, and J. Haslett. A 5GS/s voltage-to-time converter in 90 nm CMOS. In *Proceedings of IEEE European Microwave Integrated Circuits Conference*, Rome, 2009, pp. 254–257.
10. Y. Min, A. Abdullah, H. Kim, and S. Kim. A 5-bit 500-Ms/s time-domain flash ADC in 0.18 μm CMOS. In *Proceedings of IEEE International Symposium on Circuits and Systems*, Rio de Janeiro, Brazil, 2011, pp. 336–339.
11. M. Elsayed, M. Abdul-Latif, and E. Sanchez-Sinencio. A spur-frequency-boosting PLL with a −74 dBc reference-spur suppression in 90 nm digital CMOS. *IEEE Journal of Solid-State Circuits*, 48(9):2104–2117, September 2013.
12. C. Ljuslin, J. Christiansen, A. Marchioro, and O. Klingsheim. An integrated 16-channel CMOS time to digital converter. *IEEE Transactions on Nuclear Science*, 41(4):1104–1108, August 1994.
13. P. Dudek, S. Szczepanski, and J. Hatfield. A high-resolution CMOS time-to-digital converter utilizing a vernier delay line. *IEEE Journal of Solid-State Circuits*, 35(2):240–247, February 2000.
14. A. Chan and G. Roberts. A jitter characterization system using a component-invariant vernier delay line. *IEEE Transactions on VLSI Systems*, 12(1):79–95, January 2004.
15. A. Macpherson, J. Haslett, and L. Belostotski. A 5 GS/s 4-bit time-based single-channel CMOS ADC for radio astronomy. In *Proceedings of IEEE Custom Integrated Circuit Conference*, San Jose, CA, 2013, pp. 1–4.
16. J. Christiansen. Low-jitter process-independent DLL and PLL based on self-biased techniques. *IEEE Journal of Solid-State Circuits*, 31(11):1723–1732, November 1996.
17. C. Taillefer and G. Roberts. Delta-sigma A/D converter via time-mode signal processing. *IEEE Transactions on Circuits and Systems I*, 56(9):1908–1920, September 2009.
18. T. Konishi, K. Okumo, S. Izumi, M. Yoshimoto, and H. Kawaguchi. A 40-nm 640-μm² 45-dB opampless all-digital second-order MASH ΔΣ ADC. In *Proceedings of IEEE International Symposium on Circuits and Systems*, Rio de Janeiro, Brazil, 2011, pp. 518–521.
19. P. Gray, P. Hust, S. Lewis, and R. Meyer. *Analysis and Design of Analog Integrated Circuits*, 4th edn. John Wiley & Sons, New York, 2001.
20. M. Belloni, E. Bonizzoni, and F. Maloberti. A voltage-to-pulse converter for very high frequency DC-DC converters. In *Proceedings of International Symposium on Power Electronics, Electrical Devices, Automation, and Motion*, Ischia, Italy, 2008, pp. 789–791.
21. F. Yuan, A. Al-Taee, A. Ye, and S. Sadr. Design techniques for decision feedback equalization for multi-Gbps serial data links. *IET Circuit Devices & Systems*, 8(2):118–130, 2014.

22. A. Agnes, E. Bonizzoni, P. Malcovati, and F. Maloberti. An ultra-low power succes-sive approximation A/D converter with time-domain comparator. *Analog Integrated Circuits and Signal Processing*, 54:183–190, 2010.

23. H. Huang and C. Sechen. A 22 mW 227 MSps 11b self-tuning ADC based on time-to-digital conversion. In *Proceedings of IEEE Circuits and Systems Workshop*, Richardson, TX, 2009, pp. 1–4.

24. S. Mohamad, F. Tang, A. Amira, and M. Benammar. A low power temperature sen-sor based on a voltage-to-time converter cell. In *Proceedings of IEEE International Conference of Microelectronics*, Beirut, 2013, pp. 1–4.

25. T. Oh and I. Hwang. A 110 nm CMOS 0.7V input transient-enhanced digital low-drop-out regular with 99.98% current efficiency at 80 mA load. *IEEE Transactions on VLSI Systems*, July 2015, 23(7):1281–1286.

26. M. Elsayed, V. Dhanasekaran, M. Gambhir, J. Silva-Martinez, and E. Pankratz. A 0.8 ps DNL time-to-digital converter with 250 MHz event rate in 65 nm CMOS for time-mode-based $\Sigma\Delta$ modulator. *IEEE Journal of Solid-State Circuits*, 46(9):2084–2098, September 2011.

27. S. Kobenge and H. Yang. A novel low power time-mode comparator for successive approximation register ADC. *IEICE Electronics Express*, 6(16):1156–1160, 2009.

28. A. Agnes, E. Bonizzoni, P. Malcovati, and F. Maloberti. A 9.4-ENOB 1V 3.8 µ V 100 kS/s SAR ADC wit time domain comparator. In *IEEE International Solid-State Circuits Conference Digest of Technical Papers*, 2008, San Francisco, CA, pp. 246–247.

29. S. Yang, H. Zhang, W. Fu, T. Yi, and Z. Hong. A low-power 12-bit 200-kS/s SAR ADC with a differential time domain comparator. *Journal of Semiconductor*, 32(3):035002-1–035002-6, March 2011.

30. S. Lee, S. Park, H. Park, and J. Sim. A 21 fJ/conversion-step 100 kS/s 10-bit ADC with a low-noise time-domain comparator for low-power sensor interface. *IEEE Journal of Solid-State Circuits*, 46(3):651–659, March 2011.

3 Fundamentals of Time-to-Digital Converters

Fei Yuan

CONTENTS

A time-to-digital converter (TDC) maps a time variable to a digital code. TDCs are perhaps the most important building blocks of time-mode circuits. Although the deployment of TDCs in particles and high-energy physics for time-of-flight measurement in nuclear science dates back to 1970s [1,2], they have found a broad spectrum of emerging applications such as digital storage oscillators [3,4], laser range finders [5], analog-to-digital converters (ADCs) [6–8], audio signal processing [9], medical imaging [10], positron emission tomography [10], IIR and FIR filters [11,12], anti-imaging filters [13], all digital frequency synthesizers [14–18], multi-Gbps serial links [19], programmable band/channel select filters for software-defined radio [20], and laser-scanner-based perception systems [21], to name a few. Various new architectures, novel design techniques, and their CMOS realization have emerged to improve the resolution, reduce the conversion time, increase the dynamic range, or lower the power consumption of TDCs. This chapter provides a comprehensive treatment of the principles, architectures, and design techniques of TDCs with an emphasis on the critical assessment of the advantages and limitations of each class of TDCs. The chapter is organized into the following: Section 3.1 provides the classification of TDCs. The key performance indicators of TDCs are depicted in Section 3.2. Section 3.3 investigates sampling TDCs where time variables are digitized directly. These TDCs include counter TDCs, delay line TDCs, TDCs with interpolation, vernier delay line TDCs, pulse-shrinking TDCs, pulse-stretching TDCs, successive approximation TDCs (SA-TDCs), flash TDCs, and pipelined TDCs. Noise-shaping TDCs that suppress in-band quantization noise are dealt with in Section 3.4. These TDCs include gated ring oscillator (GRO) TDCs, switched ring oscillator (SRO) TDCs, gated relaxation oscillator TDCs, MASH TDCs, $\Delta\Sigma$ TDCs, and their combinations. A summary of the chapter is given in Section 3.5.

3.1 CLASSIFICATION

TDCs can be loosely classified into sampling TDCs and noise-shaping TDCs. A sampling TDC digitizes a time variable using either a high-frequency low-jitter reference clock and counting the number of the cycles of the clock within the duration of the time variable or a delay line to count the number of the stages of the delay line that the front edge of the time variable propagates before the arrival of the rear edge of the time variable directly. The asynchronization of the reference clock with the front and rear edges of the time variable gives rise to quantization errors also known as quantization noise. Sampling TDCs can be further classified into single-shot TDCs and averaging TDCs. The former digitize a time variable using a single measurement, whereas the latter digitize a time variable using the average of the multiple measurements in order to minimize the effect of the random errors of measurement and achieve a better precision. The precision of averaging TDCs is inversely proportional to the square root of the averaged results of measurement [22]. Because no noise-shaping mechanism exists to suppress quantization noise, sampling TDCs in general exhibit a high level of quantization noise as compared with their noise-shaping counterparts. Noise-shaping TDCs, on the other hand, suppress the quantization noise of TDCs using system-level techniques such as $\Delta\Sigma$ operations that are capable of moving most of in-band quantization noise to higher frequencies outside the signal band so that the displaced excessive quantization noise can be removed effectively using a decimation low-pass filter in a postprocessing step, thereby achieving a large signal-to-noise ratio (SNR).

A large number of TDCs fall into the category of sampling TDCs. These TDCs include counter TDCs, delay line TDCs, pulse-shrinking TDCs, vernier delay line TDCs, pulse-stretching TDCs, flash TDCs, SA-TDCs, pipelined TDCs, and their combinations. Sampling TDCs have the common characteristic that they digitize time variables directly in an open-loop manner with no mechanism to suppress quantization noise. As a result, the resolution of these TDCs is lower-bound by quantization noise. Since there is a corresponding digital code generated by sampling TDCs for each time variable, a one-to-one mapping between a time input variable and a corresponding output digital code exists in sampling TDCs. Such an injective relation is critical to applications such as laser distance measurement. Another attractive characteristic of sampling TDCs is their short conversion time, attributive to their open-loop architectures. Soon we will be dealing with sampling TDCs that can achieve a high resolution either via employing long delay line interpolation or pre-amplifying the time variable to be digitized prior to digitization. These crude approaches, however, are echoed with a high level of power consumption.

Noise-shaping TDCs, on the other hand, suppress in-band noise using both oversampling that spreads in-band quantization noise over oversampling frequency band and noise-shaping that moves in-band quantization noise to frequencies outside the signal band. Although the in-band noise of noise-shaping TDCs is lower than quantization noise, one-to-one mapping between input time variables and their digital output codes is lost. Compared with sampling TDCs, noise-shaping TDCs offer the key advantage of a better SNR. A number of noise-shaping TDCs such as GRO TDCs and SRO TDCs emerged recently. In these TDCs, the time variable to be quantized is the gating signal of an oscillator, typically, a ring oscillator, of a constant frequency. When the gating signal is present, the oscillator is activated and the number of the oscillation cycles of the oscillator during the duration of the time variable and the state of the delay stages of the oscillator at the end of the time variable yield the digital representation of the time variable. The continuation of the nodal voltage at the output of the delay stages of the oscillator from one sample of the time variable to the next sample of the time variable gives rise to the first-order noise-shaping of quantization noise, a characteristic only inherent to GROs and SROs.

3.2   CHARACTERIZATION

The performance of TDCs is quantified by a number of parameters; among them, resolution, precision, linearity, voltage sensitivity, temperature sensitivity, conversion time, and conversion range are perhaps the most important [23,24]. In addition to these design specifications, some figure-of-merits (FOMs) that quantify the overall performance of TDCs are also widely used when comparing the performance of TDCs of different architectures or operating on different principles. We examine them in detail in this section.

3.2.1   RESOLUTION

The resolution of a TDC is the minimum time variable that the TDC can quantize. It is determined by the architecture of the TDC. For example, the resolution of a counter TDC is set by the period of the reference clock. The resolution of a generic delay line TDC is the per-stage propagation delay of the TDC. When interpolation is used, the resolution is set by the degree of interpolation. For noise-shaping TDCs, the resolution is set by the in-band quantization noise. The imperfection of TDCs, such as the nonlinearity of the delay line of delay line TDCs arising from the mismatch of the propagation delay of the delay stages of the TDCs, dominate the resolution. Many novel techniques such as pulse-shrinking, pulse-stretching, and vernier have emerged to improve the resolution of TDCs. We shall examine these techniques in detail in this chapter.

3.2.2 PRECISION

The precision of a TDC, often known as the single-shot precision, is quantified by the standard deviation of measurement errors Δ_1 and Δ_2 in Figure 3.1 when measuring a constant time interval. When a time variable specified by START and STOP signals is measured using a low-jitter reference clock and a counter, as shown in Figure 3.1, since the reference clock is asynchronous with START and STOP, single-shot measurement errors Δ_1 and Δ_2 exist and their value is uniformly distributed in $[-T_c, T_c]$ where T_c is the period of the clock. It can be shown that the precision of averaging TDCs is inversely proportional to the square root of the averaged results of measurement [22]. To improve precision, more measurement samples are needed.

3.2.3 LINEARITY

The nonlinearity of a TDC is the deviation of the time-to-digital transfer characteristic of the TDC from that of an ideal TDC. For delay line TDCs, it is caused by the difference between the delays of the delay stages of the TDC arising from the effect of process, voltage, and temperature (PVT) variation. The nonlinearity of a delay line TDC is usually quantified by differential nonlinearity (DNL), which depicts the mismatch of the delay of adjacent delay stages, and integral nonlinearity (INL), which quantifies the accumulative effect of the mismatch of the delay of the delay stages over the entire delay line. Since the resolution of the TDC is typically measured using least significant bit (LSB), both DNL and INL are quantified using

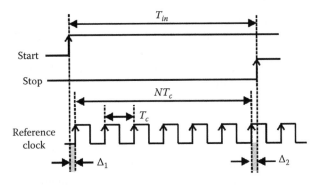

FIGURE 3.1 Counter TDCs.

LSB. Alternatively, the effect of the nonlinearity of the TDC can be depicted using Signal-to-noise plus distortion ratio (SNDR) computed over the Nyquist bandwidth of the input of the TDC. SNDR is obtained from the frequency response of the TDC by computing the ratio of the power of the signal to that of the noise and distortion tones over the Nyquist bandwidth. SNDR is widely favored over SNR when quantifying the performance of TDCs as the lower bound of the dynamic range of TDCs is often dictated by the in-band harmonic tones of the TDCs rather than the in-band quantization noise. For sampling TDCs, since no noise-shaping is present, DNL and INL are most widely used, while for noise-shaping TDCs, SNDR is preferred.

3.2.4 CONVERSION TIME

The conversion time of a TDC is the amount of the time that the TDC needs to complete the digitization of a time variable. An alternative measure of the conversion time is the throughput measured using the number of the samples that the TDC can digitize per second. For applications such as laser range measurement, conversion time is typically not of a critical concern. In fact, multiple measurements can be used in these applications to improve measurement precision. However, for applications such as $\Delta\Sigma$ modulators, conversion time directly affects the oversampling ratio and subsequently the performance of the modulators. Similarly, conversion time is also of a great importance for phase-locked loops (PLLs) with a TDC phase detector. This is because conversion time in this case directly affects the speed of the TDC phase detector and subsequently the loop dynamics of the PLLs.

3.2.5 DYNAMIC RANGE

The dynamic range of a TDC is the maximum range of the time variable that the TDC can digitize while satisfying performance specifications such as SNDR. For a generic delay line TDC, the lower bound is the per-stage delay τ of the delay line and the upper bound is the total time delay of the delay line given by $N\tau$ where N is the number of the delay stages of the TDC. Due to process spread, the linearity of the TDC deteriorates with increase in the number of the delay stages of the TDC. The value of N is determined by the allowed level of distortion. The minimum time variable that a sampling TDC can detect can be made much smaller than the per-stage delay using advanced techniques such as pulse-shrinking, vernier delay lines, and pulse-stretching so as to increase the dynamic range of the TDCs. This, however, is often at the cost of sacrificing other performances. For example, pulse-stretching TDCs trade conversion time for resolution, while vernier TDCs trade silicon and power consumption for resolution.

3.2.6 VOLTAGE SENSITIVITY

The voltage sensitivity of a delay line TDC quantifies the effect of supply voltage fluctuation on the time accuracy of the TDC. For a delay line TDC, it is typically obtained by varying the supply voltage by $\pm 10\%$ from its nominal value and measuring the resultant change of the propagation delay of the delay stage of the TDC.

For a pulse-shrinking TDC, it is measured by varying the supply voltage by ±10% from its nominal value and measuring the resultant change of the amount of the time shrinkage of the pulse-shrinking delay stage of the TDC. Because the propagation delay of a typical delay stage of TDCs is a strong function of the voltage of power and ground rails, voltage sensitivity is an important performance indicator of TDCs. As the number of the design techniques that can minimize the effect of supply voltage fluctuation and ground bouncing on the propagation delay of delay stages is rather limited, stringent constraints are usually imposed on the voltage of power and ground rails. To meet these constraints, low-dropout voltage regulators are typically mandatory in order to stabilize the supply voltage of TDCs. Further, delay-locked loops (DLLs) are widely used to minimize the effect of PVT on the propagation delay of delay lines.

3.2.7 TEMPERATURE SENSITIVITY

The temperature sensitivity of TDCs quantifies the effect of temperature variation on the time accuracy of the TDCs with a typical unit ps/°C. For a delay line TDC, it is obtained by varying temperature over the range −20°C to 80°C and measuring the resultant change of the time delay. Typical temperature sensitivity is 0.1–0.01 ps/°C. To minimize the effect of temperature variation on the propagation delay of delay line TDCs, DLL stabilization is widely used. For pulse-shrinking TDCs, delay cells with automatic temperature compensation also emerged [5]. Similar temperature compensation mechanisms were proposed for delay line TDCs [14,25].

3.2.8 POWER CONSUMPTION

Power consumption is not an explicit performance indicator of TDCs. It is, however, often a determining factor that ultimately affects the choice of the architecture, resolution, conversion time, and dynamic range of TDCs. Since most TDCs are digital circuits whose power consumption is dominated by their dynamic power consumption quantified by $P_d = \alpha f C_L V_{DD}^2$ where α is switching activity at the load capacitor, f is frequency, C_L is load capacitance, and V_{DD} is supply voltage, for a TDC with given f and V_{DD}, reducing C_L will effectively lower power consumption. As C_L typically consists of three components, namely, the output capacitance of the driving stage, the input capacitance of the driven stage, and the capacitance of the interconnect connecting the driving and driven stages, reducing device dimensions and shortening the interconnect will lower power consumption. The former, however, is echoed with worsening mismatch, which in turn, deteriorates the resolution of the TDC. Quite often, trade-offs between performance and power consumption are made.

3.3 SAMPLING TIME-TO-DIGITAL CONVERTERS

Sampling TDCs digitize time variables directly. Because no suppression of quantization noise exists, the resolution of sampling TDCs is lower-bounded by quantization noise. Noise-shaping TDCs whose in-band noise is much lower than quantization noise are investigated in Section 3.4.

3.3.1 ANALOG-BASED TDCs

Early time-to-digital conversion is performed by first converting a time variable to a voltage using a charge pump-based time-to-voltage converter (TVC), typically realized by charging or discharging a linear capacitor using a charge pump gated by the time variable and a voltage comparator that compares the voltage of the capacitor with a known reference voltage. The resultant voltage is then digitized using a conventional ADC [10,26]. The performance of analog-based TDCs is greatly affected by the linearity of the charge pump TVCs and the resolution of voltage-mode ADCs. In addition, the power consumption of these TDCs is high due to the need for ADCs. Technology scaling also has a detrimental impact on the performance of voltage-mode ADCs. As a result, the performance of analog-based TDCs scales poorly with technology.

3.3.2 COUNTER TDCs

A counter TDC quantizes a time variable T_{in} by counting the number of the cycles of a high-frequency low-jitter reference clock of a constant frequency within the duration of T_{in}, as shown in Figure 3.1. T_{in} is bordered temporally by the rising edge of START and that of STOP. The counter is enabled at the rising edge of START, synchronized with the reference clock, increments at the rising edge of the reference clock, and disabled at the rising edge of STOP. The random assertion of START and STOP results in quantization errors Δ_1 and Δ_2 that are uniformly distributed in $[-T_c, T_c]$. Note that $0 \le |\Delta_1|,|\Delta_2| \le T_c|$ and we assume that the clock is jitter-free.

Counter TDCs feature a large dynamic range that is lower-bound by the maximum quantization error T_c and upper-bound by the size of the counter only. The maximum input time variable is given by

$$T_{in} = NT_c + \Delta_1 + \Delta_2, \tag{3.1}$$

where N is the number of the cycles of the reference clock within the duration of T_{in}. In addition to a large dynamic range, counter TDCs also enjoy a good linearity as the linearity of these TDCs is only determined by the stability of the frequency of the reference clock [2]. When T_{in} is small, the timing jitter of the reference clock cannot be neglected. Both Δ_1 and Δ_2 will be affected by the jitter of the reference clock [27].

Increasing the frequency of the reference clock will increase the resolution of counter TDCs. This, however, is at the cost of more power consumption and deteriorating cross talk. Alternatively, the resolution of counter TDCs is increased if Δ_1 and Δ_2 can be further quantized with a smaller quantization error using interpolation; this is discussed in the later sections.

3.3.3 DELAY LINE TDCs

3.3.3.1 Principle

It was shown in Section 3.3.2 that the asynchronization of the reference clock with START and STOP gives rise to the quantization error of counter TDCs. The delay line TDC shown in Figure 3.2 where each delay stage has the same propagation

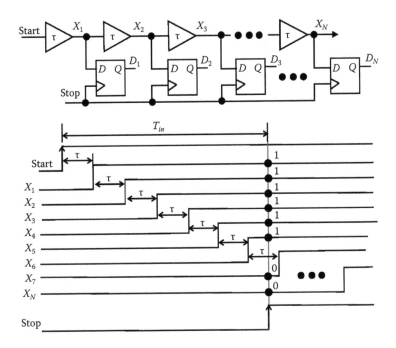

FIGURE 3.2 Delay line TDCs. The outputs of the delay stages X_1, X_2, ..., X_N are sampled simultaneously by the DFFs at the rising edge of STOP = 1. The result is a thermometer code $D_N...D_7D_6D_5D_4D_3D_2D_1 = 0...0111111$. (From Rahkonen, T. et al., Time interval measurements using integrated tapped CMOS delay lines, in *Proceedings of IEEE Mid-West Symposium on Circuits and Systems*, 1990, pp. 201–205; Aria, Y. et al., *IEEE Trans. Circuits Syst. II*, 27(3), 359, 1992; Rahkonen, T. and Kostamovaara, J., *IEEE J. Solid-State Circuits*, 28(8), 887, 1993.)

delay τ removes this drawback. In addition, it improves the resolution from the period T_c of the reference clock of counter TDCs to the per-stage delay τ of the delay line [28–30]. The operation of delay line TDCs is briefly depicted in the following: The signal START that marks the onset of time variable T_{in} propagates through the delay line, while the signal STOP that marks the end of the time variable enables the D flip-flops (DFFs) to sample the output of the delay stages at the rising edge of STOP. The resultant digital code formed by the output of the DFFs is a thermometer code and is the digital representation of T_{in}.

The conversion time of the delay line TDC in Figure 3.2 is $T_{in} + \tau_{DFF}$ where τ_{DFF} is the data-to-Q propagation delay of the DFFs. If a thermometer-to-binary converter is employed, the conversion time should also include the propagation delay of the converter. Note that the timing constraint of DFFs, specifically, the setup time and the hold time, must be satisfied in order to avoid entering metastable states, and has to be investigated shortly. If a DFF enters metastable states due to the violation of these time constraints, the amount of the time that the output of the DFF assumes a correct logic state could be significantly longer as compared with that of other DFFs that do not enter a metastable state. If the output of this DFF is read by the downstream logic before it assumes a correct logic state, an error might occur. This is one of the drawbacks of delay line TDCs, solely due to the asynchronization of STOP signal with the propagating front of START signal. In addition, although the per-stage delay of the delay line can be made small, the delay of the DFFs should be negligible as compared with the per-stage delay of the delay line. Fast DFFs are therefore critical.

3.3.3.2   Jitter and Mismatch in Delay Line TDCs

When a delay line TDC is perfectly linear, that is, the propagation delay of each stage of the delay line is identical and constant, the dynamic range of the delay line TDC is lower-bound by the per-stage delay τ of the delay line and upper-bound by the total delay of the delay line given by $N\tau$ where N is the number of the delay stages of the delay line. In reality, the jitter and mismatch of the delay stages give rise to the nonlinear behavior of the TDC. Jitter at the kth stage of the delay line consists of the contribution of the jitter of START signal that propagates from stage 1 to stage k of the delay line and that of the jitter added by stages 1, 2, ..., k. A number of sources contribute to jitter. They include switching noise, cross talk, and device noise such as thermal noise. Jitter due to switching noise and cross talk is often correlated, while that due to device noise is uncorrelated [31]. A key characteristic of jitter in a delay line is jitter accumulation, arising from the fact that any uncertainty occurring at an early transition of the delay line will affect all subsequent transitions of the delay line, as illustrated graphically in Figure 3.3. Clearly, jitter accumulation deteriorates with an increase in the number of the stages of the delay line. The worst-case jitter accumulation over the time interval ΔT is the sum of the jitter of all the transitions occurring within the interval. If we assume that the jitter of each transition is uncorrelated, then the standard deviation of the jitter at the end of ΔT interval, denoted by $\sigma\Delta_T$, is given by

$$\sigma_{\Delta T} = k\sqrt{\Delta T},\qquad(3.2)$$

where k is a circuit-specific constant. Equation 3.2 was experimentally verified in [32] and theoretically proven in [33]. It is seen from Equation 3.2 that the longer the delay line, the worse is the jitter.

The effect of mismatch among delay stages arising from process spread also manifests itself as the variation of the propagation delay of the delay stages. Similar to jitter accumulation, the longer the delay line, the worse is the effect of the mismatch on propagation delay. This is reflected by the deteriorating INL of delay line TDCs. As a result, the dynamic range of delay line TDCs does not scale linearly with the number of the delay stages. Delay line TDCs with an overly long delay line fail to provide the desirable dynamic range despite the consumption of an excessive amount of silicon and power.

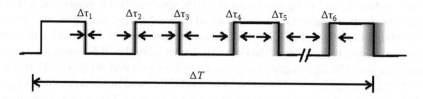

FIGURE 3.3   Jitter accumulation.

3.3.3.3 Delay-Locked-Loop Stabilized Delay Line TDCs

To minimize the undesirable effect of PVT, DLLs are widely employed to calibrate and stabilize the delay of delay stages, as shown in Figure 3.4 [30]. The DLL achieves a perfect phase alignment between START and X_N by adjusting the delay of the delay stages of the delay lines. Once the DLL is locked, we have $\tau = T_c/N$ where T_c is the period of START and N is the number of the delay stages of the TDC. It becomes evident that in the lock state, τ is independent of supply voltage fluctuation and temperature variation, and is only determined by the period of START and the number of the delay stages of the TDC. Note that since START and STOP mark the start and end of the input time variable T_{in}, the DLL is essentially locked to the input time variable to be digitized. Should the input not be a periodic signal with a constant frequency, the DLL will not be able to establish a lock state. As a result, the preceding TDC will not function properly. To get around this difficulty, delay line TDCs with DLLs locked to an external reference clock with a fixed frequency rather than the input also emerged [34]. In this case, the TDC is calibrated first by establishing a lock state in which the DLL is locked to the reference clock. Once the lock state is established, the control voltage of the delay stages of the DLL is kept unchanged. The input to be digitized is then fed to the delay line and quantized by the TDC.

3.3.3.4 Metastability in Delay Line TDCs

One critical design issue of delay line TDCs is that the sampling of the output of the delay stages by STOP might occur in the prohibited time zones, that is,

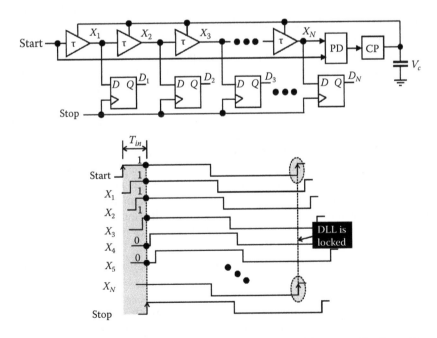

FIGURE 3.4 Delay line TDCs with DLL. The outputs of the delay stages $X_1, X_2, ..., X_N$ are sampled simultaneously by DFFs at the rising edge of STOP.

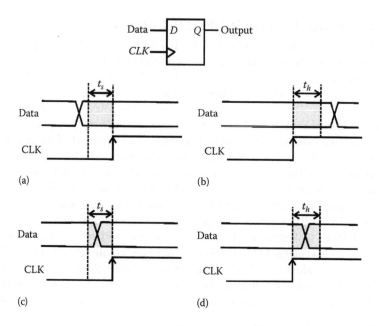

FIGURE 3.5   Setup and hold time of DFFs. (a) Sampling clock meets set-up time constraint. (b) Data meets hold-time constraint. (c) Sampling clock does not meet set-up time constraint. (d) Data does not meet hold-time constraint.

the setup time zone and hold time zone, of the DFFs due to the asynchronization nature of STOP and START signals, forcing the DFFs to enter metastable states where a long propagation delay exists and subsequently a long TDC conversion time [35]. In order for a DFF to perform properly, the input of the DFF must be stable at logic 1 or logic 0 for a minimum amount of time called *setup* time t_s before the clock of the DFF arrives. In addition, the input of the DFF must also remain unchanged for a minimum amount of time called *hold time* t_h after the arrival of the clock, as shown in Figure 3.5. Both times are needed for the DFF to establish its latch securely prior to any read-in or read-out operations. If the output of the DFFs has resolved to a valid state before the next register captures the data, the metastable output of the DFF will not negatively impact on the operation of the system. However, if the output of the DFF has not resolved to a valid logic state before the next register captures the data, the metastable output of the DFF might cause the system to fail, because the destination logic will in this case observe inconsistent logic states even though the input of this destination logic is identical [36].

A metastable state is one of three states that exist in a bistable circuit, as shown in Figure 3.6 where two cascaded inverters are used to represent the bistable circuit. States A and B are stable states. This is because in order to move away from these states, a large voltage disturbance is needed. For example, consider inverter 1, in state A, the PMOS transistor of the inverter is in triode and the NMOS transistor of the inverter is in the OFF state. To move out of state A, $v_1 > V_{IL}$ is needed. Unlike the stable states, in state M, both the PMOS and NMOS transistors of the inverter

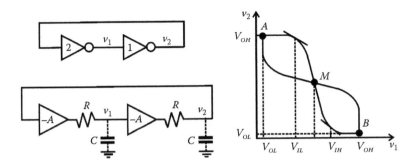

FIGURE 3.6   Metastable state of bistable circuits. States A and B are stable states, while state M is a metastable state.

are in saturation. Although in a perfectly matched and noise-free environment, if the circuit is in the metastable state, it will remain in the metastable state indefinitely, parametric imbalance and noise will cause the inverter to leave the metastable state [37,38]. Since the parametric imbalance and noise are small, a small-signal analysis can be used to quantify this process, as shown in Figure 3.6 where R represents the output resistance of the inverters and C is the load capacitance of the inverters [39,40]. It was shown in [41] that

$$v_1 = v_1(0^-)e^{\frac{A-1}{\tau}t}, \tag{3.3}$$

where

$\tau = RC$

$v_1(0^-)$ is the initial voltage of v_1, that is, the voltage at the metastable state

Equation 3.3 reveals that the regenerative mechanism formed by the positive feedback of the two cross-coupled linear amplifiers accelerates the departure from the metastable state when the small-signal loop gain exceeds unity. After departing from the metastable state, v_1 rises exponentially. This process will slow down when the voltage becomes sufficiently high such that some transistors of the inverters enter the triode mode and the loop gain drops below unity. This process will eventually terminate when the loop gain becomes sufficiently small [42,43].

If the timing constraint on the assertion of the sampling clock is violated, as shown in Figure 3.5, two unwanted results will occur: (1) The output of the DFF might not correspond to the input correctly and (2) even though the output of the DFFs corresponds to the input correctly, the DFFs might experience a long clock-to-Q delay, causing the following stage to either become metastable or fail [44], as illustrated in Figure 3.7, where the response of the DFF of Figure 3.8 is shown. It is seen that the DFF enters a metastable state from the assertion of the clock to the availability of the output. Not only the duration of the metastable state is exceedingly long, the logic state of the output cannot be predicted in advance and might not correspond to the input correctly.

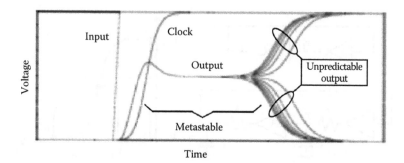

FIGURE 3.7   Response of the master–slave DFFs of Figure 3.8 with timing constraint violated. The rising edge of the input is swept with a small step to illustrate the response of the DFFs in the metastable state.

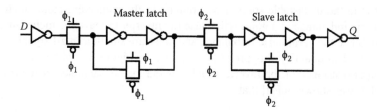

FIGURE 3.8   Master–slave DFFs.

Since the metastable state of DFFs is caused by the asynchronization of the input with the clock, to minimize the possibility that DFFs enter a metastable state, an asynchronous input can be synchronized with the clock by employing a synchronizer that consists of a sequence of DFFs clocked by the same clock. Figure 3.9a shows a synchronizer consisting of two DFFs. Although the first DFF, called synchronization register, might enter a metastable state, the second DFF, called sampling register, will never become metastable. This is because the period of the clock is typically much larger than $t_s + t_h$, as shown graphically in Figure 3.9b. Note that the input should remain unchanged for at least one clock period in order for the sampling register to obtain the correct state. Should the input change before the sampling register samples, the output of the sampling register will differ from that of the input even though no metastability will be encountered, as shown in Figure 3.9c.

3.3.3.5   D Flip-Flops

DFFs function as time samplers in delay line TDCs. DFFs can be implemented using either static or dynamic logic circuits. The former are known as static DFFs, whereas the latter are termed dynamic DFFs. As DFFs are used extensively in time-mode circuits, a short detour on reviewing the principle and implementation of DFFs from our coverage on delay line TDCs is well warranted.

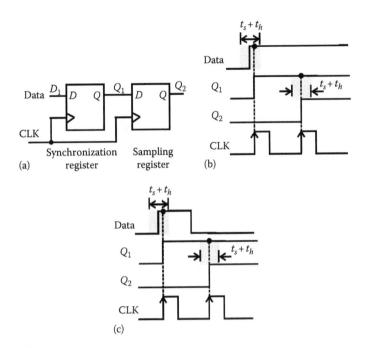

FIGURE 3.9 Synchronizer with two DFFs. (a) Configurations. (b and c) Waveforms.

Figure 3.10a and b show the implementation of static SR flip-flops using cross-coupled NOR2 gates and NAND2 gates, respectively. A static JK flip-flop can be constructed using a static SR flip-flop, as illustrated in Figure 3.10c. A static DFF can also be constructed using a static JK flip-flops, as shown in Figure 3.10d. These static flip-flops are implemented using standard logic cells and hence suffer from the drawbacks of a large number of transistors and subsequently a high level of dynamic power consumption. Figure 3.10e shows the implementation of a pseudo-static DFF. Its implementation is clearly simpler as compared with the static DFF of Figure 3.10d. Figure 3.10f shows the implementation of popular true single-phase clock (TSPC) logic implementation of positive edge-triggered dynamic DFF [45]. TSPC logic is known for its high-speed operation and low-power consumption. We briefly depict its operation here: Assume initially $D = 1$ and $\phi = 0$. C_1 is fully discharged and C_2 is fully charged. At the rising edge of $\phi = 1$, C_1 remains fully discharged while M4 turns off. The last stage becomes a static inverter and $\bar{Q} = 0$. Note that the output of this TSPC DFF is due to the charge dynamically withheld by C_2. Due to leakage, the charge of C_2 cannot be withheld indefinitely. The output of the TSPC DFF will only hold for a finite amount of time. This is one of the fundamental differences between static DFFs and dynamic DFFs. If D later becomes 0 while there is no change in the clock, C_1 will remain fully discharged and C_2 will also remain fully charged. Since there is no change in either v_{o1} or ϕ, Q will also remain unchanged. A negative edge-triggered TSPC DFF is shown in Figure 3.10g and its operation can be analyzed in a similar way.

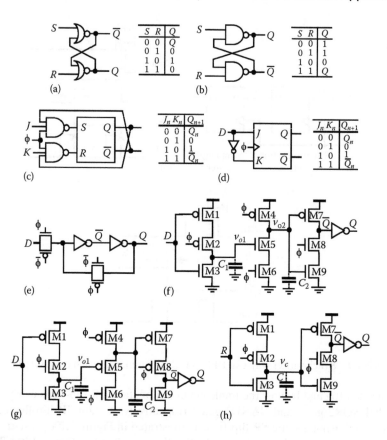

FIGURE 3.10 (a) Static SR flip-flop using cross-coupled NOR2 gates. (b) Static SR flip-flop using cross-coupled NAND2 gates. (c) Static JK flip-flop using an SR flip-flop. (d) Static DFFs using a JK flip-flop. (e) Pseudostatic DFFs. (f) Positive edge-triggered TSPC DFFs. (g) Negative edge-triggered dynamic TSPC DFFs. (h) Positive edge-triggered TSPC DFFs with reset and $D = 1$.

In many applications, $D = 1$ holds all the time. In this case, the preceding TSPC DFFs can be simplified. Figure 3.10h shows the schematic of a TSPC DFF with reset and $D = 1$ [46]. When a reset command is asserted, that is, $R = 1$, we have $v_c = 0$. This yields $Q = 0$. When the reset command is absent ($R = 0$) and $\phi = 0$, $v_c = V_{DD}$. When the rising edge of the clock arrives, that is, $\phi = 1$, v_c remains unchanged and $Q = 1$.

Although the preceding TSPC DFFs have been used in sampling time-mode variables, such as delay line TDCs, the absence of a regenerative mechanism in these DFFs is accompanied with a large set time before sampling and a large hold time after sampling. As a result, delay line TDCs with TSPC DFF samplers perform poorly as far as minimizing metastability is concerned. Samplers that utilize regeneration for speed enhancement are highly desirable. These samplers are also known as sense amplifiers, arbiters, or simply comparators. Figure 3.11a shows the schematic of the arbiter used in the coarse delay line TDC in [47]. The START signal is routed to terminal A, while STOP signal is connected to terminal B. The arbiter outputs a logic-1 when the rising edge of A leads that of B and 0 otherwise. The inclusion of a weak transistor M3 is beneficial as it balances the voltage at the drain of M1 and that of M2. Some designs do not include M3 [48]. Reset (M11 and M12 in a light color) that shunts the output to the ground can also be added, as shown in Figure 3.11a.

The operation of the arbiter is briefly depicted here: When $A = B = 0$, we have v_{o+} and v_{o-} pulled to V_{DD}. When $A = 1$ and $B = 0$, that is, $T_{in} = 1$, the cross-coupled inverter pairs formed by M4, M5, M8, and M10 are enabled and their outputs are

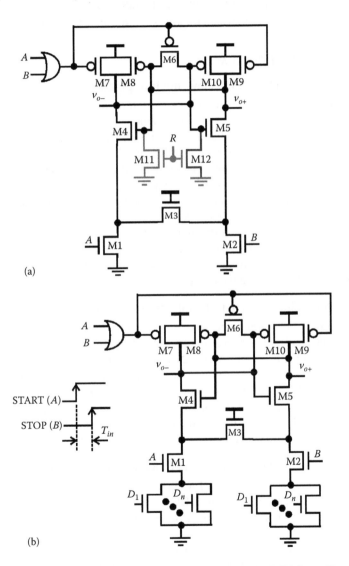

(a)

(b)

FIGURE 3.11 (a) Arbiters. (From Lee, M. and Abidi, A., *IEEE J. Solid-State Circuits*, 43(4), 769, 2008.) (b) Arbiters with digital input offset voltage compensation. (From Nose, K. et al., A 1 ps resolution jitter-measurement macro using interpolated jitter oversampling, in *IEEE International Solid-State Circuits Conference Digest of Technical Papers*, 2006, pp. 520–521.)

set by the logic states of A and B. The regeneration provided by the cross-coupled inverter pair allows a much faster state transition as compared with TSPC DFFs. The regeneration also allows the arbiter to capture a small time difference between START and STOP signals, that is, a small T_{in}, which cannot be picked up by TSPC DFFs. The regeneration also synchronizes v_o^+ and v_o^-.

Device mismatches give rise to an input offset voltage. To compensate for the effect of the input offset voltage, the source voltage of the input pair can be adjusted digitally, as shown in Figure 3.11b [49]. The digitally controlled source resistors also provide source degeneration, which improves the speed of the arbiter.

3.3.4 TDCs with Interpolation

3.3.4.1 Counter TDCs with Delay Line Interpolation

It was shown in Section 3.3.2 that the resolution of counter TDCs is lower-bound by the period of the sampling clock with quantization errors Δ_1 and Δ_2 bound by $0 \le |\Delta_1|, |\Delta_2| \le T_c$ and distributed uniformly in $[-T_c, T_c]$. To increase the resolution of counter TDCs, one needs to further digitize Δ_1 and Δ_2 with a lower level of quantization error. Since T_c is typically much larger than the per-stage delay τ of delay lines, Δ_1 and Δ_2 can be further digitized using delay line TDCs to improve resolution. This is known as interpolation. To improve the resolution of interpolation, although at the first glance that long delay lines are preferred, the linearity of delay line TDCs deteriorates with the increase in the number of the delay stages of the TDCs. With the deployment of delay line TDCs for interpolation, the dynamic range of counter TDCs is ideally improved from $T_c \le T_{in} \le NT_c$ without delay line interpolation to $\tau \le T_{in} \le NT_c$ with delay line interpolation.

3.3.4.2 Delay Line TDCs with Parallel Interpolation

The resolution of delay line interpolation can be improved using two-step interpolation without employing an overly long delay line. In the approach proposed in [50,51], the quantization errors Δ_1 and Δ_2 of the counter TDC in Figure 3.1 are first interpolated using two DLL-stabilized delay lines of M delay stages to improve the resolution to T_c/M. Each of the two consecutive stages of the delay lines is then interpolated using N-tap parallel interpolators to achieve a total of MN interpolation steps per reference clock period. The resolution is thus improved from T_c/M of one-step interpolation to $T_c/(MN)$ with only $(M + N)$ delay stages, rather than MN delay stages.

3.3.4.3 Delay Line TDCs with Vernier Delay Line Interpolation

Since the resolution of vernier delay line TDCs, to be studied shortly, is much higher as compared with that of delay line TDCs, vernier delay line TDCs are also used for interpolation to improve resolution. In References 35,52,53, the first-level interpolation is performed using multiphase sampling where the input time variable is sampled by a temporally spaced clock, while the second-level interpolation is carried out using vernier delay line TDCs. A similar approach was used in [52].

The resolution of delay line TDCs is the per-stage delay. Since one delay stage is typically realized using two cascaded static inverters, one per-stage delay is thus equal to two per-gate delays, that is, two propagation delay of static inverters.

The resolution of delay line TDCs can therefore be lowered to one gate delay using a pseudodifferential architecture where the output of each inverter is utilized [54].

3.3.4.4 Delay Line TDCs with Interstage Interpolation

To further improve resolution from per-stage delay to substage delay, interpolation between the rising edges of the output of adjacent delay stages can be utilized, as shown in Figure 3.12, where three interpolated outputs are generated from X_n and X_{n+1}. The three interpolated outputs $X_{n,k}$, $k = 1, 2, 3$, are obtained using linear interpolation.

$$X_{n,k}(t) = w_{n,k}X_n(t) + w_{n+1,k}X_{n+1}(t), \tag{3.4}$$

where

$w_{n,k}$ and $w_{n+1,k}$ are the weighting factors assigned to $X_n(t)$ and $X_{n+1}(t)$, respectively, for kth interpolated output

X_n and X_{n+1} are not Boolean quantities but rather analog signals that are displaced by τ temporally

3.3.4.5 Delay Line TDCs with Interstage Active Interpolation

The active interpolation approach proposed in [55] and shown in Figure 3.13 uses the weighted sum of the differential input voltages of v_1 and v_2 that are displaced temporally to generate the interpolated voltage that lies between v_1 and v_2, as shown in Figure 3.13. If $w = 1$, we have M1 = ON/M2 = OFF, v_o is determined by v_1 only. When $w = 0$, M1 = OFF/M2 = ON, v_o is determined solely by v_2. When $0 < w < 1$, both M1 and M2 are ON, the current of M1 and that of M2 depend upon the value of w. Assume the current of M1 and that of M2 are given by $I_{DS1} = Jw$ and $I_{DS2} = J(1 - w)$, respectively, where J is the tail current. Since

$$I_{DS3} = v_1 I_{DS1} = v_1 w J,$$

$$I_{DS4} = (1 - v_1)I_{DS1} = (1 - v_1)wJ,$$

$$I_{DS5} = v_2(1 - w)J,$$

$$I_{DS6} = (1 - v_2)(1 - w)J, \tag{3.5}$$

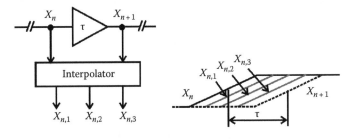

FIGURE 3.12 Interstage interpolation.

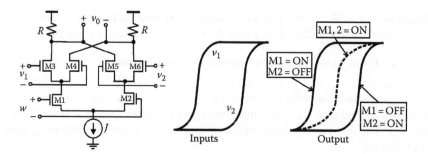

FIGURE 3.13 Interstage interpolation using the weighted sum of the voltages of adjacent stages. (From Knotts, T. et al., A 500 MHz time digitizer IC with 15.625 ps resolution, in *IEEE International Solid-State Circuits Conference on Digest of Technical Papers*, 1994, pp. 58–59.)

we have

$$v_o^+ = V_{DD} - R(I_{DS3} + I_{DS5})$$

$$= V_{DD} - R\left[v_1 w J + v_2(1-w)J\right], \tag{3.6}$$

$$v_o^- = V_{DD} - R(I_{DS4} + I_{DS6})$$

$$= V_{DD} - R\left[(1-v_1)wJ + (1-v_2)(1-w)J\right]. \tag{3.7}$$

As a result,

$$v_o = v_o^+ - v_o^-$$

$$= RJ\left[(1-2v_1)w + (1-w)(1-2v_2)\right]. \tag{3.8}$$

It is seen from Equation 3.8 that by tuning w, an interpolated output voltage v_o can be obtained.

3.3.4.6 Delay Line TDCs with Passive Interstage Interpolation

A drawback of the preceding active interpolation is its static power consumption. Interpolation can also be implemented using passive networks such as resistor networks shown in Figure 3.14 [56]. A sub-gate-delay resolution is accomplished by passive interpolation of two signals with the same switching direction but one inverter delay skew. It is seen that

$$v_{11+} = v_{2+} + \frac{3}{4}\left(v_{1+} - v_{2+}\right),$$

$$v_{12+} = v_{2+} + \frac{1}{2}\left(v_{1+} - v_{2+}\right),$$

$$v_{13+} = v_{2+} + \frac{1}{4}\left(v_{1+} - v_{2+}\right). \tag{3.9}$$

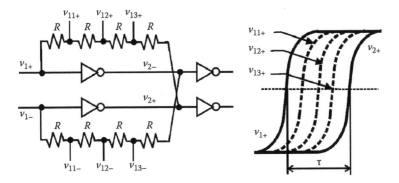

FIGURE 3.14 Interstage interpolation using resistor networks. (From Henzler, S. et al., 90 nm 4.7 ps-resolution 0.7-LSB single-shot precision and 19 pJ-per-shot local passive interpolation time-to-digital converter with on-chip characterization, in *IEEE International Solid-State Circuits Conference on Digest of Technical Papers*, 2008, pp. 548–635.)

The temporal displacement of v_{1+} and v_{2+} allows v_{11+}, v_{12+}, and v_{13+} to be spaced evenly in time, as illustrated graphically in Figure 3.14. Since the number of the level of interpolation using resistor networks is small, resolution improvement gained from resistor interpolation is rather limited.

3.3.4.7 Delay Line TDCs with Interstage Hierarchical Tree Interpolation

The phase interpolation method proposed in [57] and shown in Figure 3.15 uses hierarchical trees to increase time resolution without static power consumption. The phase interpolator is a generic differential-pair amplifier. The slopes of the outputs should be carefully controlled so that the generated interpolation points 2 and 3 will not be overlapped with points 1 and 4.

3.3.4.8 Delay line TDCs with Interstage DLL-Stabilized Delay Line TDC Interpolation

Another technique to achieve a subgate resolution is to use an array of identical DLL-stabilized delay line TDCs with per-stage delay τ_2 to interpolate the per-stage delay of the primary delay line with per-stage delay τ_1 with $\tau_1 \gg \tau_2$, as shown in

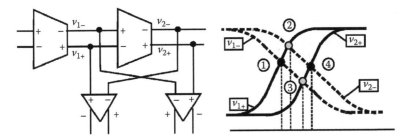

FIGURE 3.15 Interstage interpolation using hierarchical trees. (From Jang, T. et al., A highly-digital VCO-based analog-to-digital converter using phase interpolator and digital calibration, *IEEE Trans. VLSI Syst.*, 20(8), 1368, 2012.)

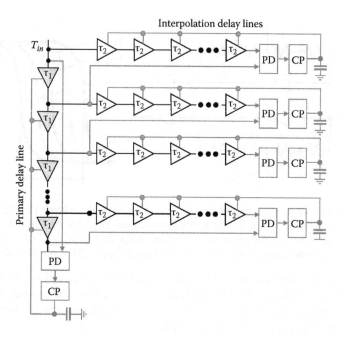

FIGURE 3.16   Interstage interpolation using DLLs. (From Christiansen, J., *IEEE J. Solid-State Circuits*, 31(7), 952, 1996.)

Figure 3.16 [58]. The time variable is measured by the primary delay line TDC. The gate delay of each delay stage of the primary delay line is interpolated by a set of identical secondary delay lines to obtain a subgate resolution. If the number of the delay stages of the primary delay line is M and that of the secondary delay lines is N, then since in the lock state of both the primary and secondary delay lines, $\tau_1 = N\tau_2$ and $T_c = M\tau_1$ hold where T_c is the period of the calibrating clock of the primary delay line, the ideal resolution of the TDC is therefore given by $T_c/(MN)$. DLL-interpolated TDCs are unattractive in terms of power and silicon consumption simply due to the need for multiple DLLs and corresponding readout circuits. They, however, provide a better immunity against PVT effect.

3.3.5   Vernier Delay Line TDCs

The resolution of delay line TDCs is lower-bounded by per-stage delay, which is set by available technology. Often TDCs with a substage-delay resolution are needed. The preceding interpolation approaches are undesirable due to their need for additional circuits in every stage. TDCs with a sub-per-stage delay resolution can be obtained using various system-level techniques. One of them is vernier technique widely used in precision length measurement. This section investigates vernier delay line TDCs.

3.3.5.1   Basic Vernier Delay Line TDCs

A vernier delay line TDC consists of two delay lines that have the same number of delay stages but slightly different per-stage propagation delays. Note that this

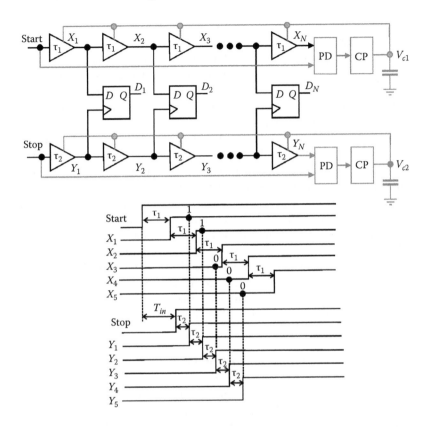

FIGURE 3.17 Vernier delay line TDCs. The output of kth delay stage X_k of START delay line is sampled by the output of kth delay stage Y_k of STOP delay line. In this example, the output is given by: $D_5D_4D_3D_2D_1 = 00011$.

difference between the per-stage propagation delay of the two delay lines is much smaller as compared with the per-stage delay of the two delay lines. It is obtained by making the dimensions of the delay stages of the two delay lines slightly different. START and STOP signals marking the start and end of an input time variable T_{in} propagate in each of the two delay lines with START leading STOP initially, as shown in Figure 3.17 [28,29,59–63]. Since $\tau_1 > \tau_2$, STOP signal propagating in STOP line, though launched late, will catch up START signal propagating in START line provided that the lines are sufficiently long and $\Delta\tau = \tau_1 - \tau_2$ is not overly small. The time instant at which a catch-up takes place is determined from

$$T_{catch} = N\tau_1 = N\tau_2 + T_{in}, \tag{3.10}$$

where N is the number of the stages that STOP signal propagates through before the catch-up takes place. For a given T_{in}, we have

$$N = \frac{T_{in}}{\tau_1 - \tau_2}. \tag{3.11}$$

Since $\tau_1 - \tau_2$ is very small, T_{in} must therefore also be small in order to have a manageable value of N.

The dynamic range of vernier delay line TDCs is upper-bound by $N(\tau_1 - \tau_2)$ and lower-bound by $\tau_1 - \tau_2$ theoretically. Similar to delay line TDCs, in order to minimize the effect of PVT on the delay of the delay lines, DLLs are normally employed to stabilize the time delay of both delay lines, as shown in Figure 3.17 (light color).

It is interesting to note that a vernier operation is essentially a fully differential operation that is capable of minimizing the effect of common-mode disturbances to the first order. The effect of supply voltage fluctuation and temperature variation on the propagation delay of the delay stages of vernier delay lines can indeed be treated as common-mode disturbances as they affect both delay lines simultaneously. As compared with delay line TDCs, vernier delay line TDCs improve resolution at the expense of more silicon and power consumption, and a reduced range.

Similar to delay line TDCs, increasing the length of the vernier delay lines worsens delay mismatches between delay stages. As a result, the dynamic range of vernier delay line TDCs does not scale linearly with the number of the delay stages of the delay lines. To improve the resolution, $\Delta\tau$ needs to be kept small. This, however, will not only require excessively long delay lines in order for STOP to catch up START before reaching the end of the line, the conversion time will also become prohibitively long.

3.3.5.2 Hierarchical Vernier Delay Line TDCs

Nose et al. showed that the conversion time of vernier delay line TDCs can be reduced while preserving the resolution by using the two-level hierarchical configuration shown in Figure 3.18 [49,64]. The architecture of hierarchical vernier delay line TDCs bears a strong resemblance to delay line TDCs with interpolation using an array of identical DLL-stabilized delay line TDCs with a smaller per-stage delay shown in Figure 3.16 [58]. A hierarchical vernier delay line TDC consists of a coarse vernier delay line TDC with a total of N stages and N fine vernier delay line TDCs. The difference between the per-stage delay of the slow and fast coarse delay lines is given by $\Delta\tau_c$, whereas that of the fine vernier delay lines is given by $\Delta\tau_f$ with $\Delta\tau_f \ll \Delta\tau_c$. The latency of the coarse vernier delay lines is $N\tau_c$. Since the digitization undertaken by the fine vernier delay line TDCs is carried out in parallel with that by the coarse vernier delay line TDC, the total latency of the hierarchical vernier delay line TDC is the same as that of the coarse vernier delay lines, that is, $N\tau_c$. If we assume that $\Delta\tau_c$ can be resolved by the fine vernier delay line TDCs with a total of M stages, $\Delta\tau_c = M\Delta\tau_f$ will hold. It should be noted that in addition to latency reduction, reducing the number of the stages of delay lines also lowers the effect of mismatches. The price paid for latency reduction is the increased silicon cost. Note that since the identical fine vernier delay line TDCs are used repeatedly, it is possible to only use one fine vernier delay line TDC to interpolate all stages of the coarse vernier delay line TDC in order to minimize silicon cost. Complex control logic that routes the input of the fine vernier delay line TDC to the output of each delay stage of the coarse vernier delay line TDC and the output of the fine vernier delay line TDC to the readout port, however, is inevitable.

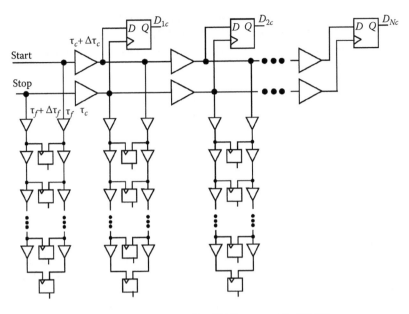

FIGURE 3.18   Hierarchical vernier delay line TDC proposed in [49,64].

3.3.5.3   Fractional Difference Conversion Vernier Delay Line TDCs

Xing et al. showed that the resolution of vernier delay line TDCs can be improved by using a fractional difference conversion mechanism [34,65,66]. Figure 3.19 shows the simplified schematic of the TDC utilizing this. The TDC employs two delay lines with their per-stage delays τ_1 and τ_2 set by two DLLs that have different number of

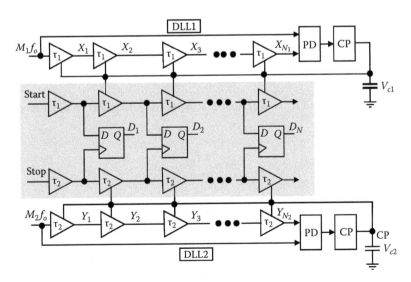

FIGURE 3.19   Fractional difference conversion vernier DLL-stabilized delay line TDC proposed in [34,65,66].

delay stages. The frequency of the inputs of the two DLLs also differs from each other and is given by $M_1 f_o$ and $M_2 f_o$ for DLL1 and DLL2, respectively, where f_o is the frequency of the reference clock and M_1 and M_2 are integers of different values. In the lock states of the two DLLs,

$$\tau_1 = \frac{1}{M_1 N_1 f_o} \tag{3.12}$$

and

$$\tau_2 = \frac{1}{M_2 N_2 f_o} \tag{3.13}$$

are established. The DLL-stabilized delay lines are mirrored to the measurement vernier delay lines that have the same number of the delay stages. In addition, the same control voltages are used for both the DLL-stabilized delay lines and the vernier delay lines so that the effect of PVT on the delay of the measurement vernier delay lines is minimized. The resolution of the TDC is given by

$$\Delta \tau = \tau_1 - \tau_2 = \frac{M_2 N_2 - M_1 N_1}{M_1 M_2 N_1 N_2} \frac{1}{f_o}. \tag{3.14}$$

If we choose $M_2 N_2 - M_1 N_1 = 1$, then

$$\Delta \tau = \frac{1}{M_1 M_2 N_1 N_2} \frac{1}{f_o}$$

$$= \frac{T_o}{M_1 M_2 N_1 N_2}. \tag{3.15}$$

If $M_1 = M_2 = 1$ and $N_1 \neq N_2$ (Note that $N_1 \neq N_2$ is required in order to ensure $\tau_1 \neq \tau_2$), we have

$$\Delta \tau = \frac{T_o}{N_1 N_2}. \tag{3.16}$$

Equation 3.16 is the resolution of conventional vernier delay line TDCs. A comparison of Equations 3.15 and 3.16 reveals that the resolution of the vernier delay line TDC with the fractional difference conversion mechanism will be $M_1 M_2$ times better as compared with that without the fractional difference conversion mechanism. The resolution $\Delta \tau$ can therefore be made sufficiently small by properly choosing the values of M_1 and M_2 without employing long delay lines, thereby minimizing silicon and power consumption and avoiding the performance degradation associated with long delay lines.

Also observed is that in order to achieve a better resolution, not only M_1 and M_2 should be made large, the difference between M_1 and M_2 should also be kept at the minimum.

3.3.5.4 Two-L Vernier Delay Line TDCs

Another way to increase the dynamic range of vernier delay line TDCs without employing overly long delay lines is to use a two-level configuration. A two-level vernier delay line TDC consists of a coarse vernier delay line TDC that has a large per-stage delay difference $\Delta \tau_c$ and a fine vernier delay line TDC that has a small per-stage delay $\Delta \tau_f$ with $\Delta \tau_f \ll \Delta \tau_c$ where the subscripts c and f specify coarse and fine vernier delay line TDCs, respectively [67]. The synchronization of the coarse and fine vernier delay line TDCs is a critical issue in ensuring the proper operation. A detailed investigation of this is available in [68].

3.3.5.5 Cyclic Vernier Delay Line TDCs

The dynamic range of two-level vernier delay line TDCs is still limited by the length of the delay lines. In order to further increase the dynamic range without employing overly long vernier delay lines, the cyclic configuration of vernier delay line TDCs emerged. Figure 3.20 shows the schematic of the cyclic vernier delay line TDC proposed by Chan and Roberts [69]. Note that the two delay lines of a conventional vernier delay line TDC are replaced with a slow cyclic delay line with a loop delay τ_s and a fast cyclic delay line with a loop delay τ_f. Since the same delay stage is used repetitively, the drawback associated with the process-spread-induced mismatch of delay stages existing in conventional vernier delay line TDCs vanishes. Cyclic delay lines are essentially ring oscillators. The delay of the AND2 gate and that of XOR2 gate should be made negligibly small as compared with τ_f. The slow cyclic loop is enabled at the assertion of START, signal while the fast cyclic loop is activated by STOP signal. Since START always leads STOP before STOP catches up START, the data input of DFF1 always leads the clock input of DFF1. As a result, Q_1 is always at logic-1, so is Q_2. The content of the counter thereby increments each time the cyclic loop completes one oscillation. When STOP catches up START, D_1 will lag CLK1. As a result, Q_1 will be at logic-0, disabling the counter. The content of the counter at this moment accurately records the number of the cycles that STOP has propagated through the delay cell since the assertion of STOP signal.

The preceding cyclic vernier delay line TDC is sensitive to PVT effect. Chen et al. showed that two PLLs that are locked to stable reference clocks can be deployed

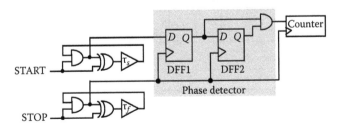

FIGURE 3.20 Cyclic vernier delay line TDC. (From Chan, A. and Roberts, G., *IEEE Trans. VLSI Syst.*, 12(1), 79, 2004.)

to minimize the effect of PVT on the delay of the delay stages [70]. Four ring oscillators, two for digitizing input time variable T_{in} bordered by START and STOP signals and the other two, are part of the PLLs that are used. The oscillators for digitizing T_{in} are the replicas of the PLL-stabilized oscillators so that the control voltages of the oscillators are stabilized by the reference clocks.

3.3.5.6 Cyclic Vernier Delay Line TDCs with Fractional Difference Conversion

Xing et al. further showed that vernier delay line TDCs with fractional difference conversion proposed in [34,65] and cyclic vernier delay line TDCs proposed in [70] can be combined to minimize PVT effect [66]. The cyclic vernier delay line TDC shown in Figure 3.21 employs two delay lines having N_1 and N_2 stages ($N_1 \neq N_2$) with their delays τ_1 and τ_2 set by two DLLs [66]

$$\tau_1 = \frac{1}{M_1 N_1 f_o} \tag{3.12}$$

and

$$\tau_2 = \frac{1}{M_2 N_2 f_o} \tag{3.13}$$

where M_1 and M_2 are integers. The delays of the two DLLs are mirrored to the delay cells in the cyclic TDC to minimize the effect of PVT on the delay of the delay cells. START and STOP signals are fed to two cyclic loops; each consists of a delay stage and a NAND2 gate. The loop delay of START loop and that of STOP loop are given by $\tau_1 + \tau_{NAND2}$ and $\tau_2 + \tau_{NAND2}$, respectively. The loops are enabled upon the arrival of the rising edge of START and STOP. Y_f samples X_f in each cycle

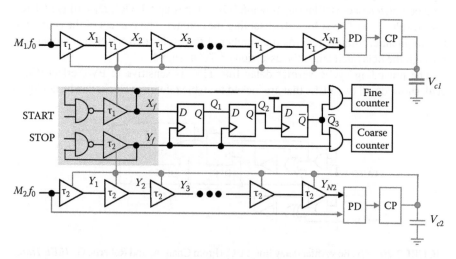

FIGURE 3.21 Cyclic vernier delay line TDC proposed in [66].

of STOP loop. A phase coincidence is detected by the phase coincidence detector (phase detector) when the sampled value changes from 1 to 0. Once this occurs, Y_f catches up X_f. Clearly the cyclic operation of the two delay loops removes the need for two long delay lines, thereby minimizing silicon consumption and mismatch-induced effect. Further, although τ_{NAND2} might be comparable to τ_1 and τ_2, since we are only concerned with the difference of the delay of START and STOP loops and assume both NAND2 gates have the same propagation delay, τ_{NAND2} plays no role in determination of the accuracy of the TDC as a vernier operation is essentially a fully differential operation.

3.3.5.7 Two-Dimensional Vernier Delay Line TDCs

As pointed out earlier, vernier delay line TDCs trade range, silicon, and power consumption for resolution. To increase conversion range, the number of the stages of Vernier delay line TDCs needs to grow exponentially with the number of the bits of the TDCs, signifying the effect of jitter and mismatch. Power and silicon consumption will also become prohibitively high. Liscidini et al. showed that the drawbacks of vernier delay line TDCs stem from the one-dimensional configuration and operation of these TDCs; specifically, not only the number of the delay stages of the fast delay line and that of the slow delay line of vernier delay line TDCs are the same, but the location at which the time comparison that determines whether a catch-up of START signal by STOP signal takes place or not is also the same [71–73]. Since a vernier delay line consists of two delay lines whose per-stage delays are slightly different, if the output of one delay stage of one of the delay lines is time-compared with that of each delay stage of the other delay line, rather than the output of the delay stage in the same location in the other delay line, a two-dimensional plane called vernier plane can be created. Figure 3.22 shows a 6×6 vernier plane with the initial time difference between START and STOP signals $\Delta\tau$ and the difference between the per-stage delay of START line and that of STOP line $\Delta\tau$. The jth horizontal axis represents the difference between the arrival time of START signal at the output of each delay stage of START line and that of the STOP signal at the output of the jth stage of the STOP line. Similarly, the ith vertical axis represents the difference of the arrival time of the STOP signal at the output of each delay stage of the STOP line and that of START signal at the output of the ith stage of START line. For example, consider $y = 0$ line. It consists of nodes $(0,0)$, $(1,0)$, $(2,0)$, ..., $(5,0)$. Since START signal leads STOP signal by $\Delta\tau$ initially and the per-stage delay of START line is $5\Delta\tau$, the time difference between the arrival time of START signal at the output of the first delay stage of START line to that of the STOP signal at the input of the first delay stage of the STOP line is $\Delta\tau + 5\Delta\tau = 6\Delta\tau$. Similarly, one can show that the time difference between the arrival time of START signal at the output of the kth delay stage of START line to that of the STOP signal at the input of the first delay stage of the STOP line is given by $(5k + 1)\Delta\tau$.

The number shown at node (x, y) of the vernier plane is the time difference between the arrival time of START signal and that of STOP signal at the node. For example, at node $(x, y) = (0, 0)$, we have $D(0, 0) = \Delta\tau$ where $D(x, y)$ denotes the time difference between the arrival time of START signal and that of STOP signal at node (x, y). Notice that at $(x, y) = (3, 4)$, we have $D(3, 4) = 0$, indicating that the output

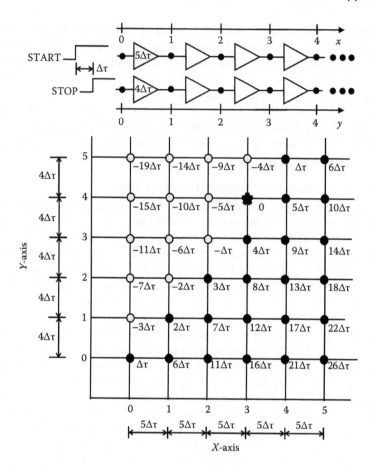

FIGURE 3.22   Two-dimensional vernier delay line TDC proposed in [71–73]. START signal leads STOP signal by $\Delta\tau$. The per-stage delay of START line is $5\Delta\tau$ and that of STOP line is $4\Delta\tau$.

of the fourth delay stage of the STOP line aligns with that of the third delay stage of START line, a catch-up therefore takes place at the node.

Now let us consider the case where the initial time difference between START and STOP signals is $5\Delta\tau$, as shown in Figure 3.23. It is seen from the figure that at (3, 5), a catch-up occurs. If we change the initial time difference between START and STOP signal from $5\Delta\tau$ to $10\Delta\tau$, as shown in Figure 3.24, a catch-up will take place at (2, 5). The preceding observations reveal that if time comparison between the output of every stage of the two delay lines is conducted, the time instant at which the STOP signal catches up START signal can be detected with the significantly reduced length of delay lines. Also observed is that the location at which a catch-up takes place is determined by the length of the delay lines, the difference between the per-stage delay of the delay lines, and the initial time difference between START and STOP signals. Further, the two delay lines do not need to have the same number of delay stages. Time comparison also does not need to be carried out at the same location of the two delay lines. Finally, since the length of both START and STOP

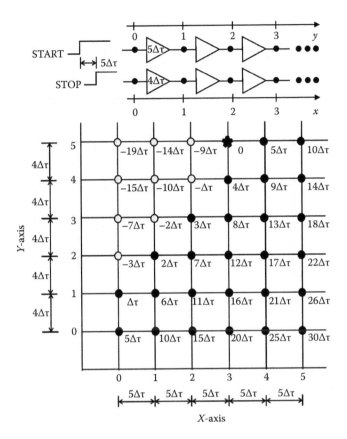

FIGURE 3.23 Vernier plane with START signal leading STOP signal by $5\Delta\tau$ initially. The per-stage delay of START and STOP lines are $5\Delta\tau$ and $4\Delta\tau$, respectively.

lines is much smaller using a two-dimensional configuration as compared with that using a one-dimensional configuration, the deleterious effect associated with jitter and mismatch that exists in long delay lines is reduced. The price paid for these gains is the increased number of time comparators and the complexity of layout.

To find out the number of time comparators needed, let the total of the delay stages of the two delay lines be M and N, as shown in Figure 3.25. It is seen that for each node of the STOP delay line, a total of M time comparators are needed. As a result, a total of $M \times N$ time comparators are needed in order to conduct time comparison between the output of each stage of the two delay lines. Since the number of time comparators connected to the output node of each delay stage is large, their loading effect must also be taken into consideration.

3.3.6 PULSE-STRETCHING TDCs

It was shown in Section 3.3.4 that the quantization errors Δ_1 or Δ_2 of counter TDCs can be further quantized using delay line–based interpolation to improve resolution from the period of the clock to per-stage delay of delay lines. Since START and

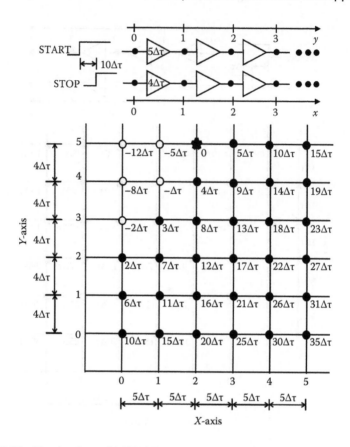

FIGURE 3.24   Vernier plane with START signal leading STOP signal by $10\Delta\tau$ initially. The per-stage delays of START and STOP lines are $5\Delta\tau$ and $4\Delta\tau$, respectively.

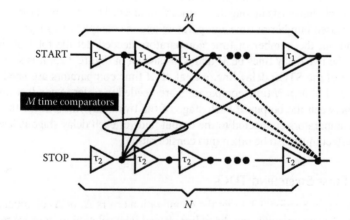

FIGURE 3.25   Two-dimensional vernier delay line TDCs.

STOP are not synchronized with the reference clock, Δ_1 or Δ_2 might be very small, although they are bounded by $0 \leq |\Delta_1|, |\Delta_2| \leq T_c|$. In this case, delay line–based interpolation will not be able to yield the desired resolution. To overcome this drawback, Δ_1 and Δ_2 can be first stretched temporally using a time stretcher also known as time amplifier. Once the resultant time variable is sufficiently large as compared with the per-stage delay of the delay line TDCs, it can then be digitized using delay line TDCs to yield the desired resolution. Clearly, the linearity of the time amplifier used for stretching is critical to ensure the overall linearity of the TDCs. The speed or bandwidth of the time amplifier is also vital if the TDCs are for high-speed applications.

3.3.6.1 Pulse-Stretching TDCs with Analog-to-Digital Conversion

Δ_1 and Δ_2 can be stretched using either analog stretching or digital stretching. The former uses a time-to-voltage converter to convert Δ_1 and Δ_2 to voltages first and then digitize the resultant voltages using ADCs [22,23,74–76]. Time-to-voltage conversion can be realized by discharging a precharged capacitor with a constant current for the duration of T_{in}. The voltage across the capacitor at the end of T_{in} is directly proportional to T_{in}. The resultant voltage is then digitized using an ADC, often a flash ADC in order to meet time constraints [10,26,75,77]. Analog stretching provides a good resolution but suffers from a poor temperature stability and high power consumption due to the use of ADCs. Further, since voltage-mode approaches are used, the performance of these TDCs inevitably scales poorly with technology.

3.3.6.2 Dual-Slope Pulse-Stretching TDCs

Digital stretching, on the other hand, stretches Δ_1 and Δ_2 to $k\Delta_1$ and $k\Delta_2$ with $k \gg 1$ using the dual-slope approach of Nutt [78] and then digitizes $k\Delta_1$ and $k\Delta_2$ using TDCs [77,79,80]. Many digital stretching methods emerged and we examine them in detail.

The dual-slope stretching method shown in Figures 3.26 and 3.27 asserts a reset (RST) command to precharge C_1 and $C_2 = MC_1$ to V_{DD} with $M \gg 1$. C_1 and C_2 are then discharged by constant currents $J_1 = NJ_2$ and J_2, respectively. v_o is set to logic-1 and will remain at logic-1 until $v_{c2} = v_{c1}$. Since v_{c2} drops slower while v_{c1} decreases

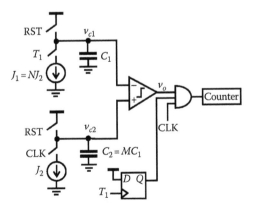

FIGURE 3.26 Dual-slope pulse stretching method. (From Park, K. and Park, J., *Rev. Sci. Instrum.*, 70(2), 1568, 1999; Kurko, B., *IEEE Trans. Nucl. Sci.*, NS-25(1), 75, 1978; Kuo, C. et al., *IET Proc. Circuits Devices Syst.*, 153(3), 247, 2006.)

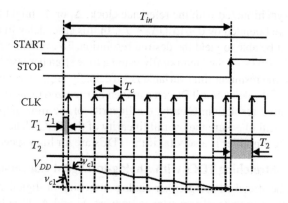

FIGURE 3.27 Waveforms of dual-slope pulse stretching method. (From Park, K. and Park, J., *Rev. Sci. Instrum.*, 70(2), 1568, 1999; Kurko, B., *IEEE Trans. Nucl. Sci.*, NS-25(1), 75, 1978; Kuo, C. et al., *IET Proc. Circuits Devices Syst.*, 153(3), 247, 2006.)

faster, it will take a total of k cycles of the reference clock to establish $v_{c1} = v_{c2}$. The number of the cycles is recorded by the counter whose content yields the digital code of the quantization error Δ_1. The same process is followed to quantize Δ_2. To determine k, from $\Delta v_{c1} = \Delta v_{c2}$ where Δv_{c1} and Δv_{c2} denote the voltage drop of C_1 and C_2 from V_{DD}, respectively, and noting that

$$\Delta v_{c1} = \frac{J_1}{C_1} T_1 \tag{3.17}$$

and

$$\Delta v_{c2} = k \frac{J_2}{C_2} T_c, \tag{3.18}$$

we arrive at

$$k = MN \left(\frac{T_1}{T_c} \right), \tag{3.19}$$

or equivalent

$$\hat{T}_1 = kT_c = (MN)T_1, \tag{3.20}$$

where $\hat{T}_1$ is the stretched version of T_1. It is evident that T_1 is stretched MN times. A notable advantage of the preceding dual-slope pulse stretching is the reduced effect of PVT as PVT affects the discharging operation of both capacitors equally and the effect can be modeled as a common-mode disturbance entering the comparator from the inverting and noninverting terminals of the comparator. Their effect is minimized by the differential operation of the comparator. One drawback of this approach is the speed penalty arising from the slow discharging of C_2. The need for a voltage comparator and two constant current sources also undermines its

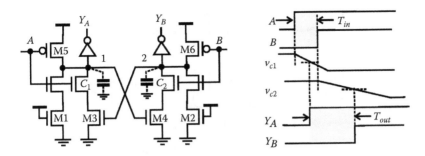

FIGURE 3.28 2× time amplifier. (From Lee, S. et al., A 1 GHz ADPLL with a 125 ps minimum-resolution sub-exponent TDC in 0.18 μm CMOS. IEEE *Journal of Solid-State Circuits*, 45(12):2827–2881, December 2010.)

compatibility with technology scaling as the performance of these analog blocks does not scale well with technology. Nevertheless, the preceding dual-slope stretching method is widely used, largely attributive to its intrinsic advantage of a low PVT sensitivity [4,77,81,82] and a large time stretch ratio.

A variant of the preceding dual-slope stretching method was proposed by Lee et al. and shown in Figure 3.28 [83]. The capacitors at nodes 1 and 2 are precharged prior to the arrival of inputs A and B. When A arrives while $B = 0$, C_1 will be discharged via two pull-down paths, one provided by M1 and the other provided by M3. When B arrives, since M4 is switched off due to the drop of v_1, only one pull-down path provided by M2 will exist to discharge C_2. v_2 therefore drops at approximately half the rate of that of v_1 provided that all transistors have the same dimension. The gain of the time amplifier is therefore approximately 2. A large gain can be obtained by cascading more stages.

Since the proper operation of the time amplifier requires that v_1 drops below the threshold voltage of M4 so that M4 can be switched off prior to the arrival of B, the application of this amplifier for a small T_{in} is rather difficult simply, because C_1 cannot be drained enough. Also, since no feedback is present, the gain of this time amplifier is subject to PVT effect. Further, the threshold voltage of the output inverters is also a strong function of PVT. As a result, precision time amplification cannot be obtained.

3.3.6.3 Delay-Locked Loop Pulse-Stretching TDCs

Rashidzadeh et al. showed that time amplification can also be achieved using DLLs, as shown in Figure 3.29, where a closed-loop approach is employed to amplify time while minimizing the effect of PVT [84]. A similar approach was used in [85]. We briefly depict its operation: Two inputs v_{in1} and v_{in2} of the same period are fed to two delay lines that have the same number of delay stages but different per-stage delays. The waveforms at nodes A and B are phase-aligned by the DLL. Because

$$\phi_A = \phi_{in1} + 2\pi \left(\frac{\tau_1}{T} \right),$$

$$\phi_B = \phi_{in2} + 2\pi \left(\frac{\tau_2}{T} \right), \tag{3.21}$$

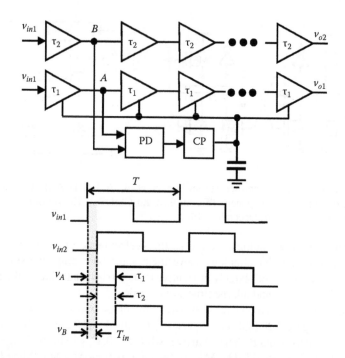

FIGURE 3.29   Time amplification using DLLs. (From Rashidzadeh, R. et al., *IEEE Trans. Instrum. Meas.*, 58(7), 2245, 2009.)

where ϕ_{in1} and ϕ_{in2} are the initial phases of v_{in1} and v_{in2}, respectively, in the lock state where $\phi_A = \phi_B$, we have

$$\phi_{in1} - \phi_{in2} = \frac{2\pi}{T}(\tau_2 - \tau_1). \tag{3.22}$$

Since the delay of the delay stage of delay line 1 (bottom) is identical and controlled by the DLL, and the overall propagation delay of delay line 1 is $N\tau_1$ while that of delay line 2 (top) is $N\tau_2$ where N is the number of the delay stages of delay lines 1 and 2, we have

$$T_{out} = N(\tau_1 - \tau_2). \tag{3.23}$$

Time amplification is evident in Equation 3.23.

We comment on this time amplifier: (1) It is a pure digital implementation. Its performance therefore scales well with technology. (2) Since the gain of the time amplifier is the number of the delay stages of the two delay lines, it is not tunable. (3) Two delay lines are needed for time amplification. It is costly in terms of silicon and power consumption. (4) The DLL must be locked in order to provide desired time amplification. This limits the use of this technique for high-speed applications. (5) No DLL is present for the second delay line. As a result, its performance is subject to the effect of PVT.

3.3.6.4   Nakura Pulse-Stretching TDCs

In References 86–88, a closed-loop time amplification scheme shown in Figure 3.30 was proposed. The two pulses whose time difference is to be amplified propagate in the opposite directions in two separate delay lines that have the same number of delay cells. The delay of the delay cells can be toggled between two different values τ_1 and τ_2 by control signal X, specifically, $\tau = \tau_2$ if $X = 1$. In this case, the transistors gated by X and $\bar{X}$ are OFF. The charging and discharging currents of the load capacitors of the two inverters are controlled by V_{b1} and V_{b2}, which set the propagation delay τ_2. When $X = 0$, the gated transistors are ON. In this case, the propagation delay is $\tau = \tau_1$, which is different from τ_2. The relation between the two delays is given by $\tau_2 = n\tau_1$ where $n > 1$ is an integer. The ratio of τ_1 to τ_2, that is, the value of n, is controlled by a DLL. The DLL also minimizes the effect of PVT.

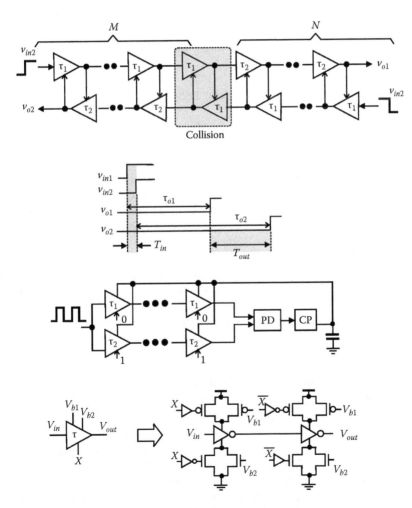

FIGURE 3.30   Time amplifier proposed by Nakura et al.

Initially when v_{in1} and v_{in2} are absent in the delay lines since the outputs of all delay stages are zero, the control voltage of all delay stages is set to 0. As a result, all delay stages have propagation delay τ_1. v_{in1} and v_{in2} are then applied to the delay lines and propagate in the delay lines. Before v_{in1} and v_{in2} meet, the delay of the delay states ahead of v_{in1} and v_{in2} is τ_1 as their control voltage is given by $X = 0$. When v_{in1} and v_{in2} collide, the delay of the cells ahead of v_{in1} and v_{in2} becomes τ_2 as the signals have already propagated through them and the control voltage of these delay stages has been set to $X = 1$. It can be shown that

$$\tau_{o1} = M\tau_1 + N\tau_2,$$
$$\tau_{o2} = N\tau_1 + M\tau_2, \tag{3.24}$$

where

τ_{o1} and τ_{o2} are the total propagation delay of the delay lines through which v_{in1} and v_{in2} propagate, respectively

M and N are the number of the delay stages that v_{in1} propagates through before and after colliding with v_{in2}

When the signals reach the end of the delay lines, we have

$$T_{out} = \tau_{o1} - \tau_{o2} = (M - N)(\tau_2 - \tau_1). \tag{3.25}$$

It becomes evident that if $M = N$, the gain of the amplifier will vanish. Also, the location of the signal collision depends upon the initial time difference T_{in}. If T_{in} is small, the collision location will be close to the center of the delay lines, resulting in a small time gain. This amplifier is clearly not suitable for amplifying small time variables. A large time gain can be provided by the amplifier only if (1) T_{in} is large, (2) $\tau_2 - \tau_1$ is large, and (3) the delay lines are sufficiently long.

Although the preceding time-stretching approaches are capable of providing time amplification using either different charging/discharging times of capacitors or the multiple duplications of delay times, they suffer from the common drawback that a large amount of time is needed for time amplification prior to digitization. As a result, they are only suitable for applications where conversion time is not a critical concern, and become less attractive for speed-constrained applications.

3.3.6.5 Regeneration Time Amplification TDCs

Abas et al. showed that the regenerative mechanism of SR latches can be used to rapidly amplify a small time variable, as shown in Figure 3.31 [89–91]. Since the underlined principle of this approach is the positive feedback provided by the regeneration of cross-coupled inverting amplifiers, time amplification using similar forms was also demonstrated [92,93]. In Figure 3.31, the cross-coupled NAND2 gates form a Set-Reset (SR) latch. It ensures that when A and B have a time displacement T_{in}, Q and $\bar{Q}$ will be set securely. The two inverters with their supply voltages provided by Q and $\bar{Q}$ ensure that when Q and $\bar{Q}$ are different, the outputs Y_A and Y_B will also be different, and when

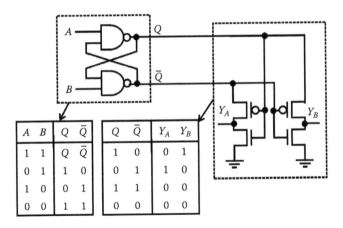

A	B	Q	$\bar{Q}$	Q	$\bar{Q}$	Y_A	Y_B
1	1	Q	$\bar{Q}$	1	0	0	1
0	1	1	0	0	1	1	0
1	0	0	1	1	1	0	0
0	0	1	1	0	0	0	0

FIGURE 3.31 RS-latch time amplifier. (From Kinniment, D. et al., *IEEE J. Solid-State Circuits*, 37(2), 202, 2002; Abas, A. et al., *IEE Electron. Lett.*, 38(23), 1437, 2002; Abas, A. et al., *IET Comput. Digit. Tech.*, 1(2), 77, 2002.)

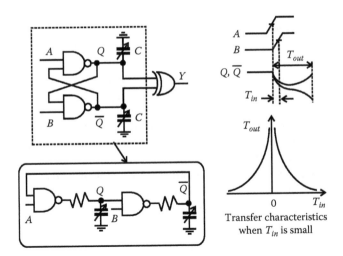

FIGURE 3.32 RS-latch time amplifier of Figure 3.31.

Q and $\bar{Q}$ are the same, the outputs Y_A and Y_B will both be at logic-0. They therefore function as a XOR2 gate. Figure 3.31 can therefore be redrawn in Figure 3.32. Since the regeneration process is controlled by time constant $\tau = RC$ where R is the output resistance of the NAND2 gate and C is the load capacitance of the NAND2 gate, by adjusting the value of C, one can control the regeneration process.

It is observed in Figure 3.32 that the RS latch time amplifier provides time amplification only when T_{in} is small, specifically when T_{in} is smaller than the regeneration time of the latch. When T_{in} is sufficiently large as compared with the regeneration time, $T_{out} \approx T_{in}$ and the gain of the time amplifier drops to unity.

A key advantage of the preceding SR latch time amplifier is its fast response and ability to amplify a small time variable. Since the gain of the time amplifier is set by

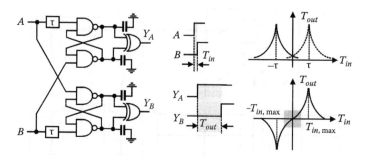

FIGURE 3.33 RS-latch time amplifier with an improved dynamic range. (From Lee, M. and Abidi, A., *IEEE J. Solid-State Circuits*, 43(4), 769, 2008.)

the regenerative characteristics of the latch, it is subject to the effect of PVT. Other drawbacks of these time amplifiers include a small input range and the poor linearity of the gain, arising from the regenerative operation where latching transistors go through all four modes of the operation of MOSFETs, namely, cutoff, subthreshold, triode, and saturation.

Lee and Abidi showed that the input range of the RS latch time amplifier in Figure 3.31 can be increased by inserting two delay units to generate unbalanced regeneration, as shown in Figure 3.33 [47]. The added time delay units shift the time transfer characteristic curves of the two RS latch time amplifiers. As a result, the input range can be greatly increased. Since the input time variable of these time amplifiers is typically small, the delay mismatch of the inserted buffers might be significant as compared with the time variable to be amplified, resulting in a nonnegligible error. This drawback can be removed by using unbalanced active charge pump loads proposed in [94].

SR latches in the preceding time amplifier can be implemented using the standard approaches given in Figure 3.34a [37] or Figure 3.34b [42]. It was shown in [47] that the gain of the SR latch time amplifier of Figure 3.32 is given by

$$A = \frac{2C}{g_m \tau},$$
(3.26)

where
  g_m is the transconductance of a NAND2 gate
  C is the load capacitance of the NAND2 gate
  τ is the input time offset

A large gain can be obtained by either increasing C, reducing g_m, or lowering τ. Since τ is directly related to the input range while C affects both the dynamic power consumption and speed of the time amplifier, lowering g_m becomes a natural choice to boost the gain without affecting power consumption and input range. Lowering g_m, however, has a detrimental impact on the strength of the feedback as the negative resistance provided by the cross-coupled transconductors of transconductance g_m

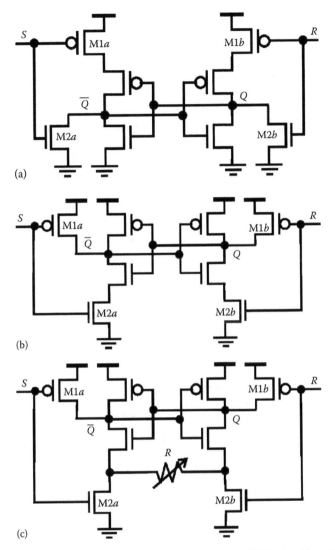

(a)

(b)

(c)

FIGURE 3.34 (a, b) SR latch with a fixed time constant. (c) SR latch with a variable time constant. (From Heo, M. et al., *IET Electron. Lett.*, 50(16), 1129, 2014.)

is given by $-2/g_m$. In [95], Heo et al. showed that when a resistor R is connected as shown in Figure 3.34c, the gain of the time amplifier becomes

$$A = \frac{C}{\left(\dfrac{g_m}{2} - \dfrac{1}{R} \right) \tau}. \tag{3.27}$$

Equation 3.27 shows that the addition of R effectively boosts the gain of the amplifier without degrading the strength of the feedback. Note that $R > 2/g_m$ is required to ensure positive feedback or regeneration.

3.3.7   PULSE-SHRINKING TDCs

3.3.7.1   Uniform Pulse-Shrinking TDCs

In the previous section, we showed that the resolution of TDCs can be improved by stretching time variables to be digitized first and then performing time-to-digital conversion on the stretched time variables. Rahkonen and Kostamovaara showed that the resolution of TDCs can also be made below per-stage delay if a delay line with a constant per-stage time delay is replaced with a pulse-shrinking delay line where the width of a propagating pulse in the line decreases stage-by-stage uniformly across the entire line, as shown in Figure 3.35a [30,96]. The pulse-shrinking delay stage consists of a current-starving inverter and a generic static inverter, as shown in Figure 3.35b. When a pulse of width T_{in} passes through the pulse-shrinking delay stage, its width is reduced due to the controlled discharge of the load capacitor. Note that the pulse is only etched away at its rising edge. The amount of pulse width shrinkage ΔT is set by both the discharge current J and the load capacitor C, and is given by

$$\Delta T = \left(V_{DD} - V_T\right)\frac{C}{J}. \tag{3.28}$$

It is seen from Equation 3.28 that the larger the discharge current J and the smaller the load capacitance C, the smaller is the amount of per-stage shrinkage. The amount of per-stage shrinkage needs to be precisely set prior to any measurement, that is,

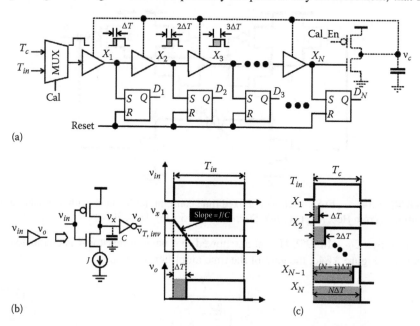

(a)

(b)   (c)

FIGURE 3.35 Pulse-shrinking TDC proposed by Rahkonen and Kostamovaara. (From Rahkonen, T. and Kostamovaara, J., *IEEE J. Solid-State Circuits*, 28(8), 887, 1993; Raisanen-Ruotsalainen, E. et al., *IEEE J. Solid-State Circuits*, 30(9), 984, 1995.) (a) DLL-stabilized uniform pulse-shrinking TDC. (b) Current-starved pulse-shrinking cell. (c) Calibration.

the TDC needs to be calibrated first. The calibration circuitry of the TDC shown in dotted lines that sets the amount of per-stage shrinkage is a DLL. To calibrate the TDC, calibration command Cal_{En} is asserted. This leads to $v_c = V_{DD}$. An external low-jitter calibration clock of known period T_c is routed to the delay line via the multiplexer by calibration command Cal. The delay of the pulse-shrinking stages is adjusted by varying the control voltage v_c set by the DLL until the pulse propagating in the line just disappears at the output of the last stage, that is, $X_N = 0$, as shown in Figure 3.35c. Once this occurs, $T_c = N\Delta T$ is established and the calibrated per-stage shrinkage is $\Delta T = T_c/N$. Note ΔT is the minimum time variable that the TDC can detect, that is, the resolution of the TDC. Clearly, if N is sufficiently large, ΔT can be made much smaller than the per-stage delay of delay line TDCs. Pulse-shrinking TDCs can therefore provide a resolution well below the per-stage delay of generic delay line TDCs.

The conversion range of the pulse-shrinking TDC is given by $\Delta T \leq T_{in} \leq N\Delta T$ or $\Delta T \leq T_{in} \leq T_c$. Although a large N is preferred from a better resolution point of view, the effect of jitter and mismatch intensifies with the increase in the number of the stages of pulse-shrinking TDCs, deteriorating the linearity of pulse-shrinking TDCs in a similar way as that in delay line TDCs. The resolution of pulse-shrinking TDCs therefore does not scale linearly with the number of the pulse-shrinking stages. As the control voltage is held by the capacitor of the DLL, it needs to be refreshed periodically. Otherwise v_c will drift away from its calibration value [97].

3.3.7.2 Cyclic Pulse-Shrinking TDCs

To increase the resolution of pulse-shrinking TDCs without employing a large number of pulse-shrinking stages, cyclic pulse-shrinking TDCs proposed by Chen et al. and shown in Figure 3.36 can be used [98,99]. The cyclic pulse-shrinking TDC consists of a delay line with stages $1, 2, \ldots, i - 1, i + 1, \ldots$ having the same dimension (homogeneous stages) and stage i having different dimensions (inhomogeneous or skewed stage), a control logic block, and a counter. The control logic is designed in such a way that when a pulse of width T_{in} is applied to the loop, the pulse will continue to circulate the loop until its width reduces to zero. The inhomogeneity of ith delay stage gives rise to a reduction in the width of the propagating pulse every time

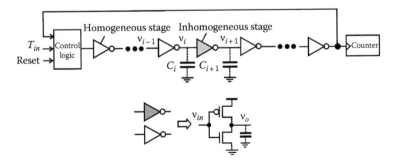

FIGURE 3.36 Cyclic pulse shrinking TDCs. (From Chen, P. and Liu, S., A cyclic CMOS time-to-digital converter with deep sub-nanosecond resolution, in *Proceedings of IEEE Custom Integrated Circuits Conference*, 1999, pp. 605–608; Chen, P. et al., *IEEE Trans. Circuits Syst. II*, 47(9), 954, 2000.)

it completes a round trip. A counter is used to record the number of the round trips that the pulse completes before it diminishes. Since the number of the round trips that the pulse completes is directly proportional to T_{in}, when the pulse vanishes, the content of the counter yields the digital representation of T_{in}.

To quantify the amount of the shrinkage of the pulse width at the interfaces between the inhomogeneous stage i and its neighboring homogeneous stages $i - 1$ and $i + 1$, we assume that the input to each delay stage is a step input. The high-to-low and low-to-high propagation delays of the output voltage of $(i - 1)$th stage, denoted by $\tau_{PHL,i-1}$ and $\tau_{PLH,i-1}$, respectively, can be obtained using approaches given in standard textbooks on digital integrated circuits, such as [100], and the results are given here

$$\tau_{PHL,i-1} = \frac{2C_iV_T}{k_{n,i-1}\left(V_{DD} - V_T\right)^2} + \frac{C_i}{k_{n,i-1}\left(V_{DD} - V_T\right)} \ln\left(\frac{3V_{DD} - 4V_T}{V_{DD}}\right), \qquad (3.29)$$

and

$$\tau_{PLH,i-1} = \frac{2C_iV_T}{k_{p,i-1}(V_{DD} - V_T)^2} + \frac{C_i}{k_{p,i-1}(V_{DD} - V_T)} \ln\left(\frac{3V_{DD} - 4V_T}{V_{DD}}\right). \qquad (3.30)$$

where we have assumed that NMOS and PMOS transistors have the same threshold voltage, that is, $V_{Tn} = |V_{Tp}| = V_T$, $k_{n,i-1} = \frac{1}{2}\mu_nC_{ox}(W/L)_{i-1}$, and $k_{p,i-1} = \frac{1}{2}\mu_pC_{ox}(W/L)_{i-1}$. In order to have the same rising and falling profiles, $k_{n,i-1} = k_{p,i-1}$ is required. This leads to the well-known design constraint on static CMOS inverters

$$\frac{(W/L)_n}{(W/L)_p} = \frac{\mu_p}{\mu_n}. \qquad (3.31)$$

It is seen from Figure 3.37 that when Equation 3.31 is satisfied, there is no reduction in the width of the pulse when the pulse passes through the delay stage. Homogeneous delay stages are designed in this way. Therefore, no reduction in pulse width occurs when a pulse passes through homogeneous stages.

If we purposely make $k_{n,i-1} \neq k_{p,i-1}$, then $\tau_{PHL,i-1} \neq \tau_{PLH,i-1}$. For example, if we make $\tau_{PLH,i-1} > \tau_{PHL,i-1}$, as shown in Figure 3.37, the width of the pulse at the output of the delay stage will be smaller as compared with that at the input of the delay stage due to unequal rising and falling slopes. The amount of the reduction in the width of the propagating pulse from $(i - 1)$th stage to ith stage is obtained from

$$\Delta\tau_{i-1} = \tau_{PLH,i-1} - \tau_{PHL,i-1}$$

$$= \alpha C_i\left(\frac{1}{k_{p,i-1}} - \frac{1}{k_{n,i-1}}\right), \qquad (3.32)$$

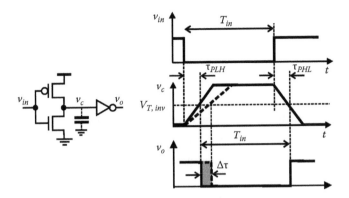

FIGURE 3.37 Pulse shrinking of cyclic pulse shrinking TDCs. (From Chen, P. and Liu, S., A cyclic CMOS time-to-digital converter with deep sub-nanosecond resolution, in *Proceedings of IEEE Custom Integrated Circuits Conference*, 1999, pp. 605–608; Chen, P. et al., *IEEE Trans. Circuits Syst. II*, 47(9), 954, 2000.)

where

$$\alpha = \frac{2V_T}{\left(V_{DD} - V_T\right)^2} + \frac{1}{V_{DD} - V_T} \ln\left(\frac{3V_{DD} - 4V_T}{V_{DD}}\right) \tag{3.33}$$

is a constant. Similarly, one can show that the amount of the shrinkage of the width of the propagating pulse from ith stage to $(i + 1)$th stage is given by

$$\Delta\tau_i = -\alpha C_{i+1}\left(\frac{1}{k_{p,i}} - \frac{1}{k_{n,i}}\right). \tag{3.34}$$

The total amount of the pulse shrinkage in one cycle is obtained by summing up $\Delta\tau_{i-1}$ and $\Delta\tau_i$ in Equation 3.35

$$\Delta\tau = \Delta\tau_{i-1} + \Delta\tau_i$$

$$= \alpha\left[C_i\left(\frac{1}{k_{p,i-1}} - \frac{1}{k_{n,i-1}}\right) - C_{i+1}\left(\frac{1}{k_{p,i}} - \frac{1}{k_{n,i}}\right)\right]. \tag{3.35}$$

If we choose $L_i = L_{i-1} = L_{i+1}$ and $W_i = \beta W_{i-1} = \beta W_{i+1}$ for both NMOS and PMOS transistors, then $k_{n,i} = \beta k_{n,i-1}$, and $k_{p,i} = \beta k_{p,i-1}$. Further, although the capacitance at node i consists of three components, namely, the output capacitance of $(i - 1)$th stage $C_{out,i-1}$, the capacitance of the interconnect connecting $(i - 1)$th and ith stages $C_{int,i}$, and the input capacitance of ith stage $C_{in,i}$, that is,

$$C_i = C_{out,i-1} + C_{int,i} + C_{in,i}, \tag{3.36}$$

to simplify analysis, we assume that the capacitance at node i mainly comes from the input capacitance of ith stage

$$C_i \approx C_{in,i} = C_{ox}\left[(WL)_{n,i} + (WL)_{p,i}\right]. \tag{3.37}$$

Note that when an MOS transistor is in triode, the capacitance looking to the gate of the transistor is given by $C_g = C_{gs} + C_{gd} = C_{ox}(WL)$ where $C_{gs} = \frac{1}{2}C_{ox}(WL)$ and $C_{gd} = \frac{1}{2}C_{ox}(WL)$ are gate-source and gate-drain capacitances, respectively. When the MOS transistor is in cut-off, the capacitance looking to its gate is gate-substrate capacitance given by $C_g = C_{ox}(WL)$. We thus have $C_i = \beta C_{i-1} = \beta C_{i+1}$. Equation 3.35 can then be simplified to

$$\Delta\tau = \alpha C_{i-1}\left(\frac{1}{k_{p,i-1}} - \frac{1}{k_{n,i-1}}\right)\left(\beta - \frac{1}{\beta}\right). \tag{3.38}$$

It is seen from Equation 3.38 that as long as $\beta \neq 1$ and $k_{n,i-1} \neq k_{p,i-1}$, we will have $\Delta\tau \neq 0$.

The preceding cyclic pulse-shrinking TDC exhibits a perfect linearity as the amount of cycle-to-cycle pulse shrinkage remains unchanged. This is one of the key advantages of Chen pulse-shrinking TDC. Chen cyclic pulse-shrinking TDC does not need to be calibrated periodically. This is because the amount of pulse shrinkage is only set by the physical dimensions of the delay stages. The all-digital nature of Chen cyclic pulse-shrinking TDC also makes it an ideal candidate for applications where power consumption is of a critical concern. One drawback of Chen cyclic pulse-shrinking delay line TDC is that an input pulse can be applied to the TDC only after the previous one has vanished completely. The use of only one inhomogeneous stage, though yielding a high resolution, is at the cost of a long conversion time. One might suggest using multiple pulse-shrinking stages to speed up the conversion process; this, however, will deteriorate resolution unless counters are added at the output of every pulse-shrinking stage. This, of course, is at the cost of high power and silicon consumption. Another drawback of the preceding generic cyclic pulse-shrinking TDC is the absence of PVT compensation.

3.3.7.3 Cyclic Pulse-Shrinking TDCs with Temperature Compensation

To minimize the effect of temperature, Chen et al. showed that the propagation delay stages of pulse-shrinking TDCs can be temperature-compensated so that the propagation delay is less sensitive to temperature variation [14,26]. The effect of temperature on CMOS devices has been extensively studied. An exhaustive collection of these studies can be found in [101]. The key temperature-dependent parameters of MOSFETs are the surface mobility μ of minority charge carriers that varies with temperature as per Equation 3.39

$$\mu = \mu_o \left(\frac{T}{T_o} \right)^m,$$ (3.39)

where
$-1.2 \leq m \leq -2.0$
T_o is a reference temperature, typically room temperature (300 K)
μ_o is the surface mobility of minority charge carriers at T_o, and the threshold voltage that varies with temperature as per Equation 3.40

$$V_T(T) = V_T(T_o) + \alpha_{V_T}(T - T_o),$$ (3.40)

where -0.5 mV/K $\leq \alpha_{V_T} \leq -3.0$ mV/K. It is seen from Equations 3.39 and 3.40 that both the surface mobility of minority charge carriers and threshold voltage drop when temperature rises. The former decreases the channel current, while the latter increases the channel current.

Examining Chen temperature-compensated inverter shown in Figure 3.38, we notice that when temperature rises, the decreasing threshold voltage of M1–M3 will give rise to an increase in the charge and discharge currents of the load capacitor. At the same time, the rising temperature will also result in a decrease in the surface mobility of minority charge carriers and subsequently the charge and discharge currents of the load capacitor. If both effects cancel out completely, the charge and discharge currents of the load capacitor will be independent of temperature. The amount of pulse shrinkage in this case will become temperature independent.

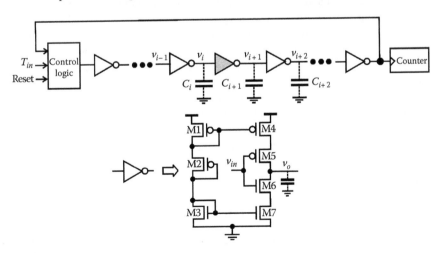

FIGURE 3.38 Cyclic pulse shrinking TDCs with temperature compensation. (From Chen, C. et al., *IEEE Trans. Nucl. Sci.*, 52(4), 834, 2005.)

To find out the condition upon which the temperature-independent charge and discharge currents of the load capacitor can be obtained, let us consider i_{SD4}, the current of M4. To simplify analysis, we neglect the effect of channel length modulation. Because

$$i_{SD4} \approx \frac{1}{2}\mu_{p,o}\left(\frac{T}{T_o}\right)^m C_{ox}\left(\frac{W}{L}\right)_4 \left[v_{SG4} - V_T(T_o) - \alpha_{V_T}(T - T_o)\right]^2, \qquad (3.41)$$

where $\mu_{p,o}$ is the surface mobility of holes at reference temperature T_o. Differentiating i_{SD4} with respect to temperature and setting the result to 0 yield the optimal value of v_{SG4}, denoted by v^*_{GS4}, at which perfect thermal compensation takes place

$$v^*_{SG4}(T) = V_T(T_o) + \alpha_{V_T}(T - T_o) + \left(\frac{2\alpha_{V_T}}{m}\right)T, \qquad (3.42)$$

Equation 3.42 is the condition for the first-order temperature compensation of i_{DS4}. For a given temperature, the desired $v^*_{SG4}(T)$ is computed from Equation 3.42 first. The value of v_{SG4} is then adjusted by varying the dimensions of M1 and M3 such that v_{SG4} is as close as possible to v^*_{SG4}. The results reported in [5] show that the variation of the resolution of thermally compensated pulse-shrinking TDC is only $\pm6\%$ over temperature range $0°C–100°C$.

In the preceding Chen cyclic pulse-shrinking TDC, the delay line is homogeneous. The inhomogeneous elements are the logic gates in the control logic that are also part of the loop. Such an arrangement allows the uniform layout of the homogeneous stages that are identical. Since thermal compensation is performed in every element of the loop, that is, both homogeneous delay elements and inhomogeneous control logic gates. The power and silicon consumption of the temperature compensated cyclic pulse-shrinking TDC of Figure 3.38 is significantly higher as compared with that of the generic cyclic pulse-shrinking TDC of Figure 3.36. Chen et al. pointed out that since pulse-shrinking is only due to the insertion of the inhomogeneous stage and we are only interested in the amount of pulse shrinkage, only the inhomogeneous stage needs to be thermally compensated. No thermal compensation is needed for homogeneous stages [102]. The exclusion of the homogeneous elements from thermal compensation greatly reduces silicon and power consumption. It has been demonstrated by Chen et al. that the cyclic pulse-shrinking TDC with selective thermal compensation achieves $\pm5.5\%$ resolution variation over temperature range $0°C–100°C$, comparable to $\pm6\%$ resolution variation of the TDC with every element of the loop temperature-compensated, while consuming a significantly less amount of power and silicon.

3.3.8   Successive Approximation TDCs

SA is an effective means to perform analog-to-digital conversion with low power consumption. The principle of SA-ADC can also be applied to time-to-digital conversion. A successive approximation TDC (SA-TDC) typically consists of a time comparator, a successive approximation register (SAR), and a digital-to-time

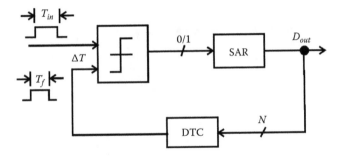

FIGURE 3.39 Architecture of SA-TDC. (From Mantyniemi, A. et al., *IEEE J. Solid-State Circuits*, 44(11), 3067, 2009.)

converter (DTC), as shown in Figure 3.39 [103]. The DTC maps a digital code D_{out} to a time variable T_f with its value proportional to D_{out}. T_{in} and T_f are compared by the time comparator. The time comparator yields 1 if $T_{in} > T_f$ and 0 otherwise. The output of the SAR is adjusted by the output of the time comparator accordingly until the difference $\Delta T = T_{in} - T_f$ drops below the minimum time variable that the time comparator can detect. Once this occurs, time-to-digital conversion is completed and the output of the TDC is given by the SAR.

The key component of SA-TDCs is the DTC. Digital-to-time conversion can be realized using a capacitor array shown in Figure 3.40 [103–108]. The load capacitor of the current-starved inverter is digitally selected by the output of the SAR. The selected load capacitor is charged to V_{DD} prior to the arrival of D_{in} by the driving inverter and discharged by the constant current J upon the arrival of D_{in}. The width of the output pulse is adjusted by the incoming digital code that selects the capacitance of the load capacitor. Although capacitor-array DTCs provide a large dynamic range and a good linearity, they are silicon and power-greedy especially when the number of bits is large.

Digital-to-time conversion can also be carried out using a delay line shown in Figure 3.41 [109,110]. The input code from the SAR selects the location of the output of the delay line using a multiplexer. Delay line DTCs enjoy a good linearity and low power consumption but suffer from a small dynamic range. To improve resolution, vernier delay lines can also be used, as shown in Figure 3.42 [110]. The output of the

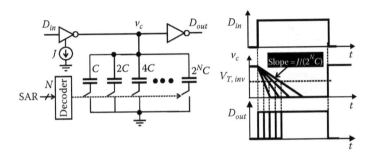

FIGURE 3.40 Digital-to-time conversion using a capacitor array.

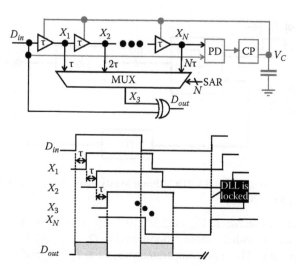

FIGURE 3.41 Digital-to-time conversion using a DLL-stabilized delay line. (From Roberts, G. and Ali-Bakhshian, M., *IEEE J. Solid-State Circuits*, 57(3), 153, 2010; Li, S. and Salthouse, C., Digital-to-time converter for fluorescence lifetime imaging, in *Proceedings of IEEE International Instrumentation and Measurement Technical Conference*, 2012, pp. 894–897.) Assume the output of SAR selects X_3. The resultant time variable is 3τ.

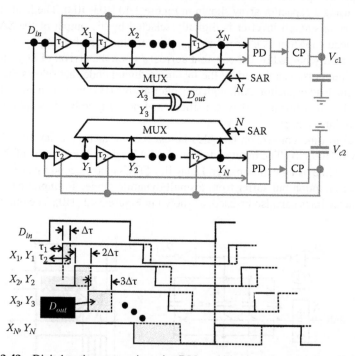

FIGURE 3.42 Digital-to-time conversion using DLL-stabilized vernier delay lines. (From Li, S. and Salthouse, C., Digital-to-time converter for fluorescence lifetime imaging, in *Proceedings of IEEE International Instrumentation and Measurement Technical Conference*, 2012, pp. 894–897.) Assume the output of SAR selects X_3. The resultant time variable is $3\Delta\tau = 3(\tau_2 - \tau_1)$.

vernier delay line DTC is the multiple of the difference between the per-stage delays of the two delay lines. Clearly, although the resolution of the vernier delay line DTC is higher than that of the corresponding delay line DTC, its upper-bound is lower.

3.3.9   Flash TDCs

A flash TDC digitizes a time variable by comparing the edge of STOP to the temporally displaced edge of START in a similar way as a delay line TDC, as shown in Figure 3.43a [111,112]. In delay line TDCs, the displaced edges of START are connected to the data node of the arbiters, that is, samplers, while in delay line flash TDCs, they are routed to the clocking node of the arbiters. The sampling of delay line TDCs takes place only at the rising edge of STOP, while flash TDCs sample STOP at each rising edge of the output of the delay stages. Although the per-stage delay of delay lines scales well with technology, the resolution of delay lines is lower-bound by the mismatch of the delay stages, and is limited to approximately 20 ps [54]. To improve resolution, the vernier flash TDC shown in Figure 3.43b can be employed. Similar to delay line TDCs, the resolution of the flash TDC of Figure 3.43a is per-stage delay τ, while that of the vernier TDC of Figure 3.43b is the difference between the per-stage delays of the two delay lines, that is, $\tau_2 - \tau_1$.

The resolution of the flash TDC in Figure 3.43a can be further improved if the delay line is removed, as shown in Figure 3.43c [113]. This TDC operates based on the time offset of the arbiters caused by device mismatches, and are therefore termed sampling offset TDC. Since the offset time of the arbiters differs, typically by 2–30 ps [111], this TDC needs to be calibrated prior to its operation [112]. The preceding flash TDCs employ balanced arbiters, that is, arbiters with a zero offset time between their two inputs. Alternatively, a flash TDC can be constructed with unbalanced arbiters with a gradually increased offset time, as shown in Figure 3.43d [114,115]. Flash TDCs with parallel delay elements shown in Figure 3.43e also emerged recently [116,117]. As compared with flash TDCs with cascaded delay stages, mismatch-induced delay error accumulation, which is present in delay lines and sets the upper-bound of INL, does not exist in flash TDCs with parallel delay elements.

3.3.10   Pipelined Time-to-Digital Converters

Pipelined TDCs that provide a better throughput also emerged [118–120]. As compared with other types of TDCs, pipelined TDCs typically require more hardware simply due to the pipelined operation of these TDCs. In applications such as bioelectronics where power consumption is of a critical concern or wireless communications where a stringent constraint is typically imposed on SNDR, pipelined TDCs are less attractive for these applications. Readers are referred to the cited references for details on pipelined TDCs.

3.4   NOISE-SHAPING TIME-TO-DIGITAL CONVERTERS

It was shown in Section 3.3 that sampling TDCs quantize a time variable either using a high-frequency low-jitter reference clock or counting the number of the delay stages of the delay line that START signal propagates through before the arrival of the STOP signal. The resolution of counter TDCs is lower-bound by the period of the clock. Increasing the clock frequency lowers the quantization noise, however, at the cost of increased dynamic power consumption and worsens cross talk. For delay line TDCs, the resolution is lower-bound by per-stage delay of the delay lines. A sub-per-stage delay resolution can be obtained using interstage interpolation, either passive or active with a limited resolution improvement. A sub-per-stage delay resolution can

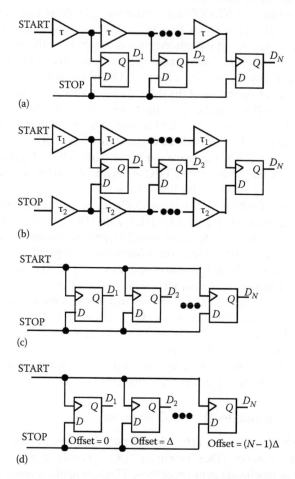

(a)

(b)

(c)

(d)

FIGURE 3.43 Flash TDCs. (a) Delay line flash TDC. (From Levine, P. and Roberts, G., A calibration technique for a high-resolution flash time-to-digital converter, in *Proceedings of IEEE International Symposium on Circuits and Systems*, vol. 1, 2004, pp. 253–256; Levine, P. and Roberts, G., *IEEE Proc. Comput. Digit. Tech.*, 152(3), 415, 2005.) (b) Vernier flash TDC. (c) Sampling offset flash TDC. (From Gutnik, V. and Chandrakasan, A., On-chip picosecond time measurement, in *Symposium on VLSI Circuits Digest of Technical Papers*, 2000, pp. 52–53.) (d) Flash TDC with unbalanced arbiters. (From Minas, N. et al., A high resolution flash time-to-digital converter taking into account process variability, in *Proceedings of IEEE International Symposium on Asynchronous Circuits and Systems*, 2007, pp. 163–174; Yamaguchi, T. et al., A CMOS flash TDC with 0.84–1.3 ps resolution using standard cells, in *Proceedings of IEEE Radio-Frequency Integrated Circuits*, 2012, pp. 527–530. *(Continued)*

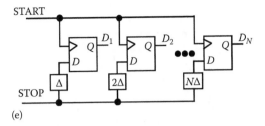

START

D_1 D_2 D_N

Δ 2Δ $N\Delta$

STOP

(e)

FIGURE 3.43 (*Continued*) Flash TDCs. (e) Flash TDC with parallel delay elements. (From Zanuso, M. et al., Time-to-digital converter with 3-ps resolution and digital linearization algorithm, in *Proceedings of IEEE ESSCIRC*, 2010, pp. 262–265; Yao, C. et al., A high-resolution time-to-digital converter based on parallel delay elements, in *Proceedings of IEEE International Symposium on Circuits and Systems*, 2012, pp. 3158–3161.)

also be obtained using vernier delay line TDCs, pulse-stretching TDCs, and pulse-shrinking TDCs that trade power, silicon, speed, or range for resolution. A common characteristic of sampling TDCs is that they improve resolution by lowering quantization noise uniformly across the entire spectrum rather than over a specific frequency range of the spectrum of the TDCs. As a result, they are costly both in terms of power and silicon consumption.

Since in applications such as all digital PLLs where a TDC functions as a phase detector, we are only interested in the performance of the TDC over the loop bandwidth of the PLLs while the noise transfer function from the phase detector to the output of the PLL has a low-pass characteristic, system-level approaches such as noise-shaping obtained from $\Delta\Sigma$ operations can be utilized to reduce the in-band quantization noise of TDCs well below that of sampling TDCs by moving excessive quantization noise to higher frequencies outside the signal band, which can then be removed effectively by the loop dynamics of the PLLs. Better phase noise can therefore be obtained. We term these TDCs noise-shaping TDCs. Several noise-shaping techniques that provide the frequency-dependent shaping of the quantization noise of TDCs emerged recently. In this section, we examine them in detail.

3.4.1 GATED RING OSCILLATOR TDCs

Ring oscillators are widely used in a broad range of applications, attributive to their attractive characteristics such as all-digital realization, and therefore full compatibility with technology scaling, low silicon and power consumption, a high oscillation frequency, and a broad frequency tuning range. GROs are a special class of ring oscillators of constant oscillation frequencies whose ON state (oscillation) and OFF state (no oscillation) are controlled by a time variable called gating signal. The gating signal is a pulse of width T_{in}. It starts the oscillator when $T_{in} = 1$ and stops the oscillator when $T_{in} = 0$. The TDC utilizing GRO is shown in Figure 3.44. The time variable to be digitized is the gating signal of the GRO [7]. The operation of this GRO-based TDC is briefly depicted as follows: The gating signal T_{in} activates the ring oscillator whose oscillation frequency is constant when it is at logic-1 and stops the oscillator when it is at logic-0. The gating signal only switches the oscillator on or off and has no effect on the frequency of the oscillator. Since the frequency of the

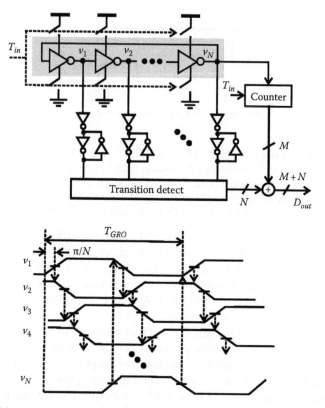

FIGURE 3.44 Gated ring oscillator TDC. (From Straayer, M. and Perrott, M., *IEEE J. Solid-State Circuits*, 44(4), 1089, 2009.) The number of the cycles of the oscillator is recorded by the counter and the number of threshold-crossings per oscillation cycle of the oscillator is determined by transition detection logic.

oscillator is constant, the number of the oscillation cycles of the oscillator when the gating signal is at logic-1 is directly proportional to the duration of T_{in}. It therefore provides the digital representation of T_{in}.

The number of the oscillation cycles of the oscillator when the gating signal is at logic-1 is recorded by a counter. The counter functions as a phase quantizer with a quantization error of 2π. To lower the phase quantization error from 2π to the per-stage phase shift of the oscillator, π/N, where N is the number of the stages of the oscillator, the output of each stage of the oscillator needs to be sampled using a counter. To minimize the possibility of the counting error caused by sampling the output of the delay stage of the oscillator at its state transitions, cross-coupled voltage buffers shown in Figure 3.44 are often used at the output of the delay stages to sharpen the transition edge of the output [7,121].

The resolution of GRO-based TDCs is lower-bound by the oscillation frequency of the ring oscillator of the TDCs, which is inversely proportional to the number of the delay stages of the oscillator. The quantization error of GRO-based TDCs, given by π/N where N is the number of the stages of the oscillator, is also inversely proportional to the number of the delay stages of the oscillator. The larger the number of the

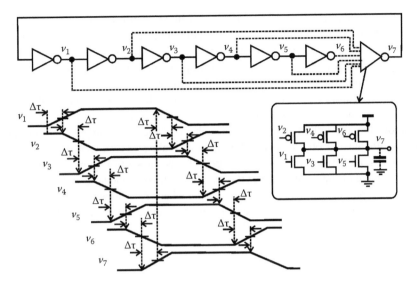

FIGURE 3.45 Multipath ring oscillator TDC. Each load capacitor has multiple charging and discharging paths controlled by the output of selected delay stages.

stages of the oscillator, the lower is the quantization noise and worse the resolution. In order to lower the quantization noise and at the same time to improve the resolution of GRO-based TDCs, the multipath technique proposed by Lee et al. is an effective means to reduce the per-stage delay of ring oscillators without reducing the number of the delay stages of the oscillators [122]. References 123,124 provide the examples of multipath ring oscillators and performance comparison with their single-path counterparts. We use Figure 3.45 to illustrate how multipath can significantly speed up the oscillation of a ring oscillator without lowering the number of the delay stages of the oscillator. Let us focus on the rising edge of stage 7. The rising of v_7 is caused by the charging of the load capacitor C_7. If only the generic inverters are used as the delay stages of the oscillator, the load capacitor C_7 will be charged by v_6, which arrives much later as compared with v_2 and v_4. This observation suggests that if the outputs of stages 2 and 4 are routed to the input of stage 7, as shown in Figure 3.45, it will provide two additional precharging paths for C_7 before the arrival of v_6, thereby speeding up the charging process of C_7. The same holds for the charging process of C_7. Similar approaches apply to the other delay stages of the oscillator as well.

For a ring oscillator with a large number of delay stages, the choice of precharging or predischarging inputs for each delay stage should also be chosen after due consideration of the goal of achieving a balance between performance improvement and routing cost. Also, the number of the pull-up and pull-down paths for each stage do not need to be equal. For example, in [7], although the oscillator has 47 delay stages, only 2 pull-up paths and 3 pull-down paths are provided for each delay stage. Further, the locations from which the pull-up and pull-down signals originate do not need to be the same. Since PMOS transistors are slower as compared with NMOS transistors, the locations from which the pull-up signals originate should be further away as compared with that from which the pull-down signals originate. For example, in [7], the pull-up signals are 11 and 13 stages away while the pull-down signals are only 1, 5, and 9 stages away. Finally, the strengths of the pull-up operation and that of pull-down operation, that is, the width of the pull-up and pull-down transistors of different stages, need not be the same. Large pull-up and pull-down transistors should be used for precharging/predischarging signals with long distances.

When the gating signal $T_{in} = 0$ is asserted, the delay stages of the oscillator are isolated from both the supply voltage and ground rails. As a result, the oscillation of the oscillator is halted and the output voltage of the delay stage of the oscillator remains unchanged provided that the effect of leakage, cross talk from neighboring devices, and switching noise are negligible, as shown in Figure 3.46. Since the output voltage of the delay stage of the oscillator remains unchanged during $T_{in} = 0$, the residual phase of $(k-1)$th sampling period where $T_{in} = 1$, denoted by $e_f(k-1)$, is carried over to kth sampling period in its entirety and becomes the initial phase of kth sampling period, denoted by $e_i(k)$, that is,

$$e_i(k) = e_f(k-1), \tag{3.43}$$

as illustrated graphically in Figure 3.46. Note that the carry-over of the residual phase is due to the continuity of the charge of the load capacitor of the delay stage. The net phase accumulation in kth sampling period is thereby given by

$$\phi(k) = K_{vco}v_{in} + \left[e_f(k) - e_i(k)\right], \tag{3.44}$$

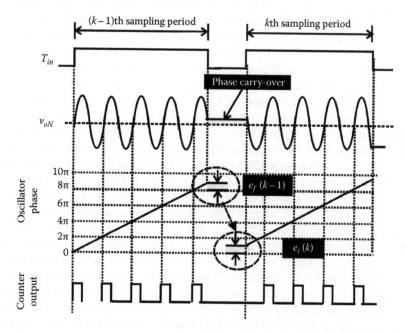

FIGURE 3.46 Gated ring oscillator TDCs. The residual phase of $(k-1)$th sampling period is carried over to kth sampling period as the initial phase of kth sampling period.

where K_{vco} is the voltage-to-phase gain of the oscillator. If we let $e(k)$ be the residual phase, that is, the quantization noise in phase k, that is, $e(k) = e_f(k)$, we have

$$e_i(k) = e_f(k-1) = e(k-1). \tag{3.45}$$

As a result,

$$\phi(k) = K_{vco}v_{in} + \left[e(k) - e(k-1)\right]. \tag{3.46}$$

Its frequency response is given by

$$\Phi(z) = K_{vco}V_{in}(z) + \left(1 - z^{-1}\right)E(z). \tag{3.47}$$

The noise transfer function, $NTF(z) = \Phi(z)/E(z)$, is given by

$$NTF(z) = 1 - z^{-1}. \tag{3.48}$$

Equation 3.48 shows that the quantization noise is first-order shaped, specifically, quantization noise at low frequencies is pushed to high frequencies.

In addition to the first-order noise-shaping of quantization noise, the effect of delay element mismatch is also first-order shaped. This is because the effect of mismatch arising from process spread manifests itself as the variation of the propagation delay or equivalently the phase of delay stages, which is subject to the effect of the phase continuity from one sampling period to another. The randomness of the initial phase of each sampling period also effectively scrambles quantization error across different sampling periods so that it can be first-order shaped.

Since the output nodes of the oscillator are floating during the OFF state ($T_{in} = 0$) of the oscillator, the noise-shaping that originates from the continuity of the phase of the oscillator during the OFF state of the oscillator is rather vulnerable to the effect of the leakage of p–n junctions at the output nodes and charge redistribution during the switching of the gating transistors [125]. This causes the output phase to vary during the OFF state, resulting in a phase error known as skew error. Another source of skew error is the gating delay of GROs arising from the fact that GROs cannot be started or stopped instantaneously by the gating signal [126]. This is particularly true if the oscillation frequency of the oscillator in the ON state is high. Although the effect of charge sharing can be mitigated if the gating transistors are made small, this, however, is at the cost of limiting oscillation frequency due to the rising channel resistance of the gating transistors [127]. The floating output nodes of the delay stages of the oscillator in the OFF state also make the phase of the oscillator susceptible to disturbances external to the oscillators such as switching noise and cross talk [121]. GROs also exhibit dead-zone behavior if the period of the gating signal T_{in} is in the vicinity of an integer multiple of the period of the oscillator [126].

3.4.2 VERNIER GATED RING OSCILLATOR TDCs

It was shown in the last section that GRO TDCs offer an improved resolution obtained from the first-order noise-shaping of quantization noise. It was shown in early sections of the chapter that vernier delay line TDCs provide a much better raw resolution, that is, the resolution without noise-shaping, as compared with that of delay line TDCs. It is therefore natural to utilize the noise-shaping characteristics of GRO TDCs and the fine resolution of vernier delay line TDCs simultaneously to further improve resolution. The simplified schematic of the vernier GRO TDC proposed by Lu et al. is shown in Figure 3.47 [128–130]. The TDC consists of two gated ring oscillators that have slightly different oscillation frequencies. START signal is applied to the slow GRO, identified by SGRO, while STOP signal is fed to the fast GRO, identified by FGRO. When START is asserted, SGRO starts to oscillate. FGRO will start to oscillate when STOP arrives. Since the rising edge of x_j, $j = 1, 2, ..., N$, leads that of y_j before STOP catches up START, we have $D_j = 0$, $j = 1, 2, ..., N$. Once the rising edge of jth stage of FGRO catches up with that of SGRO, $D_j = 1$, a flag signal is set. Once this flag is set, the time-to-digital conversion is completed and the resultant digital code is read from the output of the samplers.

It is important to note that in vernier delay line TDCs shown in Figure 3.17, DFFs are employed as the samplers. As pointed out earlier that when the rising edge of START signal and that of STOP signal are close to each other, the timing constraint of DFFs will be violated. As a result, the DFFs will enter a metastable state. To prevent this from happening, an SR latch is placed before the DFF, as shown in Figure 3.47. The SR latch functions as a time amplifier that stretches the time

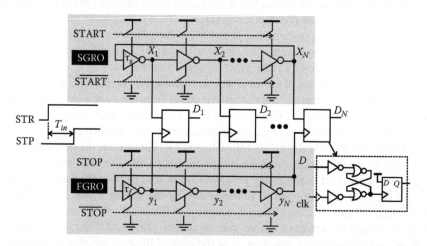

FIGURE 3.47   Vernier gated ring oscillator TDCs. (From Lu, P. and Andreani, P., A high-resolution vernier gated-ring-oscillator TDC in 90-nm CMOS, in *Proceedings of NORCHIP*, 2010, pp. 1–4; Lu, P. et al., A 90 nm CMOS gated-ring-oscillator-based vernier time-to-digital converter for DPLLs, in *Proceedings of IEEE ESSCIRC*, 2011, pp. 459–462; Lu, P. et al., A 90 nm CMOS digital PLL based on vernier-gated-ring oscillator time-to-digital converter, in *Proceedings of IEEE International Symposium on Circuits and Systems*, 2012, pp. 2593–2596.) *Abbreviations*: SGRO, Slow GRO; FGRO, Fast GRO.

difference between the rising edge of START and that of STOP sufficiently so that
no violation of the timing constraints of DFFs will occur [47].

3.4.3   Two-Step Gated Ring Oscillator TDCs

It was shown earlier that interpolation is an effective means to improve the resolution
of TDCs. In a counter TDC with interpolation, a time variable is first quantized by
the counter TDC. The quantization error of the counter TDC is then quantized using
a delay line TDC whose resolution is much higher than that of the counter TDC.
In delay line TDCs with interpolation, passive or active interpolation techniques
are used to interpolate the outputs of the adjacent delay stages of the TDC so as to
lower resolution from the per-stage delay of delay line TDCs to a sub-per-stage delay.
Constrained by circuit complexity, resolution improvement obtained from interstage
interpolation is rather moderate.

Lee and Abidi showed that the resolution of delay line TDCs can be greatly
improved using a two-stage TDC consisting of a coarse TDC and a fine TDC, both
realized using delay line TDCs [47]. Figure 3.48 shows conceptually the principle of
two-step TDCs. A two-step TDC consists of a coarse TDC of resolution ΔT_1 and a
fine TDC of resolution ΔT_2. A time amplifier of gain A is used to amplify ΔT_1 to the
full scale of the fine TDC that has a total of N stages. The time amplifier is designed
in such a way that its input range is ΔT_1 and its output range is $N\Delta T_2$. The overall
resolution of the two-step TDC is given by $\Delta T_1/N$. Clearly, the performance of the
time amplifier directly affects the performance of the TDC. We will come back to
this point shortly.

Figure 3.49 shows the architecture of Lee-Abidi two-step TDC [47]. The coarse
TDC is a delay line TDC that digitizes an input time variable T_{in} directly in a similar
way as that of delay line TDCs studied earlier. The resolution of the coarse TDC is
the per-stage delay τ. Time difference between the rising edge of the output of jth

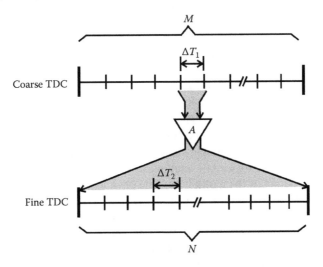

FIGURE 3.48   Two-step time-to-digital conversion.

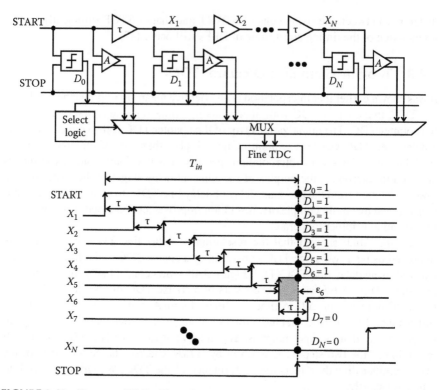

FIGURE 3.49 Two-step TDCs. (From Lee, M. and Abidi, A., *IEEE J. Solid-State Circuits*, 43(4), 769, 2008.)

delay stage X_j and that of STOP signal is the quantization error of stage j, denoted by ϵ_j. Note that $|\epsilon_j| \leq \tau$. In the example shown in Figure 3.49, $\epsilon_j > 0$ for $j = 0, 1, 2, ..., 6$ and $\epsilon_j < 0$ for $j = 7, 8, ..., N$. Also, ϵ_6 is the smallest positive quantization error. Each quantization error ϵ_j is amplified by a shifted RS latch time amplifier studied earlier in Figure 3.33. Since the output of the shifted RS latch time amplifier is at logic-0 when $|T_{in}| > T_{in,\max}$, when ϵ_j is large, the output of time amplifier j will settle to a valid logic state. When ϵ_j is small, the time amplifier will stretch ϵ_j by the gain of the time amplifier so that the resultant time variable is large enough and can be digitized by the fine TDC with a sufficiently high resolution. The output of the time amplifiers is multiplexed by a 1–N multiplexer. The output of the multiplexer is conveyed to the fine TDC. The selection command of the multiplexer is generated from the output of the delay stages. In the example shown in Figure 3.49, since $D_{0,1,2,...,6} = 1$ and $D_{7,8,...,N} = 0$, the selection logic of the multiplexer should route the output of sixth time amplifier to the output of the multiplexer. One obvious way to achieve this is to use XOR2 gates as $D_i \otimes D_j = 0$ for $i, j = 0, 1, 2, ..., 6$, $D_6 \otimes D_7 = 1$, and $D_i \otimes D_j = 0$ for $i, j = 7, 8, ..., N$ where $\otimes$ is the exclusive-OR operator.

To further improve the resolution of two-step TDCs, Chung et al. showed that the fine TDC of a two-step TDC can be implemented using a GRO TDC so as to utilize its first-order noise-shaping characteristic [48]. Vernier delay line TDCs are another candidate for the fine TDC.

A number of factors critically affect the performance of the preceding two-step TDC. In what follows we examine them in detail. An important source of timing error is the metastable state of the arbiters. In Figure 3.49, we show that the smallest positive quantization error ϵ_6 is routed to the output of the multiplexer for further digitization. Since min $\{\epsilon_j\} \leq \tau, j = 0, 1, 2, ..., N$, min $\{\epsilon_j\}$ might be overly small such that the timing constraint of the arbiter is violated. If this occurs, the arbiter will enter a metastable state and the output of the arbiter will no longer be min $\{\epsilon_j\}$ but rather an erroneous value that is much larger than the residue. The device mismatch of time amplifiers is another source of timing error. The device mismatch of time amplifiers shifts the zero-crossing of the gain curve of the amplifiers, manifesting itself as an input offset time, similar to the input offset voltage of voltage amplifiers. As the dynamic range of the fine TDC is typically much smaller as compared with that of the coarse TDC, not only the input offset time sets the minimum input time variable that the fine TDC can detect, it might also drive the fine TDC into overrange. Another source of timing error is the kickback effect associated with the arbiters of the two-step TDC [131]. Kickback injects charge into both the delay and STOP lines, giving rise to the variation of the voltage of the lines and subsequently timing errors. Moreover, mismatch between the time amplifiers also affects the performance of the two-step TDC. As the time amplifiers stretch the residue of the coarse TDC and the number of the time amplifiers is the same as that of the delay stages, mismatch between the time amplifiers will yield different outputs even though the inputs of the time amplifier are identical. Finally, the number of time amplifiers used in [47] is the same as the number of the delay stages, giving rise to both high power and silicon consumption and mismatches. Mandai and Charbon showed that the number of the time amplifiers can be reduced from the number of the delay stages to only one so as to remove the preceding drawbacks [88].

3.4.4   TWO-DIMENSIONAL VERNIER GATED RING OSCILLATOR TDCs

As pointed out earlier, one-dimensional vernier delay line TDCs trade conversion range for resolution. To increase conversion range, the number of the stages of vernier delay line TDCs needs to grow exponentially. The drawback of one-dimensional vernier delay line TDCs stems from the fact that not only the number of the delay stages of the fast delay line and that of the slow delay line of vernier delay line TDCs are the same, the location at which time comparison that determines whether a catch-up of the START signal by STOP signal takes place or not is also identical. The drawback of one-dimensional vernier delay line TDCs can be mitigated using two-dimensional vernier delay line TDCs. Since the comparison of the output of every stage of the two delay lines is conducted in two-dimensional vernier delay line TDCs, the time instant at which STOP signal catches up the START signal can be detected with the significantly reduced length of delay lines. Not only the cost associated with the exceedingly long delay lines of one-dimensional vernier delay line TDCs is reduced, the deleterious effect–associated jitter and mismatch that exist in overly long delay lines is also minimized. No improvement in resolution, however, is made from one-dimensional vernier delay line TDCs to two-dimensional vernier delay line TDCs.

It was shown in Figure 3.47 vernier delay line TDCs can be replaced with vernier GRO TDCs to reduce the cost, minimize the effect of jitter and mismatch, scramble the effect of mismatch via variable initial phase, and lower quantization noise obtained from first-order noise-shaping. Although two-dimensional vernier delay line TDCs achieve the same resolution as that of one-dimensional vernier delay line TDCs with shortened delay lines, Lu et al. showed that the resolution of two-dimensional vernier delay line TDCs can be improved by utilizing the first-order noise-shaping of GROs while reducing the number of delay stages. Two vernier delay lines of a two-dimensional vernier delay line TDC can be replaced by two gated ring oscillators so as to convert a two-dimensional vernier delay line TDC to a two-dimensional vernier GRO TDCs [132,133]. The principle of two-dimensional vernier GRO TDCs is the same as that of two-dimensional vernier delay line TDCs except that two-dimensional vernier GRO TDCs provide first-order noise-shaping on quantization noise.

Figure 3.50 shows the typical configuration of a two-dimensional vernier GRO TDC. The gating signals denoted by FGRO and SGRO, respectively, are generated from the START and STOP signals with SGRO leading FGRO by T_{in}. SGRO activates the

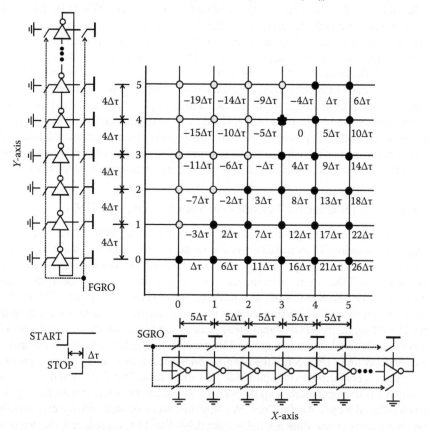

FIGURE 3.50 Two-dimensional vernier gated ring oscillator TDC. (From Lu, P. et al., A 90 nm CMOS gated-ring-oscillator-based 2-dimensional vernier time-to-digital converter, in *Proceedings of IEEE NOCHIP*, 2012, pp. 1–4; Lu, P. et al., A 2-D GRO vernier time-to-digital converter with large input range and small latency, in *Proceedings of IEEE Radio-Frequency Integrated Circuits*, 2013, pp. 151–154.)

slow oscillator, while FGRO starts the fast oscillator. The time instant at which FGRO catches up SGRO is detected by the time comparators placed at every node of the vernier plane. Once this occurs, the time-to-digital conversion is completed and the resultant digital code can be determined by reading out the output of the time comparators.

3.4.5 GATED RELAXATION OSCILLATOR TDCs

Although ring oscillators are known for their high oscillation frequency, a broad frequency tuning range, low power and silicon consumption, and full compatibility with technology scaling, they suffer from high sensitivity to temperature variation and supply voltage fluctuation. Relaxation oscillators, on the other hand, exhibit a low sensitivity to supply voltage fluctuation, a characteristic critically important for passive wireless microsystems such as implanted or embedded sensors whose operational power is harvested from radio-frequency waves [101,134]. Cao et al. showed that the first-order noise-shaping characteristic possessed by both GRO TDCs and switched ring oscillator TDCs is also intrinsic to relaxation oscillator TDCs [135,136]. We use Figure 3.51 to briefly show this. The relaxation oscillator consists

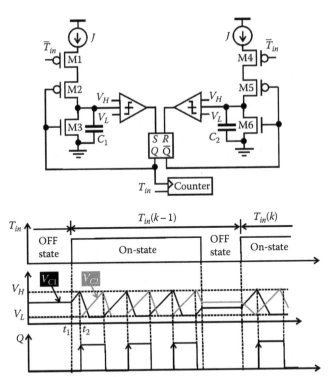

FIGURE 3.51 Relaxation oscillator TDCs. V_L and V_H are lower and upper threshold voltages of the comparators. (From Cao, Y. et al., A 0.7 mW 11b 1-1-1 MASH $\Delta\Sigma$ time-to-digital converter, in *IEEE International Solid-State Circuits Conference on Digest of Technical Papers*, 2011, pp. 480–481; Cao, Y. et al., A 0.7 mW 13b temperature-stable MASH $\Delta\Sigma$ TDC with delay-line assisted calibration, in *Proceedings of IEEE Asian Solid-State Circuits Conference*, 2011, pp. 361–364.)

of six transistors M1–M6, all operating in an ON/OFF mode, that is, they function as switches: two comparators that compare the voltage of linear capacitors C_1 and C_2 with reference voltages V_H and V_L, an SR flip-flop that sharpens the output of the comparators and drives transistors M2, M3, M5, and M6, and a pair of current sources that supply a constant current J. M1, M2, M4, and M5 control the charging process of the capacitors, while M3 and M6 manage the discharging process of the capacitors. Since the current of the current sources is constant while the capacitors are linear, v_{c1} and v_{c2} rise with time linearly. The discharging process of C_1 and C_2, however, follows a RC discharge profile where R is the sum of the channel resistances of M3 and M6. Whether the relaxation oscillator will oscillate or not is controlled by T_{in}. When $T_{in} = 1$, the oscillator is enabled. It is disabled when $T_{in} = 0$. T_{in} thus functions as a gating signal in a similar way as that in GROs. The counter records the number of the oscillation cycles of the oscillator when $T_{in} = 1$ and the content of the counter at the end of the gating signal yields the digital representation of T_{in}.

Consider $t = t_1$, assume C_1 is being charged while C_2 is discharged. v_{c1} rises linearly with time while $v_{c2} = 0$. At t_2 where $v_{c1} = V_H$, $Q = 1$ and $\bar{Q} = 0$. C_1 starts to discharge and $v_{c1} = 0$ when it is fully discharged while C_2 starts to be charged. When $v_{c2} = V_H$, $Q = 0$ and $\bar{Q} = 1$. This process repeats and oscillation is sustained. Since the charge of the capacitors is held unchanged when $T_{in} = 0$, that is, in the OFF state of the oscillator, the residual phase of the oscillator in $(k - 1)$th sampling phase is the initial phase of kth sampling phase, yielding the first-order noise-shaping, in a similar way as that of GRO TDCs.

As pointed out earlier, GRO TDCs suffer from the drawback of phase variation in the OFF state caused by leakage and charge redistribution. Gating error arising from the inability of the oscillators to start oscillation immediately after T_{in} is asserted also undermines the noise-shaping capability of GRO TDCs. Unlike GROs whose load capacitance, which is made of the output capacitance of the driving stage, the input capacitance of the load stage, and the capacitance of the interconnect connecting the driving and driven stages, is comparable to parasitic capacitances, the capacitance of C_1 and C_2 is much larger as compared with parasitic capacitances of transistors. As a result, the effect of charge redistribution and leakage becomes negligible. One downside of gated relaxation oscillator TDCs is the low frequency of relaxation oscillators, which limits the oversampling ratio of the TDCs. Note that one can increase the frequency of relaxation oscillators by either increasing the current of the current sources or lowering the capacitance of the capacitors. The former increases the power consumption, whereas the latter signifies the effect of the parasitic capacitances of transistors via leakage and charge redistribution.

3.4.6 SWITCHED RING OSCILLATOR TDCS

It was shown in Section 3.4.1 that the first-order noise-shaping of GRO TDCs is vulnerable to the effect of leakage and charge redistribution. To eliminate this drawback, Konishi et al. showed that freezing the output state of a GRO in the off state of the oscillator can be replaced by another oscillation state of the oscillator in which the output voltage of the delay stage of the oscillator is securely defined. Specifically, the ring oscillator operates between two oscillation states that have different

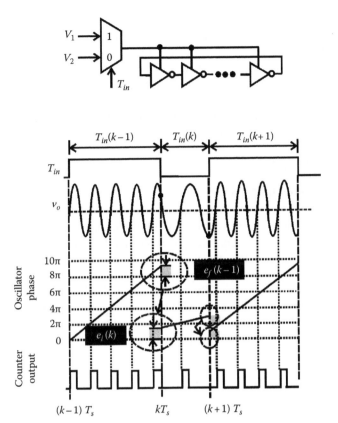

FIGURE 3.52   Switched ring oscillator TDCs.

oscillation frequencies, as shown in Figure 3.52. The duration of kth oscillation state is controlled by $T_{in}(k)$. The two different frequencies of the oscillator are obtained by applying two different control voltages to the oscillator. We term this oscillator switched ring oscillator (SRO), as opposed to the GROs that only have one control voltage and operate between an OFF state and an ON state [125–127,137,138]. GROs can therefore be viewed as a special case of SROs with the frequency of one of the two states of the oscillator set to 0.

As the state of a SRO is represented by the voltage of the delay stages of the oscillator, the charge of the load capacitors of the delay stages of the oscillator at the end of $(k-1)$th sampling phase is carried over in its entirety to the following kth sampling phase, similar to the continuity of the phase of the output of GROs except without phase-holding, the first-order noise-shaping characteristic of GROs is therefore also inherent to SROs. Unlike a GRO, since the output voltage of the delay stages of a SRO before and after a switching instant at which the oscillation state of the oscillator changes is well defined, floating nodes existing in GROs do not exist in SROs. As a result, the inimical effect associated with the floating nodes of GROs such as leakage and charge redistribution is eliminated completely. As mentioned earlier, GRO TDCs suffer from gating delay arising from the fact that GROs cannot start or stop oscillation instantaneously after the application or removal of the gating signal. The higher the frequency of the oscillator in the ON state, the severer is the effect of the gating delay. Unlike GRO TDCs, since the oscillator in a SRO TDC oscillates in both states, the effect of gating delay is reduced [126]. Similar to the dead zone characteristic of GRO TDCs, SRO TDCs also exhibit a dead zone characteristic if f_H, the frequency of the oscillator in the fast state, is an integer multiple of f_L, the frequency of the oscillator in the slow state. The dead zone can be removed if f_H and f_L are asynchronous with each other.

3.4.7   MASH TDCs

3.4.7.1   Fundamental of MASH

MASH is an effective means widely used in design of ADCs to obtain high-order noise-shaping without sacrificing stability [139]. This technique can also be used to construct high-order TDCs. For example, a 1-1 MASH TDC can be constructed by cascading two TDCs, namely, a signal TDC and a residue TDC. The signal TDC digitizes input time variables directly. The quantization error of the signal TDC is extracted and fed to the residue TDC for further digitization so as to obtain a lower quantization error. We use the 1-1 MASH shown in Figure 3.53 to illustrate the principle of the MASH operation. The input time variable T_{in} is fed to the signal TDC and digitized by the TDC. The output of the signal TDC forms the upper portion of the digital code of T_{in} given by

$$D_{out1}(z) = H_1(z)T_{in}(z) + N_1(z)E_1(z), \tag{3.49}$$

where

   $H_1(z)$ and $N_1(z)$ denote the signal transfer function and quantization noise transfer function of the signal TDC, respectively
   $E_1(z)$ is the quantization error of the signal TDC

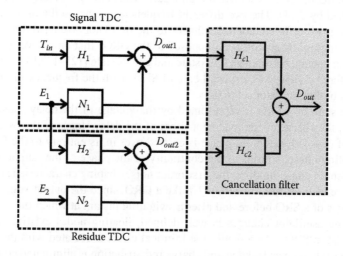

FIGURE 3.53   1-1 MASH TDCs.

The quantization error of the signal TDC is extracted and then fed to the residue TDC as its input for further quantization. The output of the residue TDC constitutes the lower portion of the digital code of T_{in} and is obtained from

$$D_{out2}(z) = H_2(z)E_1(z) + N_2(z)E_2(z), \qquad (3.50)$$

where

$H_2(z)$ and $N_2(z)$ denote the signal transfer function and quantization noise transfer function of the residue TDC, respectively

$E_2(z)$ denotes the quantization error of the residue TDC

Although it seems that if we add D_{out1} and D_{out2}, the outputs of the signal and residue TDCs directly, we would get the complete digital code of T_{in}, a detail analysis shows that if we did this, the quantization error of the signal TDC $E_1(z)$ would also appear at the output of the TDC. Since the signal TDC has to handle the full range of the input time variable T_{in}, it is normally realized using a coarse TDC that typically has a large dynamic range but a poor quantization error. Unlike the signal TDC, the input of the residue TDC is the quantization error of the signal TDC. The quantization error of the residue TDC is typically much smaller as compared with the input of the signal TDC. As a result, the residue TDC can be implemented using a fine TDC that typically has a smaller dynamic rang but a much finer resolution. E_1 is therefore much larger as compared with E_2. As a result, the quantization error of the MASH TDC is dominated by that of the signal TDC.

It becomes evident from the preceding observation that in order to minimize the overall quantization error of the MASH TDC, the quantization error of the signal TDC needs to be removed from the output of the MASH TDC. This can be accomplished in a postprocessing step taking place in the digital domain by a cancellation filter. The output of the signal TDC and that of the residue TDC are fed to the cancellation filter, specifically, H_{c1} and H_{c2} implemented digitally in the digital domain. The combined output is given by

$$D_{out} = H_{c1}(z)H_1(z)T_{in}(z) + \left[H_{c1}(z)N_1(z) - H_{c2}(z)H_2(z) \right]E_1(z)$$

$$-H_{c2}(z)N_2(z)E_2(z). \qquad (3.51)$$

To remove $E_1(z)$ from D_{out}, we impose the noise cancellation condition

$$H_{c1}(z)N_1(z) - H_{c2}(z)H_2(z) = 0. \qquad (3.52)$$

A natural choice to satisfy Equation 3.52 is given by

$$H_{c1}(z) = kH_2(z),$$
$$H_{c2}(z) = kN_1(z), \qquad (3.53)$$

where k is a constant, often $k = 1$. Equation 3.51 thus becomes

$$D_{out} = kH_1(z)H_2(z)T_{in}(z) - kN_1(z)N_2(z)E_2(z). \tag{3.54}$$

If the noise-shaping provided by both the signal TDC and residue TDC is first-order, that is,

$$N_1(z) = N_2(z) = 1 - z^{-1}, \tag{3.55}$$

while there is no attenuation on the signal, that is,

$$H_1(z) = H_2(z) = z^{-1}, \tag{3.56}$$

Equation 3.54 becomes

$$D_{out} = kz^{-2}T_{in}(z) - k\left(1 - z^{-1}\right)^2 E_2(z). \tag{3.57}$$

Equation 3.57 reveals that although both the signal and residue TDCs only offer first-order noise-shaping, the 1-1 MASH TDC provides second-order noise-shaping on the quantization noise of the residue TDC, removes the quantization noise of the signal TDC completely, and provides no attenuation on the input. Since the quantization noise of the residual TDC is typically much smaller as compared with that of the signal TDC, the additional order of noise-shaping provided by the MASH further lowers the effect of the quantization noise of the residue TDC. It should be emphasized that the preceding results are based on the condition of the satisfaction of the cancellation condition given in Equation 3.52.

Let us now examine the postprocessing of the cancellation filter implemented in the digital domain. It is evident from Equation 3.51 that if the cancellation condition Equation 3.52 is not completely satisfied due to errors in estimating H_2 and N_1 needed for constructing H_{c1} and H_{c2} as per Equation 3.53, the quantization error of the signal TDC $E_1(z)$ will leak to the output. To demonstrate this, we assume that

$$H_{c1} = k(H_2 + \Delta H),$$
$$H_{c2} = k(N_1 + \Delta N), \tag{3.58}$$

where ΔH and ΔN denote the errors in estimation of H_2 and N_1, respectively; it can be shown that the amount of the leakage of $E_1(z)$ to the output of the TDC is given by $(N_1\Delta H - H_2\Delta N)E_1(z)$. Since a perfect estimation of H_2 and N_1 is not possible in reality, the quantization error of the signal TDC $E_1(z)$ will leak to the output of the TDC. To minimize the leakage of $E_1(z)$ to the output of the TDC, $E_1(z)$ itself should be minimized. The signal TDC should therefore be chosen taking into due consideration both a large dynamic range that can handle T_{in} and a low quantization error.

Let us now examine the profile of the leakage of the quantization error of the signal TDC. We continue to assume that the noise-shaping provided by both the signal TDC and residue TDC is first-order, that is, $N_1(z) = N_2(z) = 1 - z^{-1}$, while there is no attenuation on the signal, that is, $H_1(z) = H_2(z) = z^{-1}$. Since

$$H_{c1}(z) = kH_2 = kz^{-1} \tag{3.59}$$

and

$$H_{c2}(z) = kN_1 = k\left(1 - z^{-1}\right), \tag{3.60}$$

$H_{c1}(z)$ has an all-pass characteristic, while $H_{c2}(z)$ provides first-order noise-shaping. As a result, ΔN, the error in estimation of $N_1(z)$, dominates ΔH, the error in estimation of $H_2(z)$. This is because the former passes through $H_{c1}(z)$ without attenuation, while the latter is suppressed by $H_{c2}(z)$ to the first order.

Having studied the fundamentals of MASH, in the next few sections, we examine several recently reported MASH TDCs.

3.4.7.2 MASH Gated Ring Oscillator TDCs

It was shown in Section 3.4.5 that GRO TDCs provide first-order noise-shaping on quantization error. These TDCs also offer a number of desirable characteristics such as fast conversion and subsequently a large oversampling ratio, low power and silicon consumption, and of course, full compatibility with technology scaling. An 1-1 MASH TDC can be realized with both coarse and fine TDCs implemented using GRO TDCs so as to take the advantage of the first-order noise-shaping provided by each of the TDCs to achieve overall second-order noise-shaping. Figure 3.54a and b show the 1-1 MASH GRO TDC proposed by Konishi et al. [46,140]. The input of the TDC, T_{in1}, is applied to the first GRO, hereafter referred to as the coarse GRO. The coarse GRO is activated when $T_{in1} = 1$ and halted when $T_{in1} = 0$. The output of the coarse GRO is first sharpened by a voltage buffer and then fed to a quantization noise propagator implemented using a resettable DFF [140]. The quantization noise propagator ideally extracts the quantization noise of the coarse TDC and propagates it to the fine TDC. The rising edge of the output of the quantization noise propagator T_{in2} is aligned with the first rising edge of the output of the coarse GRO, v_{o1}, and its falling edge is aligned with the falling edge of T_{in1} by routing v_{o1} to the clock input of the DFF and connecting $\overline{T}_{in1}$ to the reset input of the DFF.

If there is a sufficiently large time gap between the rising edge of v_{o1} and that of $\overline{T}_{in1}$, no violation of the timing constraints of the DFF will occur. The quantization noise propagator will function properly, as shown in Figure 3.54b. If the rising edge of v_{o1} is very close to that of $\overline{T}_{in1}$, the timing constraints of the DFF will not be met. In this case, the DFF will enter a metastable state and the location of the rising edge of T_{in2} will become unpredictable, as illustrated graphically in Figure 3.54c. Since T_{in2} is the gating signal of the fine TDC, the uncertainty of the rising edge of T_{in2} will give rise to an uncertainty in the activation time of GRO2 and subsequently the initial phase of GRO2, thereby deteriorating the performance of the TDC. Okuno et al. showed that

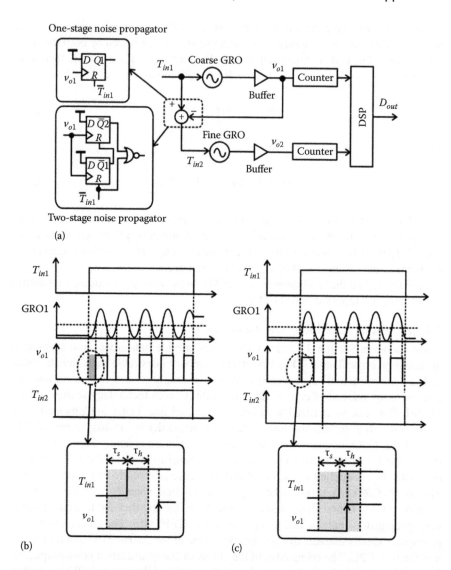

FIGURE 3.54 (a) 1-1 MASH gated ring oscillator TDC [46,140]. GRO1 is the waveform of the output voltage of the coarse gate ring oscillator and v_{o1} is the waveform of the voltage of the output of the voltage buffer whose input is GRO1. (b) Waveforms with a one-stage noise propagator. (c) Waveforms with a two-stage noise propagator.

this uncertainty can be removed using the two-stage quantization noise propagator shown in Figure 3.54a [46]. It is seen that although the rising edge of Q_1 is unpredictable, since the duration of the metastable state of the first DFF is typically less than the period of GRO1, the second DFF will be triggered at the second rising edge of v_{o1} at which Q_1 has already assumed a stable state. It is worthy to note that the principle of the preceding two-stage quantization noise propagator is similar to the synchronization technique used to combat the metastability of delay line TDCs studied earlier.

It is often preferred to have GROs to oscillate as fast as possible so as to improve resolution; the time gap between the rising edge of T_{in1} and the second rising edge of v_{o1} of the two-stage quantization noise propagator might still not be sufficiently large such that the timing constraints of the DFF might still be violated. In order to prevent the quantization noise propagating DFF from entering a metastable stage, a three-stage quantization noise propagator that cascades 3 DFFs in a similar way as that of the two-stage noise propagator in Figure 3.54c was also used [127].

3.4.7.3 MASH Switched Relaxation Oscillator TDCs

It was shown in Section 3.4.5 that GRO TDCs are vulnerable to the effect of leakage and charge redistribution, and SRO TDCs eliminate these drawbacks inherently. GRO TDCs also suffer from gating delay. Since the oscillator in a switched relaxation oscillator TDC oscillates in both states, the effect of gating delay is reduced. If the switched oscillators are realized using switched relaxation oscillators, the large integrating capacitors of these oscillators also minimize the effect of leakage and charge redistribution, as detailed in Section 3.4.6. MASH TDCs can therefore be implemented using switched relaxation oscillator TDCs to take the advantages of these desirable properties.

Figure 3.55 shows the 1-1 MASH TDC proposed by Cao et al. with both the coarse and fine TDCs implemented using switched relaxation oscillator TDCs [135,136,141,142]. The input time variable T_{in1} is digitized by the signal TDC implemented using a switched relaxation oscillator TDC. The signal TDC provides first-order noise-shaping on its quantization noise. The input of the residue TDC, T_{in2}, is generated using a resettable DFF with its clock input as the first rising edge of v_{o1} and its reset input as the falling edge of T_{in1}. This quantization error propagation method was introduced by Konishi et al. in [46,127,140] and studied in detail in Figure 3.54 earlier. To minimize the effect of the metastability of DFFs, two resettable DFFs can be cascaded, also detailed in Figure 3.54 [46].

The input of the residue TDC is digitized by another switched relaxation oscillator that is identical to that of the signal TDC. The cascaded MASH configuration ensures that not only the quantization noise of the signal TDC is eliminated completely via a postprocessing step taking place in the digital domain when the cancellation condition is satisfied, the MASH TDC also provides second-order noise-shaping on the quantization noise of the residue TDC.

The performance of this MASH TDC is affected by a number of factors. Let us examine them in detail. It is seen from Figure 3.55 that the output of the quantization noise propagator T_{in2} contains not only the quantization error of the signal TDC, it also contains the multiple cycles of the relaxation oscillator of the signal TDC. As a result, the time variable of the input of the residue TDC is rather not small. The large input range constraint imposed on the residue TDC prevents us from choosing TDCs that have a high resolution as these TDCs often only have a small dynamic range. This will inevitably have a detrimental effect on the overall performance of the MASH TDC.

In the MASH TDC of Figure 3.55, both the signal and residue TDCs are implemented using identical switched relaxation oscillator TDCs. Relaxation oscillators typically have a low oscillation frequency, mainly due to the deployment of large integration capacitors to minimize the effect of the parasitic capacitances of the active devices of the oscillator and small charging and discharging currents to minimize

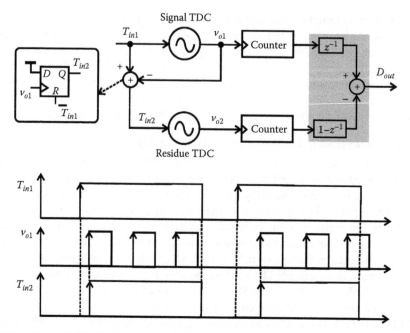

FIGURE 3.55 1-1 MASH TDC proposed by Cao et al. Both the coarse and fine TDCs are implemented using identical switched relaxation oscillator TDCs. (From Cao, Y. et al., A 0.7 mW 11b 1-1-1 MASH $\Delta\Sigma$ time-to-digital converter, in *IEEE International Solid-State Circuits Conference on Digest of Technical Papers*, 2011, pp. 480–481; Cao, Y. et al., A 0.7 mW 13b temperature-stable MASH $\Delta\Sigma$ TDC with delay-line assisted calibration, in *Proceedings of IEEE Asian Solid-State Circuits Conference*, 2011, pp. 361–364; Cao, Y. et al., *IEEE J. Solid-State Circuits*, 47(9), 2093, 2012; Cao, Y. et al., *IEEE Trans. Nucl. Sci.*, 59(4), 1382, 2012.)

the power consumption of the oscillator. The low oscillation frequency of relaxation oscillators limits the resolution of switched relaxation oscillator TDCs. This unique characteristic of switched relaxation oscillator TDCs differs from that of SRO TDCs where the ring oscillators typically oscillate at much higher frequencies [138].

The need for voltage comparators in construction of relaxation oscillators does not scale well with technology, largely due to the need for a precision reference voltage typically obtained using band-gap configurations and input offset voltage compensation circuitry.

3.4.7.4 MASH Gated Switched-Ring Oscillator TDCs

The relaxation oscillator of the signal TDC and that of the residue TDC of the preceding MASH switched relaxation oscillator TDC in Figure 3.55 oscillate at the same frequency. Since the input of the residue TDC is typically smaller as compared with that of the signal TDC, to improve the resolution of the residue TDC, it is desirable to have the oscillator of the residue TDC to oscillate at a frequency that is higher than that of the oscillator of the signal TDC. It was also shown that the input of the residue TDC of the preceding MASH switched relaxation oscillator TDC is only smaller than the input of the signal TDC by one oscillation period of the relaxation ring oscillator, the maximum. The large input of the residue TDC hence severely

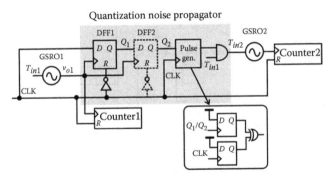

FIGURE 3.56 1-1 MASH gated switched ring oscillator TDC. (From Yu, W. et al., *IEEE Trans. Circuits Syst. I*, 61(8), 2281, 2014; Yu, W. et al., *IEEE Trans. Circuits Syst. I*, 60(4), 856, 2013; Yu, W. et al., A 148 fs_{rms} integrated noise 4 MHz bandwidth second-order $\Delta\Sigma$ time-to-digital converter with gated switched-ring oscillator, in *Proceedings of IEEE Custom Integrated Circuits Conference*, 2013, pp. 1–4.) The quantization noise propagator extracts the quantization noise of the signal TDC and feeds it to the residue TDC.

limits the choice of TDCs for the residue TDC. It becomes evident that in order to improve the performance of MASH TDCs, not only the input of the residue TDC should be smaller, preferably only the quantization noise of the signal TDC, the frequency of the oscillator of the residue TDC should also be higher as compared with that of the signal TDC. Figure 3.56 shows the 1-1 MASH TDC proposed by Yu et al. that provides these desirable characteristics [138,143,144].

The 1-1 MASH TDC consists of two gated SROs that oscillate at different frequencies with the oscillator of the residue oscillating at a higher frequency. The quantization error of the signal TDC is extracted and fed to the residue TDC by a quantization noise propagator. In the MASH switched relaxation oscillator TDC studied in Figure 3.55, the counters that are driven by the switched relaxation oscillators function as phase quantizers with a quantization error of 2π.

In the MASH TDC proposed by Yu et al. of Figure 3.56, both the number of the oscillation cycle of the SROs and the state transition of the output of the delay stages of the oscillators are recorded, as shown in Figure 3.57. The quantization noise of the TDC is therefore reduced from 2π of the TDC in Figure 3.55 to π/N in Figure 3.56 where N is the number of the delay stage of the oscillators. To capture the state transition of the output of the delay stages, the first-order digital differentiators shown in Figure 3.57 are employed. It is seen from the truth table given in the figure that the first-order digital differentiators can indeed capture the transition of the output of the delay stages of the oscillators. In order to perform digital differentiation, a clocking signal is employed to drive the DFFs that sample and store the incoming data. The exclusive-OR gates perform digital differentiation.

On the arrival of the sampling clock, both counters are reset. The content of the counters per sampling period yields the digital representation of T_{in1} as the content of the counters increments at both $T_{in1} = 1$ and $T_{in1} = 0$. The oscillator of the signal TDC oscillates at a higher frequency when $T_{in1} = 1$ and a lower frequency when $T_{in1} = 0$. Counter 1 consists of two parts: a counter whose content increments once the oscillator of the signal TDC completes an oscillation cycle, and state transitions detection logic

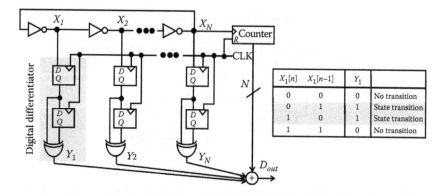

FIGURE 3.57 The counter records the number of the oscillation cycle of the oscillator per sampling period, while the first-order digital differentiators detect the state transition of the output of the delay stages of the oscillator per sampling period.

that records the number of the state transitions of the output of the delay stages of the oscillator per oscillation cycle. Since the counters are reset at the arrival of the sampling clock, T_{in1} must be smaller than the period of the sampling clock. Similarly, the counter of the residue TDC increments once the oscillator of the residue TDC completes an oscillation cycle and the state transition detection logic records the number of the state transition of the stages of the oscillator of the residue TDC per oscillation cycle.

To extract the quantization error of the signal TDC, which is the pulse bordered by the rising edge of T_{in1}, the input of the signal TDC, and the first rising edge of v_{o1}, the output of the oscillator of the signal TDC, a DFF is used with its data input from the sampling clock and clock input from v_{o1}, as shown in Figure 3.56. The output of DFF1 is set to logic-1 at the first rising edge of v_{o1}. The output of DFF1 and the sampling clock are fed to a pulse generator, which can be implemented using two DFFs and an OR2 gate. The AND2 gate generates the quantization error of the signal TDC. As pointed out by Ye et al. in [143]: since the assertion of T_{in1} is random, the phase difference between the sampling clock and T_{in1} may be small. As a result, the exacted quantization error pulse of the signal TDC may be too narrow for the oscillator of the residue TDC to digitize it accurately. To resolve this issue, another DFF can be employed, as shown in Figure 3.56. Note that the addition of the second DFF is similar to the synchronization technique used to combat the metastability of delay line TDCs studied earlier. The waveforms with the addition of the second DFF are shown in Figure 3.58. It is seen that the pulse width of the input of the residue TDC is the sum of the quantization error of the signal TDC and one oscillation period of the oscillator of the signal TDC. Since the added 2π phase offset is static, it can be removed later without difficulty.

3.4.8 ΔΣ TDCs

Although oversampling ΔΣ modulators are effective in achieving a good SNDR by means of spreading in-band quantization noise over the oversampling frequency range, noise-shaping, and the suppression of both quantization noise and nonlinearities residing in the forward path by the loop gain, their time-mode realization is

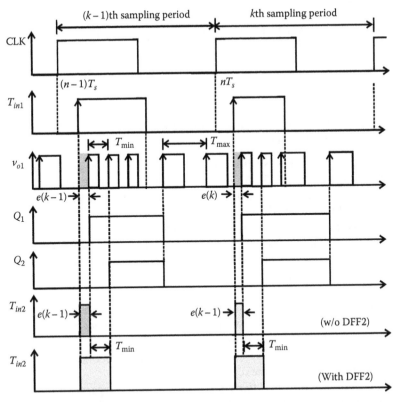

FIGURE 3.58 Waveforms of 1-1 MASH gated switched ring oscillator TDC. (From Yu, W. et al., *IEEE Trans. Circuits Syst. I*, 61(8), 2281, 2014; Yu, W. et al., *IEEE Trans. Circuits Syst. I*, 60(4), 856, 2013; Yu, W. et al., A 148 *fs$_{rms}$* integrated noise 4 MHz bandwidth second-order $\Delta\Sigma$ time-to-digital converter with gated switched-ring oscillator, in *Proceedings of IEEE Custom Integrated Circuits Conference*, 2013, pp. 1–4.) The ring oscillator of the signal TDC oscillates at $f_{max} = 1/T_{min}$ when $T_{in1} = 1$ and $f_{min} = 1/T_{max}$ when $T_{in1} = 0$.

rather difficult mainly due to the absence of a time integrator with a large in-band gain. A time integrator performs an integration operation on a time variable T_{in} over a given time duration and outputs a time variable T_{out}, i.e.,

$$T_{out}(t) = \int_0^t T_{in}(\tau)d\tau$$

Integrators play a cruel role in providing a large in-band loop gain critical to suppressing quantization noise, forward-path nonlinearities, and noise-shaping. Figure 3.59a shows the generic configuration of a time-mode $\Delta\Sigma$ modulator utilizing a time integrator. It is seen that if a digitally realized time integrator is available, the entire $\Delta\Sigma$ modulator will be digital, that is, an all-digital $\Delta\Sigma$ modulator.

Due to the unavailability of time integrators, most time-mode $\Delta\Sigma$ TDCs reported recently are realized using a partial time-mode partial voltage-mode approach; specifically, integrators are implemented using operational amplifiers (OTA) whose

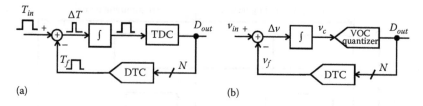

(a)   (b)

FIGURE 3.59   (a) Time-mode $\Delta\Sigma$ modulators with a TDC quantizer. (b) Pseudotime-mode $\Delta\Sigma$ modulators with a VCO quantizer.

input and output are voltages while quantizers are VCO-based multibit quantizers, as shown in Figure 3.58b [145]. Although VCO-based multibit quantization provides many attractive advantages over their voltage-mode counterparts, the performance of this pseudotime-mode $\Delta\Sigma$ modulator does not scale well with technology simply due to the poor scalability of the performance of the operational amplifiers of the integrator with technology. This modulator is also not attractive for low-power applications due to the difficulties in lowering the power consumption of operational amplifiers while fulfilling ever-stringent performance constraints on slew rate, bandwidth, and settling time.

Recently, a time-difference accumulator functioning as a time integrator was proposed by Hong et al. [18]. The core of the time difference accumulator is a time register capable of withholding a time variable indefinitely and releasing the stored time variable upon the assertion of a triggering signal [146]. Figure 3.60 shows the simplified schematic of the time register.

The time register consists of two gated delay cells and two generic static inverters. The gated delay cell is composed of a 2-to-1 multiplexer, a DFF, and a three-state inverter that has logic-1, logic-0, and hold states. The gated delay cell has two control signals, namely, Tr (trigger) and Hold. The former enables the gated delay cell so that its output can be set by its input, whereas the latter disabled the gated delay cell so

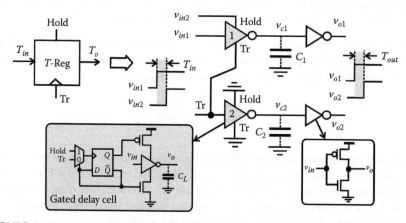

FIGURE 3.60   Time register. (From Hong, J. et al., A 0.004 mm² 250 μW $\Delta\Sigma$ TDC with time-difference accumulator and a 0.012 mm² 2.5 mW bangbang digital PLL using PRNG for low-power SoC applications, in *IEEE International Conference on Solid-State Circuits Digest of Technical Papers*, 2012, pp. 240–242.) *Abbreviation*: T-Reg, Time register.

that the output of the delay cell retains its current logic state indefinitely. In the following, we examine the operation of the gated delay cell.

Hold operation: Consider $Q = 0$ and $\bar{Q} = 1$. The inverter is enabled and the output of the inverter is set by the input of the inverter. The multiplexer in this case selects Hold as its output and the data input of the DFF is set to 1. If Hold = 1, the clock input of the DFF is at logic-1. As a result, $Q = 1$ and $\bar{Q} = 0$. The inverter is disabled and the output of the gated delay cell holds indefinitely.

Trigger operation: Now let us consider $Q = 1$ and $\bar{Q} = 0$. The inverter is disabled and the output of the inverter withholds. The multiplexer selects Tr as its output and the data input of the DFF is set to logic-0. If Tr is asserted, that is, Tr = 1, the clock input of the DFF is set to 1. As a result, $Q = 0$ and $\bar{Q} = 1$. The inverter is enabled and the output of the inverter held in the holding state is read out.

The preceding investigation shows that when Hold = 1, the output of the gated delay cell withholds its current logic state indefinitely provided that leakage is neglected. When Tr = 1, the output of the gated delay cell is the state when Hold = 1.

Having studied the operation of the gate delay cell, let us now look into the operation of the time register. The operation of the time register is briefly depicted as follows: Assume initially $v_{cl} = V_{DD}$. When $v_{in1} = 1$ while $v_{in2} = 0$ and Tr = 0, that is, $t_1 \leq t < t_2$, gated delay cell 1 is enabled. v_{cl} starts to drop with the discharge of C_1. If $v_{cl} < V_{T,inv}$, $v_{o1} = 0$ remains unchanged. When $v_{in2} = 1$ arrives, gated delay cell 1 enters a hold state and v_{cl} remains unchanged (Figure 3.61).

When the triggering signal Tr arrives, gated delay cell 1 ends its hold state and is reactivated. As a result, C_1 continues to discharge and v_{cl} starts to drop again with time. When v_{cl} drops below the threshold voltage of the downstream inverter $V_{T,inv}$, $v_{o1} = 1$ is set. Since gated delay cells 1 and 2 are identical, they have the same delay τ_d. As a result, we have $(t_2 - t_1) + (t_5 - t_4) = \tau_d$. Since $t_2 - t_1 = T_{in}$, $(t_6 - t_5) + (t_5 - t_4) = \tau_d$, and $t_6 - t_5 = T_{out}$, we have $T_{out} = T_{in}$. The preceding investigation shows that the time variable T_{in} is stored by the time register upon the arrival of v_{in2} and read out upon the arrival of the triggering signal Tr.

It becomes evident from the preceding analysis that in order to ensure the proper operation of the time register, $v_{cl} > V_{T,inv}$ is required at $t = t_2$, that is, the initial voltage drop during $T_{in} = 1$ must not be overly large. Since the discharge process of C_1 is governed by the time constant $R_{eq,n}C_1$ where $R_{eq,n}$ is the equivalent resistance of the two series-connected NMOS transistors of the gated delay cell, the discharge process is rather uncontrolled. One might suggest that the discharge of C_1 can be controlled by replacing the NMOS transistor in the gated delay cell with a current source. The inclusion of a current source undermines the goal of achieving an all-digital time integrator. Alternatively, the discharge process of the load capacitor can be digitally controlled, as shown in Figure 3.62. It is seen that by sizing transistors as per $(W/L)_1 = W/L$, $(W/L)_2 = 2(W/L)$, ..., $(W/L)_N = 2^N(W/L)$, the discharge resistance can be digitally controlled, so is the discharge process of the load capacitor.

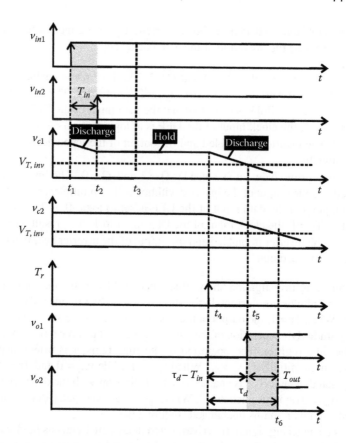

FIGURE 3.61 Waveforms of critical nodes of time register. τ_d is the delay of the gated delay cells. Specifically, τ_d is the amount of the time for the output of the inverter of the gated delay cell to drop from V_{DD} to $V_{T,inv}$.

The preceding time register can be utilized to construct a time subtracter or adder, as shown in Figure 3.63 [146]. It is seen that although the time register in Figure 3.60 needs two gated delay cells, the time adder only needs two, rather than four gated delay cells. To perform time addition, one needs to reverse the order of the second time input so that $T_{out} = T_{in1} - (-T_{in2}) = T_{in1} + T_{in2}$.

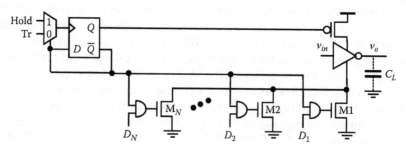

FIGURE 3.62 Gated delay cell with digitally controlled discharge resistance.

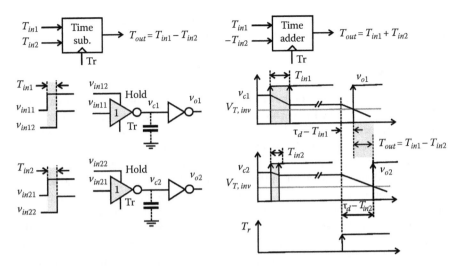

FIGURE 3.63 Waveforms of critical nodes of time adder and subtracter.

It should be noted that although the addition of two time variables can also be achieved by first converting the time variables into two voltages using time-to-voltage converters, the resultant voltages are added in the voltage-mode, the sum can then be converted back to a time variable using a VTC. Alternatively, the two time variables can first be converted into two digital variables using two TDCs, the resultant two digital variables are summed. The result is then converted back to a time variable using a DTC. Clearly, the preceding alternatives are not attractive due to multiple conversions.

The preceding time adder was utilized in [18] to construct a time accumulator, as shown in Figure 3.64. It uses a time-interleaved approach to perform addition using two back-to-back connected time adders to perform time accumulation. The difference between the input time variable T_{in} and the feedback time variable T_f in Figure 3.59b, denoted by ΔT, is accumulated over one sampling period by the accumulator. The accumulator functions as a time integrator. The output of the accumulator is digitized by the TDC and the output of the TDC is converted to the feedback time variable T_f using a DTC. Since the time accumulator is a first-order integrator, it will

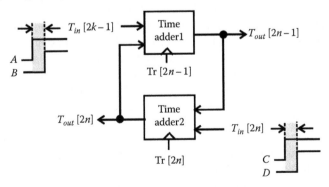

FIGURE 3.64 Time accumulator utilizing time adders and time registers.

provide 20 dB/decade noise-shaping if a sampling TDC is used. If a GRO TDC is used, a 40 dB/decade noise-shaping will be obtained.

3.5   SUMMARY

A comprehensive treatment of the principles, architectures, and design techniques of TDCs was provided. The key performance matrices of TDCs such as resolution, precision, nonlinearity, voltage and temperature sensitivities, conversion time, and conversion range were examined. Sampling TDCs that digitize time variables directly were studied with an emphasis on their advantages and limitations. Counter TDCs enjoy a large dynamic range but suffer from a high level of quantization noise and a low conversion speed. Resolution can be improved from the period of the reference clock in counter TDCs to the per-stage delay of delay lines if delay line interpolation is used to digitize the quantization errors of counter TDCs. Although the longer the delay line, the better the resolution of interpolation, the nonlinearity of the delay lines arising from the mismatch of the propagation delay of delay stages deteriorates with the increase in the length of the delay lines. To overcome this drawback, two-level delay line interpolation can be used. Resolution can be further improved to below one buffer delay by using pulse-shrinking delay lines, vernier delay lines, pulse stretching TDCs, or offset-sampling flash TDCs. Interstage interpolation using either active or passive networks also lowers resolution to per-stage delay.

Delay line TDCs provide a better resolution and enjoy low power consumption but suffer from a small dynamic range. Pulse-stretching TDCs enjoy a low level of power consumption but suffer from a long conversion time. The same holds for pulse-shrinking TDCs. Vernier delay line TDCs offer a good resolution at the cost of a higher level of silicon and power consumption, and a smaller range. Cyclic vernier delay line TDCs remove the drawbacks of conventional vernier delay line TDCs. Sampling offset flash TDCs provide the best resolution among all flash TDCs. SA-TDCs offer a high resolution and a large conversion range but suffer from a long conversion time and a high level of power consumption.

Noise-shaping TDCs that achieve a better resolution over a specific frequency range were also investigated. Unlike sampling TDCs, GRO TDCs offer an intrinsic characteristic of first-order noise-shaping. The noise-shaping of these TDCs, however, is undermined by charge leakage, charge redistribution, switching noise, and cross talk in the phase-holding state. SRO TDCs outperform GRO TDCs by eliminating the effect of charge leakage, charge redistribution, switching noise, and cross talk in the phase-holding stage of GRO TDCs. Switched relaxation oscillator TDCs offer the advantage of a low sensitivity to supply voltage fluctuation and the first-order noise-shaping of quantization noise. Time-mode $\Delta\Sigma$ TDCs that utilize time accumulators as time integrators remove the need for voltage-mode integrators whose performance scales poorly with technology. Mechanisms to achieve high-order time integration critical to time-mode $\Delta\Sigma$ modulators are yet to be developed.

The power consumption of sampling TDCs is generally higher as compared with that of noise-shaping TDCs. This is because the former use power-greedy complex configurations such as vernier delay lines or multilevel interpolation to improve resolution over the entire spectrum, while the latter improve resolution only over a specific frequency range by lowering in-band noise and distortion using feedback. Noise-shaping TDCs are therefore more attractive for low-power applications especially those that utilize time integrators. To digitize high-frequency time variables, sampling TDCs are the preferred choice.

REFERENCES

1. T. Yoshiaki and A. Takeshi. Simple voltage-to-time converter with high linearity. *IEEE Transactions on Instrumentation and Measurement*, 20(2):120–122, May 1971.

2. D. Porat. Review of sub-nanosecond time-interval measurements. *IEEE Transactions on Nuclear Science*, NS-20:36–51, September 1973.

3. K. Park and J. Park. 20 ps resolution time-to-digital converter for digital storage oscillator. *Proceedings of IEEE Nuclear Science Symposium*, 2, 876–881, Toronto, ON, 1998.

4. P. Chen, C. Chen, and Y. Shen. A low-cost low-power CMOS time-to-digital converter based on pulse stretching. *IEEE Transactions on Nuclear Science*, 53(4):2215–2220, August 2006.

5. C. Chen, P. Chen, C. Hwang, and W. Chang. A precise cyclic CMOS time-to-digital converter with low thermal sensitivity. *IEEE Transactions on Nuclear Science*, 52(4):834–838, August 2005.

6. M. Straayer and M. Perrott. A 12-bit, 10-MHz bandwidth, continuous-time $\Delta\Sigma$ ADC with a 5-bit, 950-MS/s VCO-based quantizer. *IEEE Journal of Solid-State Circuits*, 43(4):805–814, April 2008.

7. M. Straayer and M. Perrott. A multi-path gated ring oscillator TDC with first-order noise shaping. *IEEE Journal of Solid-State Circuits*, 44(4):1089–1098, April 2009.

8. M. Park and M. Perrott. A single-slope 80 Ms/s ADC using two-step time-to-digital conversion. In *IEEE International Symposium on Circuits and Systems*, Taipai, Taiwan, 2009, pp. 1125–1128.

9. B. Yousefzadeh and M. Sharifkhani. An audio band low voltage CT-$\Delta\Sigma$ modulator with VCO-based quantizer. In *Proceedings of IEEE International Conference on Electronics, Circuits and Systems*, Beirut, 2011, pp. 232–235.

10. B. Swann, B. Blalock, L. Clonts, D. Binkley, J. Rochelle, E. Breading, and K. Baldwin. A 100-ps time-resolution CMOS time-to-digital converter for positron emission tomography imaging applications. *IEEE Journal of Solid-State Circuits*, 39(11):1839–1852, November 2004.

11. M. Guttman and G. Roberts. K-locked-loop and its application in time mode ADC. In *Proceedings of IEEE International Symposium on Integrated Circuits*, Singapore, 2009, pp. 101–104.

12. V. Ravinuthula. Time-mode circuits for analog computations. PhD dissertation, University of Florida, Gainesville, FL, 2006.

13. S. Aouini, K. Chuai, and G. Roberts. Anti-imaging time-mode filter design using a PLL structure with transfer function DFT. *IEEE Transactions on Circuits and Systems I*, 59(1):66–79, January 2012.

14. P. Chen, C. Chen, C. Tsai, and W. Lu. A time-to-digital-converter-based CMOS smart temperature sensor. *IEEE Journal of Solid-State Circuits*, 40(8):1642–1648, August 2005.

15. T. Tokairin, M. Okada, M. Kitsunezuka, T. Maeda, and M. Fukaishi. A 2.1-to-2.8 GHz low-phase-noise all-digital frequency synthesizer with a time-windowed time-to-digital converter. *IEEE Journal of Solid-State Circuits*, 45(12):2582–2590, December 2010.

16. C. Hsu, M. Straayer, and M. Perrott. A low-noise wide-BW 3.6-GHz digital $\Delta\Sigma$ fractional-N frequency synthesizer with a noise-shaping time-to-digital converter and quantization noise cancellation. *IEEE Journal of Solid-State Circuits*, 43(12): 2776–2786, December 2008.

17. A. Ghaffari and A. Abrishamifar. A novel wide-range delay cell for DLLs. In *Proceedings of IEEE International Electrical and Computer Engineering Conference*, Dhaka, 2006, pp. 497–500.
18. J. Hong, S. Kim, J. Liu, N. Xing, T. Jang, J. Park, J. Kim, T. Kim, and H. Park. A 0.004 mm² 250 μW ΔΣ TDC with time-difference accumulator and a 0.012 mm² 2.5 mW bang-bang digital PLL using PRNG for low-power SoC applications. In *IEEE International Conference on Solid-State Circuits Digest of Technical Papers*, San Francisco, CA, 2012, pp. 240–242.
19. M. Rashdan, A. Yousif, J. Haslett, and B. Maundy. A new time-based architecture for serial communication links. In *Proceedings of IEEE International Conference on Electronics, Circuits, and Systems*, Yasmine Hammamet, Tunisia, 2009, pp. 531–534.
20. J. Kim, W. Yu, and S. Cho. A digital-intensive multi-mode multi-band receiver using a sinc² filter-embedded VCO-based ADC. *IEEE Transactions on Microwave Theory and Techniques*, 60(10):3254–3262, October 2012.
21. J. Jansson, V. Koskinen, A. Mantyniemi, and J. Kostamovaara. A multichannel high-precision CMOS time-to-digital converter for laser-scanner-based perception systems. *IEEE Transactions on Instrumentation and Measurement*, 61(9):2581–2590, September 2012.
22. J. Kostamovaara and R. Myllyla. Time-to-digital converter with an analog interpolation circuit. *Review of Scientific Instruments*, 57(11):2880–2885, November 1986.
23. J. Kalisz. Review of methods for time interval measurements with picosecond resolution. *Metrologia*, 41:17–32, 2004.
24. P. Napilitano, F. Alimenti, and P. Carbone. A novel sample-and-hold-based time-to-digital converter architecture. *IEEE Transactions on Instrumentation and Measurement*, 59(5):1019–1026, May 2010.
25. P. Chen, C. Hwang, and W. Chang. A precise cyclic CMOS time-to-digital converter with low thermal sensitivity. *IEEE Transactions on Nuclear Science*, 52(4):834–838, August 2005.
26. E. Raisanen-Ruotsalainen, T. Rahkonen, and J. Kostamovaara. An integrated time-to-digital converter with 30 ps single-shot precision. *IEEE Journal of Solid-State Circuits*, 35(10):1507–1510, October 2000.
27. S. Henzler. *Time-to-Digital Converters*. Springer, New York, 2010.
28. T. Rahkonen, J. Kostamovaara, and S. Saynajakangas. Time interval measurements using integrated tapped CMOS delay lines. In *Proceedings of IEEE Mid-West Symposium on Circuits and Systems*, Champaign, IL, 1990, pp. 201–205.
29. Y. Aria, T. Matsumura, and K. Endo. A CMOS four-channel × 1K memory LSI with 1-ns/b resolution. *IEEE Transactions on Circuits and Systems II*, 27(3):359–364, March 1992.
30. T. Rahkonen and J. Kostamovaara. The use of stabilized CMOS delay lines in the digitization of short time intervals. *IEEE Journal of Solid-State Circuits*, 28(8):887–894, August 1993.
31. A. Hajimiri, S. Limotyrakis, and T. Lee. Jitter and phase noise in ring oscillators. *IEEE Journal of Solid-State Circuits*, 34(6):790–804, June 1999.
32. J. McNeil. Jitter in ring oscillators. *IEEE Journal of Solid-State Circuits*, 32(6):870–879, June 1997.
33. F. Herzal and B. Razavi. A study of oscillator jitter due to supply and substrate noise. *IEEE Transactions on Circuits and Systems II*, 46(1):56–62, January 1999.
34. N. Xing, W. Shin, D. Jeong, and S. Kim. A PVT-insensitive time-to-digital converter using fractional difference vernier delay lines. In *Proceedings of IEEE International SOC Conference*, Belfast, 2009, pp. 43–46.
35. C. Hwang, P. Chen, and H. Tsao. A high-precision time-to-digital converter using a two-level conversion scheme. *IEEE Transactions on Nuclear Science*, 51(4):1349–1352, August 2004.

36. Altera Corporation. Understanding metastability in FPGAs. White Paper, 2009, pp. 1–6.
37. T. Kacprzak and A. Albicki. Analysis of metastable operation in RS CMOS flip-flops. *IEEE Journal of Solid-State Circuits*, 22(1):57–64, February 1987.
38. W. Noije, W. Liu, and J. Navarro. Precise final state determination of mismatched CMOS latches. *IEEE Journal of Solid-State Circuits*, 30(5):607–611, May 1995.
39. L. Marino. General theory of metastable operation. *IEEE Transactions on Computers*, 30(2):107–115, February 1981.
40. T. Kacprzak. Analysis of oscillatory metastable operation of a RS flip-flop. *IEEE Journal of Solid-State Circuits*, 23(1):260–266, February 1988.
41. H. Veendrick. The behavior of flip-flop used as synchronizers and prediction of their failure rate. *IEEE Journal of Solid-State Circuits*, 15(2):169–176, April 1980.
42. L. Kim and R. Dutton. Metastability of CMOS latch/flip-flop. *IEEE Journal of Solid-State Circuits*, 25(4):942–950, 1990.
43. K. Jeppson. Comments on the metastable behavior of mismatched CMOS latches. *IEEE Journal of Solid-State Circuits*, 31(2):275–277, February 1996.
44. R. Ginosar. Metastability and synchronizers: A tutorial. *IEEE Design and Test of Computers*, September/October 2011, pp. 23–35.
45. J. Rabaey. *Digital Integrated Circuits—A Design Perspective*, 1st edn. Prentice Hall, Upper Saddle River, NJ, 1996.
46. K. Okuno, T. Konishi, S. Izumi, M. Yoshimoto, and H. Kawaguchi. A 62 dB SNDR second-order gated ring oscillator TDC with two-stage dynamic D-type flipflops a quantization noise propagator. In *Proceedings of IEEE New Circuits and Systems*, Montreal, Que, 2012, pp. 289–292.
47. M. Lee and A. Abidi. A 9b, 1.25 ps resolution coarse-fine time-to-digital converter in 90 nm CMOS that amplifies a time residue. *IEEE Journal of Solid-State Circuits*, 43(4):769–777, April 2008.
48. S. Chung, K. Hwang, W. Lee, and L. Kim. A high resolution metastability-independent two-step gated ring oscillator TDC with enhanced noise shaping. In *Proceedings of IEEE International Symposium on Circuits and Systems*, Paris, 2010, pp. 1300–1303.
49. K. Nose, M. Kajita, and M. Mizuno. A 1 ps resolution jitter-measurement macro using interpolated jitter oversampling. In *IEEE International Solid-State Circuits Conference Digest of Technical Papers*, San Francisco, CA, 2006, pp. 520–521.
50. E. Raisanen-Ruotsalainen, T. Rahkonen, and J. Kostamovaara. Integrated time-to-digital converters based on interpolation. *Analog Integrated Circuits and Signal Processing*, 15:49–57, 1997.
51. A. Mantyniemi, T. Rahkonen, and J. Kostamovaara. An integrated digital CMOS time-to-digital converter with sub-gate-delay resolution. *Analog Integrated Circuits and Signal Processing*, 22:61–70, 1999.
52. H. Huang, S. Wu, and Y. Tsai. A new cycle-time-to-digital converter with two level conversion scheme. In *Proceedings of IEEE International Symposium on Circuits and Systems*, New Orlean, 2007, pp. 2160–2163.
53. N. Narku-Tetteh, A. Titriku, and S. Palermo. A 15b, sub-10 ps resolution, low dead time, wide range two-stage TDC. In *Proceedings of IEEE Mid-West Symposium on Circuits and Systems*, College Station, TX, 2014, pp. 13–16.
54. R. Staszewski, S. Vemulapalli, P. Vallur, J. Wallberg, and P. Balsara. 1.3 V 20 ps time-to-digital converter for frequency synthesis in 90 nm CMOS. *IEEE Transactions on Circuits and Systems II*, 53(3):220–224, March 2006.
55. T. Knotts, D. Chu, and J. Sommer. A 500 MHz time digitizer IC with 15.625 ps resolution. In *IEEE International Solid-State Circuits Conference on Digest of Technical Papers*, 1994, pp. 58–59.

56. S. Henzler, S. Koeppe, W. Kamp, and D. Schmitt-Landsiedel. 90 nm 4.7 ps-resolution 0.7-LSB single-shot precision and 19 pJ-per-shot local passive interpolation time-to-digital converter with on-chip characterization. In *IEEE International Solid-State Circuits Conference on Digest of Technical Papers*, 2008, pp. 548–635.

57. T. Jang, J. Kim, Y. Yoon, and S. Cho. A highly-digital VCO-based analog-to-digital converter using phase interpolator and digital calibration. *IEEE Transactions on VLSI Systems*, 20(8):1368–1372, August 2012.

58. J. Christiansen. An integrated high resolution CMOS timing generator based on an array of delay locked loop. *IEEE Journal of Solid-State Circuits*, 31(7):952–957, July 1996.

59. R. Baron. The vernier time-measuring technique. *Proceedings of the IRE*, 45(1):21–30, January 1957.

60. C. Gray, W. Liu, A. Wilhelmus, N. Van, A. Thomas, and K. Ralph. A sampling technique and its CMOS implementation with 1Gb/s bandwidth and 25 ps resolution. *IEEE Journal of Solid-State Circuits*, 29(3):340–349, March 1994.

61. C. Ljuslin, J. Christiansen, A. Marchioro, and O. Klingsheim. An integrated 16-channel CMOS time to digital converter. *IEEE Transactions on Nuclear Science*, 41(4):1104–1108, August 1994.

62. P. Dudek, S. Szczepanski, and J. Hatfield. A high-resolution CMOS time-to-digital converter utilizing a vernier delay line. *IEEE Journal of Solid-State Circuits*, 35(2):240–247, February 2000.

63. C. Huang, P. Chen, and H. Tsao. A high-resolution and fast conversion time-to-digital converter. *Proceedings of IEEE International Symposium on Circuits and Systems*, Bangkok, 1:37–40, 2003.

64. K. Nose, M. Kajita, and M. Mizuno. A 1-ps resolution jitter-measurement macro using interpolated jitter oversampling. *IEEE Journal of Solid-State Circuits*, 41(12):2911–2920, December 2008.

65. N. Xing, W. Shin, D. Jeong, and S. Kim. High-resolution time-to-digital converter utilizing fractional difference conversion scheme. *IET Electronics Letters*, 46(6):398–399, March 2010.

66. N. Xing, J. Woo, W. Shin, H. Lee, and S. Kim. A 14.6 ps resolution, 50 ns input-range cyclic time-to-digital converter using fractional difference conversion method. *IEEE Transactions on Circuits and Systems I*, 57(12):3064–3072, December 2010.

67. G. Li and H. Chou. A high resolution time-to-digital converter using two-level vernier delay line technique. In *Proceedings of IEEE Nuclear Science Symposium on Conference Record*, Honolulu, 2007, pp. 276–280.

68. J. Jansson, A. Mntyniemi, and J. Kostamovaara. Synchronization in a multi-level CMOS time-to-digital converter. *IEEE Transactions on Circuits and Systems I*, 56(8):1622–1634, August 2009.

69. A. Chan and G. Roberts. A jitter characterization system using a component-invariant vernier delay line. *IEEE Transactions on VLSI Systems*, 12(1):79–95, January 2004.

70. P. Chen, C. Chen, J. Zheng, and Y. Shen. A PVT insensitive vernier-based time-to-digital converter with extended input range and high accuracy. *IEEE Transactions on Nuclear Science*, 54(2):294–302, April 2007.

71. A. Liscidini, L. Vercesi, and R. Castello. Time to digital converter based on a 2-dimension vernier architecture. In *Proceedings of IEEE Custom Integrated Circuits Conference*, San Jose, CA, 2009, pp. 45–48.

72. A. Liscidini, L. Vercesi, and R. Castello. Two-dimensions vernier time to digital converter. *IEEE Journal of Solid-State Circuits*, 45(8):1504–1512, August 2010.

73. L. Vercesi, L. Fanori, F. Bernardinis, A. Liscidini, and R. Castello. A dither-less all digital PLL for cellular transmitters. *IEEE Journal of Solid-State Circuits*, 47(8):1908–1920, August 2012.

74. J. Kalisz, M. Pawlowski, and R. Pelka. Error analysis and design of the Nutt time-interval digitiser with picosecond resolution. *Journal of Physics E: Scientific Instruments*, 20:1330–1341, 1987.

75. K. Maatta and J. Kostamovaara. A high-precision time-to-digital converter for pulsed time-of-flight laser radar applications. *IEEE Transactions on Instrumentation and Measurement*, 47(2):521–536, April 1998.

76. P. Keranen, K. Maatta, and J. Kostamovaara. Wide-range time-to-digital converter with 1 ps single-shot precision. *IEEE Transactions on Instrumentation and Measurement*, 60(9):3162–3172, September 2011.

77. K. Park and J. Park. Time-to-digital converter of very high pulse stretching ratio for digital storage oscilloscope. *Review of Scientific Instruments*, 70(2):1568–1574, February 1999.

78. R. Nutt. Digital time intervalometer. *Review of Scientific Instruments*, 39:1342–1345, 1968.

79. B. Kurko. A picosecond resolution time digitizer for laser ranging. *IEEE Transactions on Nuclear Science*, NS-25(1):75–80, February 1978.

80. C. Kuo, S. Chen, and S. Liu. Magnetic-field-to-digital converter using PWM and TDC techniques. *IET Proceedings of Circuits Devices & Systems*, 153(3):247–252, June 2006.

81. B. Dehlaghi, S. Magierowski, and L. Belostotski. Highly-linear time-difference amplifier with low sensitivity to process variations. *IET Electronics Letters*, 47(13):743–745, June 2011.

82. H. Kwon, J. Lee, J. Sim, and H. Park. A high-gain wide-input-range time amplifier with an open-loop architecture and a gain equal to current bias ratio. In *Proceedings of IEEE Asian Solid-State Circuits Conference*, 2011, pp. 325–328.

83. S. Lee, Y. Seo, H. Park, and J. Sim. A 1 GHz ADPLL with a 125 ps minimum-resolution sub-exponent TDC in 0.18 μm CMOS. *IEEE Journal of Solid-State Circuits*, Jeju, Korea, 45(12):2827–2881, December 2010.

84. R. Rashidzadeh, R. Muscedere, M. Ahmadi, and W. Miller. A delay generation technique for narrow time interval measurement. *IEEE Transactions on Instrumentation and Measurement*, 58(7):2245–2252, July 2009.

85. C. Lin and M. Syrzycki. Pico-second time interval amplification. In *Proceedings of IEEE International SOC Design Conference*, Seoul, 2010, pp. 201–204.

86. T. Nakura, S. Mandai, M. Ikeda, and K. Asada. Time difference amplifier using closed-loop gain control. In *Symposium on VLSI Circuits Digest of Technical Papers*, Kyoto, 2009, pp. 208–209.

87. S. Mandai, T. Nakura, M. Ikeda, and K. Asada. Cascaded time difference amplifier using differential logic delay cell. In *Proceedings of International SOC Design Conference*, Busan, Korea, 2009, pp. 299–304.

88. S. Mandai and E. Charbon. A 128-channel, 8.9-ps LSB, column-parallel two-stage TDC based on time difference amplification for time-resolved imaging. *IEEE Transactions on Nuclear Science*, 59(5):2463–2470, October 2012.

89. D. Kinniment, A. Bystrov, and A. Yakovlev. Synchronization circuit performance. *IEEE Journal of Solid-State Circuits*, 37(2):202–209, February 2002.

90. A. Abas, A. Bystrov, D. Kinnimnt, O. Maevsky, G. Russell, and A. Yakovlev. Time difference amplifier. *IEE Electronics Letters*, 38(23):1437–1438, November 2002.

91. A. Abas, G. Russell, and D. Kinniment. Embedded high-resolution delay measurement system using time amplification. *IET Computers and Digital Techniques*, 1(2):77–86, March 2002.

92. M. Oulmane and G. Roberts. CMOS time amplifier for femto-second resolution timing measurement. In *Proceedings of IEEE International Symposium on Circuits and Systems*, Vancouver, BC, 2004, pp. 509–512.

93. B. Tong, W. Yan, and X. Zhou. A constant-gain time-amplifier with digital self-calibration. In *Proceedings of IEEE International ASIC Conference*, Changsha, China, 2009, pp. 1133–1136.

94. A. Alahmadi, C. Russell, and A. Yakovlev. Time difference amplifier design with improved performance parameters. *IET Electronics Letters*, 48(10):562–563, May 2012.

95. M. Heo, D. Kwon, and M. Lee. Low-power programmable high-gain time difference amplifier with regeneration time control. *IET Electronics Letters*, 50(16):1129–1131, July 2014.

96. E. Raisanen-Ruotsalainen, T. Rahkonen, and J. Kostamovaara. A low-power CMOS time-to-digital converter. *IEEE Journal of Solid-State Circuits*, 30(9):984–990, September 1995.

97. K. Karadamogou, N. Paschalidis, E. Sarris, N. Stamatopoulos, G. Kottaras, and V. Paschalidis. An 11-bit high-resolution and adjustable-range CMOS tie-to-digital converter for space science instruments. *IEEE Journal of Solid-State Circuits*, 39(1): 214–222, January 2004.

98. P. Chen and S. Liu. A cyclic CMOS time-to-digital converter with deep sub-nanosecond resolution. In *Proceedings of IEEE Custom Integrated Circuits Conference*, San Diego, CA, 1999, pp. 605–608.

99. P. Chen, S. Liu, and J. Wu. A CMOS pulse-shrinking delay element for time interval measurement. *IEEE Transactions on Circuits and Systems II*, 47(9):954–958, September 2000.

100. S. Kang and Y. Leblebici. *CMOS Digital Integrated Circuits*, 3rd edn. McGraw-Hill, Boston, MA, 2003.

101. F. Yuan. *CMOS Circuits for Passive Wireless Microsystems*. Springer, New York, 2010.

102. C. Chen and H. Chen. A low-cost CMOS smart temperature sensor using a thermal-sensing and pulse-shrinking delay line. *IEEE Sensors Journal*, 14(1):278–284, January 2014.

103. A. Mantyniemi, T. Rahkonen, and J. Kostamovaara. A CMOS time-to-digital converter (TDC) based on a cyclic time domain successive approximation interpolation method. *IEEE Journal of Solid-State Circuits*, 44(11):3067–3078, November 2009.

104. G. Nagaraj, S. Miller, B. Stengel, G. Cafaro, T. Gradishar, S. Olson, and R. Hekmann. A self-calibrating sub-picosecond resolution digital-to-time converter. In *Proceedings of IEEE International Microwave Symposium*, Honolulu, 2007, pp. 2201–2204.

105. Y. Choi, S. Yoo, and H. Yoo. A full digital polar transmitter using a digital-to-time converter for high data rate system. In *Proceedings of IEEE International Symposium on Radio-Frequency Integration Technology*, Singapore, 2009, pp. 56–59.

106. S. Al-Ahdab, A. Mantyniemi, and J. Kostamovaara. A 12-bit digital-to-time converter (DTC) for time-to-digital converter (TDC) and other time domain signal processing applications. In *Proceedings of IEEE NORCHIP*, Tampere, Finland, 2010, pp. 1–4.

107. S. Al-Ahdah, A. Mantyniemi, and J. Kostamovaara. A 12-bit digital-to-time converter (DTC) with sub-ps-level resolution using current DAC and differential switch for time-to-digital converter (TDC). In *Proceedings of IEEE Instrumentation and Measurement Technology Conference*, Graz, Austria, 2012, pp. 2668–2671.

108. S. Al-Ahdab, A. Mantyniemi, and J. Kostamovaara. Review of a time-to-digital converter (TDC) based on cyclic time domain successive approximation interpolator method with sub-ps-level resolution. In *Proceedings of IEEE Nordic-Mediterranean Workshop on Time-to-Digital Converters*, Perugia, Italy, 2013, pp. 1–5.

109. G. Roberts and M. Ali-Bakhshian. A brief introduction to time-to-digital and digital-to-time converters. *IEEE Journal of Solid-State Circuits*, 57(3):153–157, March 2010.

110. S. Li and C. Salthouse. Digital-to-time converter for fluorescence lifetime imaging. In *Proceedings of IEEE International Instrumentation and Measurement Technical Conference*, Minneapolis, MN, 2012, pp. 894–897.

111. P. Levine and G. Roberts. A calibration technique for a high-resolution flash time-to-digital converter. In *Proceedings of IEEE International Symposium on Circuits and Systems*, vol. 1, Vancouver, BC, 2004, pp. 253–256.

112. P. Levine and G. Roberts. High-resolution flash time-to-digital conversion and calibration for system-on-chip testing. *IEEE Proceedings of Computers and Digital Techniques*, 152(3):415–426, May 2005.

113. V. Gutnik and A. Chandrakasan. On-chip picosecond time measurement. In *Symposium on VLSI Circuits Digest of Technical Papers*, Honolulu, HI, 2000, pp. 52–53.

114. N. Minas, D. Kinniment, K. Heron, and G. Russell. A high resolution flash time-to-digital converter taking into account process variability. In *Proceedings of IEEE International Symposium on Asynchronous Circuits and Systems*, Berkeley, CA, 2007, pp. 163–174.

115. T. Yamaguchi, S. Komatsu, M. Abbas, K. Asada, N. Maikhanh, and J. Tandon. A CMOS flash TDC with 0.84–1.3 ps resolution using standard cells. In *Proceedings of IEEE Radio-Frequency Integrated Circuits*, Montreal, Que, 2012, pp. 527–530.

116. M. Zanuso, S. Levantino, A. Puggelli, C. Samori, and A. Lacait. Time-to-digital converter with 3-ps resolution and digital linearization algorithm. In *Proceedings of IEEE ESSCIRC*, Seville, Spain, 2010, pp. 262–265.

117. C. Yao, F. Jonsson, J. Chen, and L. Zheng. A high-resolution time-to-digital converter based on parallel delay elements. In *Proceedings of IEEE International Symposium on Circuits and Systems*, Seoul, 2012, pp. 3158–3161.

118. Y. Seo, J. Kim, H. Park, and J. Sim. A 0.63 ps resolution, 11 b pipelined TDC in 013 μm CMOS. In *Symposium on VLSI Circuits Digest of Technical Papers*, Honolulu, 2012, pp. 152–153.

119. J. Kim, Y. Seo, Y. Suh, H. Park, and J. Sim. A 300-MS/s 1.76-ps-resolution 10b asynchronous pipelined time-to-digital converter with on-chip digital background calibration in 0.13 μm CMOS. *IEEE Journal of Solid-State Circuits*, 48(2):516–526, February 2013.

120. K. Kim, W. Yu, and S. Cho. A 9 b 1.12 ps resolution 2.5 b/stage pipelined time-to-digital converter in 65 nm CMOS using time-register. In *Symposium on VLSI Circuits Digest of Technical Papers*, Honolulu, 2013, pp. 136–137.

121. K. Hwang and L. Kim. An area efficient asynchronous gated ring oscillator TDC with minimum GRO stages. In *Proceedings of IEEE International Symposium on Circuits and Systems*, Paris, 2010, pp. 3973–3976.

122. S. Lee, B. Kim, and K. Lee. A novel high-speed ring oscillator for multiphase clock generation using negative skewed delay scheme. *IEEE Journal of Solid-State Circuits*, 32(2):289–291, February 1997.

123. C. Park and B. Kim. A low-noise, 900-MHz VCO in 0.6 μm CMOS. *IEEE Journal of Solid-State Circuits*, 34(5):586–591, May 1999.

124. F. Yuan. A modified Park-Kim voltage-controlled ring oscillator for multi-Gbps serial links. *Analog Integrated Circuits and Signal Processing*, 47(3):345–353, June 2006.

125. A. Elshazly, S. Rao, B. Young, and P. Hanumolu. A 13b 315 $f_{s,rms}$ 2 mW 500 MS/s 1 MHz bandwidth highly digital time-to-digital converter using switched ring oscillators. In *IEEE International Solid-State Circuits Conference on Digest of Technical Papers*, 2012, pp. 464–465.

126. A. Elshazly, S. Rao, B. Young, and P. Hanumolu. A noise-shaping time-to-digital converter using switched-ring oscillators—Analysis, design, and measurement techniques. *IEEE Journal of Solid-State Circuits*, 49(5):1184–1197, May 2014.

127. T. Konishi, K. Okumo, S. Izumi, M. Yoshimoto, and H. Kawaguchi. A 61 dB SNDR 700 μm second-order all-digital TDC with low-jitter frequency shift oscillator and dynamic flipflops. In *Symposium on VLSI Circuits Digest of Technical Papers*, Honolulu, 2012, pp. 190–191.

128. P. Lu and P. Andreani. A high-resolution vernier gated-ring-oscillator TDC in 90-nm CMOS. In *Proceedings of NORCHIP*, Tampere, Finland, 2010, pp. 1–4.

129. P. Lu, P. Andreani, and A. Liscidini. A 90 nm CMOS gated-ring-oscillator-based vernier time-to-digital converter for DPLLs. In *Proceedings of IEEE ESSCIRC*, Helsinkl, Finland, 2011, pp. 459–462.

130. P. Lu, Y. Wu, and P. Andreani. A 90 nm CMOS digital PLL based on vernier-gated-ring-oscillator time-to-digital converter. In *Proceedings of IEEE International Symposium on Circuits and Systems*, Seoul, 2012, pp. 2593–2596.

131. T. Carusone, D. Johns, and K. Martin. *Analog Integrated Circuit Design*, 2nd edn. John Wiley & Sons, New York, 2012.

132. P. Lu, P. Andreani, and A. Liscidini. A 90 nm CMOS gated-ring-oscillator-based 2-dimensional vernier time-to-digital converter. In *Proceedings of IEEE NOCHIP*, Copenhagen, 2012, pp. 1–4.

133. P. Lu, P. Andreani, and A. Liscidini. A 2-D GRO vernier time-to-digital converter with large input range and small latency. In *Proceedings of IEEE Radio-Frequency Integrated Circuits*, Seattle, WA, 2013, pp. 151–154.

134. F. Yuan and N. Soltani. A low-voltage low VDD sensitivity relaxation oscillator for passive wireless microsystems. *IET Electronics Letters*, 45(21):1057–1058, October 2009.

135. Y. Cao, P. Leroux, W. De Cock, and M. Steyaert. A 0.7 mW 11b 1-1-1 MASH $\Delta\Sigma$ time-to-digital converter. In *IEEE International Solid-State Circuits Conference on Digest of Technical Papers*, 2011, pp. 480–481.

136. Y. Cao, P. Leroux, W. De Cock, and M. Steyaert. A 0.7 mW 13b temperature-stable MASH $\Delta\Sigma$ TDC with delay-line assisted calibration. In *Proceedings of IEEE Asian Solid-State Circuits Conference*, Jeju, Korea, 2011, pp. 361–364.

137. T. Konishi, K. Okumo, S. Izumi, M. Yoshimoto, and H. Kawaguchi. A 51 dB SNDR DCO-based TDC using two-stage second-order noise shaping. In *Proceedings of IEEE International Symposium on Circuits and Systems*, Seoul, 2012, pp. 3170–3173.

138. W. Yu, K. Kim, and S. Cho. A 148 fs_{rms} integrated noise 4 MHz bandwidth second-order $\Delta\Sigma$ time-to-digital converter with gated switched-ring oscillator. *IEEE Transactions on Circuits and Systems I*, 61(8):2281–2289, August 2014.

139. R. Schreier and G. Temes. *Understanding Delta-Sigma Data Converters*. John Wiley & Sons, Hoboken, NJ, 2005.

140. T. Konishi, K. Okumo, S. Izumi, M. Yoshimoto, and H. Kawaguchi. A 40-nm 640-μm² 45-dB opamp-less all-digital second-order MASH $\Delta\Sigma$ ADC. In *Proceedings of IEEE International Symposium on Circuits and Systems*, Rio de Janeiro, Brazil, 2011, pp. 518–521.

141. Y. Cao, W. De Cock, M. Steyaert, and P. Leroux. 1-1-1 MASH $\Delta\Sigma$ time-to-digital converters with 6 ps resolution and third-order noise-shaping. *IEEE Journal of Solid-State Circuits*, 47(9):2093–2106, September 2012.

142. Y. Cao, W. De Cock, M. Steyaert, and P. Leroux. Design and assessment of a 6 ps-resolution time-to-digital converter with 5 MGy gamma-dose tolerance for LIDAR application. *IEEE Transactions on Nuclear Science*, 59(4):1382–1389, August 2012.

143. W. Yu, J. Kim, K. Kim, and S. Cho. A time-domain high-order MASH $\Delta\Sigma$ ADC using voltage-controlled gated-ring oscillator. *IEEE Transactions on Circuits and Systems I*, 60(4):856–866, August 2013.

144. W. Yu, K. Kim, and S. Cho. A 148 fs_{rms} integrated noise 4 MHz bandwidth second-order $\Delta\Sigma$ time-to-digital converter with gated switched-ring oscillator. In *Proceedings of IEEE Custom Integrated Circuits Conference*, San Jose, CA, 2013, pp. 1–4.

145. M. Gande, N. Maghari, T. Oh, and U. Moon. A 71 dB dynamic range third-order $\Delta\Sigma$ TDC using charge pump. In *Symposium on VLSI Circuits Digest of Technical Papers*, Honolulu, 2012, pp. 168–169.

146. S. Kim. Time domain algebraic operation circuits for high performance mixed-mode system. MS thesis, Korean Advanced Institute of Science and Technology, Daejeon, South Korea, 2010.

4 A Novel Three-Step Time-to-Digital Converter with Phase Interpolation

Kang-Yoon Lee

CONTENTS

4.1 INTRODUCTION

A time-to-digital converter (TDC) is similar to an analog-to-digital converter (ADC), except that, instead of quantizing voltage or current, the TDC quantizes time intervals between two rising edges. Originally developed for nuclear experiments to locate single-shot events [1], the TDC is now being used in many applications such as laser range finders, space science instruments, and measurement devices.

Recently, it has been employed to measure phase in all-digital phase-locked loops (ADPLLs) [2]. The TDC replaces a charge pump, the only true analog component in conventional PLLs, and allows the output word to drive a digital loop filter. An all-digital PLL brings with it the advantages of programmability and easy calibration. As the minimum feature size decreases, the supply voltage will also decrease. While an analog PLL needs to be redesigned as the process is changed, a digital PLL can be easily translated into a new process.

However, just as in any digital replacement of an analog function, the TDC creates quantization noise, which dominates the loop's in-band phase noise in a digital PLL [2].

To improve the frequency resolution and tuning range performance of the ADPLL, the performance of the TDC is very important. A conventional analog PLL uses a charge-pump circuit. However, an analog PLL gets worse in jitter performance because of the charge-pump feed-through and mismatches.

Figure 4.1 shows the block diagram of example ADPLL [3]. It uses a TDC instead of a charge-pump. The use of the TDC eliminates the current sources and the RC filter found in analog PLLs. The TDC is used to compare the DCO output phase to a reference phase. It needs to have a high resolution in order to improve the phase noise of the digital PLL [2]. The relationship between the resolution (ΔT_{RES}) of TDC and in-band phase noise of ADPLL is shown in Equation 4.1.

$$\text{In-band phase noise} = 10\log\left[\frac{(2\pi)^2}{12}\left(\frac{\Delta t_{inv}}{T_V}\right)^2 \cdot \frac{1}{f_R}\right] \tag{4.1}$$

where T_V and f_R are the period of the DCO clock and frequency of the reference clock, respectively. In order to meet the phase noise requirements, the resolution of TDC should be fine based on Equation 4.1.

The delay chain of buffers [1,4] and the vernier delay line [5] are well-known methods to realize a TDC. In the delay chain shown in Figure 4.2, the rising edge of the F_{DCO} signal propagates through the chain of buffers; when the rising edge of the F_{REF} signal arrives, a flip-flop samples the output of each buffer and produces a thermometer code that locates the relative time interval. However, this simple scheme cannot resolve the time interval better than a single buffer delay.

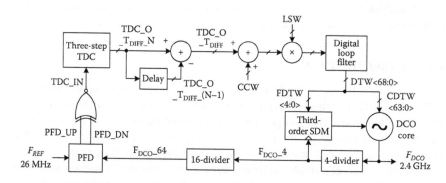

FIGURE 4.1   Block diagram of example ADPLL.

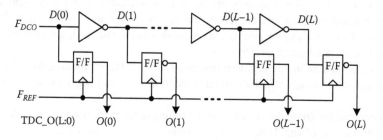

FIGURE 4.2   Block diagram of conventional TDC with a chain of inverters.

On the other hand, using the delay difference between unequal buffers, the vernier delay line can resolve more finely, but its area increases linearly with the resolution, and the devices must match more tightly. This leads to high power consumption. Calibrating the time offsets in the buffers and comparators can improve the resolution [6], but the measurement range is limited because of the calibration complexity. The TDC can be designed with a delay time that is less than that of the inverter by using a phase-interpolator (PI) [7]. However, it is sensitive to process variation.

A coarse–fine ADC improves resolution by amplifying the residue between the input and closest coarse level, then quantizing the amplified residue again with the same coarse resolution. So, this brings us to the concept of time amplification. A coarse–fine two-step TDC becomes feasible if a time amplifier (TA) can be realized [8].

In this chapter, we will cover the high resolution, multistep TDC architectures based on the TAs for ADPLL.

Section 4.2 describes the operating principle of TDC architectures and its proposed realization. Using the TA, a coarse–fine two-step, three-step, and cyclic TDC architectures and its detailed circuit implementations are elaborated in Section 4.2. Experimental results are discussed in Section 4.3. Finally, Section 4.4 summarizes and concludes this chapter.

4.2 TIME-TO-DIGITAL CONVERTER ARCHITECTURES

4.2.1 COARSE–FINE TWO-STEP TDC

In order to improve the resolution, two-step TDC architecture with TA is adopted [8,9]. Figure 4.3 shows the conventional two-step TDC architecture. A two-step TDC improves the resolution by amplifying the residue between the input and closest coarse level and then, by quantizing the amplified residue again with the same coarse resolution. Both coarse TDC and fine TDC use the thermometer-to-binary (T2B) and have the same number of resolutions.

The coarse TDC stage converts the time difference between the two inputs into digital bits. Moreover, the residual time difference is also amplified by using the TA and it is transferred to the fine TDC stage. As a final step, fine TDC converts the amplified time difference into fine TDC codes.

Figure 4.4 shows the TA used in Figure 4.3. The small-signal gain of the TA is calculated by Equation 4.2 [8,9].

$$\text{Small signal gain of the TA} = \frac{2C}{g_m \times T_{off}} \tag{4.2}$$

where T_{off} is limited by the delay of the inverter.

4.2.2 THREE-STEP TDC WITH PHASE INTERPOLATOR

In order to achieve the wide input range and high resolution compared with two-step TDC, three-step TDC with PI is proposed in [3]. The conceptual diagram of the proposed three-step TDC architecture is shown in Figure 4.5. It is composed

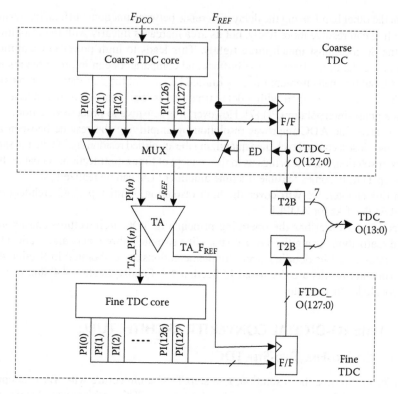

FIGURE 4.3   Block diagram of conventional two-step TDC.

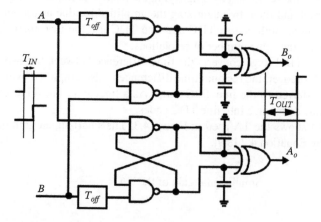

FIGURE 4.4   Time amplifier.

of the first coarse TDC, the second coarse TDC, and the fine TDC with a PI and a TA. The time difference between the edge of each output of delay line $D(n)$, and the edge of TDC_Fall, generates every possible time residue. The edge detector (ED) in the first coarse TDC and the second coarse TDC, determines which edge of $D(n)$ is closest to the TDC_Fall. The multiplexer (MUX) then passes that

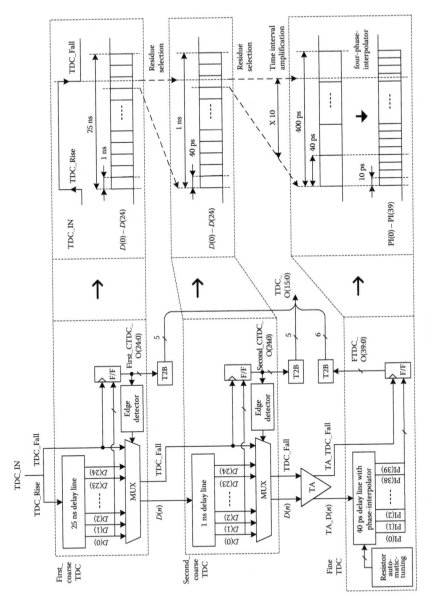

FIGURE 4.5 The conceptual diagram of the proposed three-step TDC.

residue to the fine TDC through the TA. The three-step TDC improves the resolution by amplifying the residue between the input and closest coarse level, then quantizing the amplified residue again with the same coarse resolution. Both the first coarse TDC, the second coarse TDC and the fine TDC use T2B, resolving 5, 5, and 6 bits, respectively.

The ΔT_{RES}, which is the time resolution of the proposed three-step TDC, is 1 ps. The quantization noise due to the finite resolution of the TDC, ΔT_{RES}, affects the in-band phase noise of the ADPLL of Figure 4.1. Therefore, the in-band phase-noise, at the output of the ADPLL due to the TDC timing quantization, can be calculated using Equation 4.1.

Substituting $\Delta T_{RES} = 1$ ps, $F_{REF} = 26$ MHz, $F_{DCO} = 2.4$ GHz, $T_V = 417$ ps, we obtain in-band phase noise of -121.37 dBc/Hz. Equation 4.1 was validated experimentally within 1 dB of measurement error for ΔT_{RES} spanning 0.89–1.12 ps through varying the TDC supply voltage. The next generations of deep-submicron CMOS processes can only bring reductions in ΔT_{RES}, so the phase noise performance will be further improved.

Figure 4.6a through c shows the detailed block diagram of the proposed three-step TDC. The first coarse TDC in Figure 4.6a is composed of a delay cell that consists of two cascaded inverters, a PI with a resistor automatic-tuning circuit, a F/F, an ED, and a MUX. The second coarse TDC is similar to the first coarse TDC in Figure 4.6b. The fine TDC in Figure 4.6c is composed of a delay cell that consists of two cascaded inverters, a PI with a resistor automatic-tuning circuit, and a F/F. The F/F is used as a comparator whose output determines the thermometer codes of the TDC. By locating the transition from 1 to 0, the ED logic identifies the critical residue to be sent to the TA and the fine TDC for fine conversion.

Figure 4.7 shows the block diagram of the proposed resistor automatic-tuning and PI that is composed of resistor tuning arrays (RTA) using passive resistors. The passive resistors usually vary by about 15% due to process variations. The variation of passive resistors in the PI can be compensated by adjusting component values for regular time intervals. The RTA(i) in the PI is controlled by the resistor automatic-tuning circuit to achieve regular time intervals. Assume there are two rising signal transitions with two inverter delay TDs. A new signal PI(i) defined by

$$PI(i) = D(0) + a_i \cdot \{D(0) - D(1)\}, \quad 0 < a_i < 1 \tag{4.3}$$

lies between the two generating signals $D(0)$ and $D(1)$ in the transition region [7]. Together with a comparator latch that detects the midlevel crossing, the new signal can be used to quantize the time interval between $D(0)$ and $D(1)$. A passive voltage divider, as shown in Figure 4.7, is connected between $D(0)$ and $D(1)$ and is used to generate the interpolated signals defined by Equation 4.3.

Figure 4.8a shows the RTA, which is composed of a main resistor ($R0$) and subresistors ($R1$–$R4$). It is controlled by the $R_{TUNE}(3{:}0)$ signal from the resistor automatic-tuning circuit shown in Figure 4.8b. In Figure 4.8b, V_{TUNE} is generated by I_{REF} and the replica resistor of $R0$–$R4$. It is then compared with the reference voltage, V_{REF}. $R_{TUNE}(3{:}0)$ is then controlled based on the result from the digital controller. When the

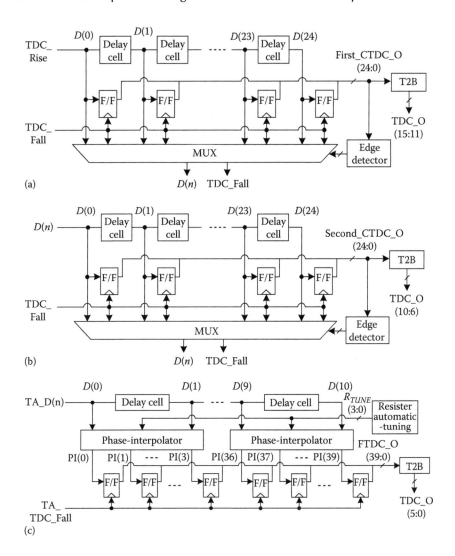

FIGURE 4.6 The detailed block diagram of the proposed three-step TDC: (a) the first coarse TDC, (b) the second coarse TDC, and (c) the fine TDC.

resistor is changed due to process variation, it is restored by the negative feedback of the resistor automatic-tuning circuit.

Figure 4.9 shows the timing diagram of the proposed resistor automatic-tuning method. If the resistor of the PI is increased by process variation, the V_{TUNE} is higher than the V_{REF}. In this case, the switch control bits $R_{TUNE}(3:0)$ are decreased. The resistor tuning is completed when the V_{TUNE} crosses the V_{REF}.

Figure 4.10 shows the schematic of the proposed TA. It is composed of a latch with a large output capacitance and delay cells with separate delay times (T_{off}, $T_{off} + \alpha$).

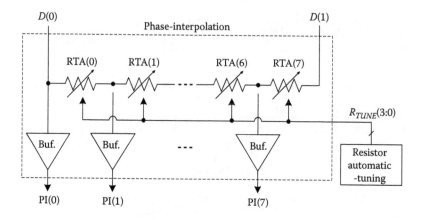

FIGURE 4.7 The block diagram of the proposed resistor automatic-tuning and PI.

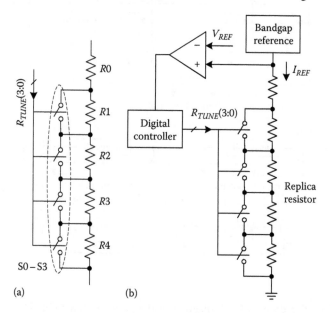

FIGURE 4.8 The schematic of the proposed (a) RTA and (b) resistor automatic-tuning circuit of PI.

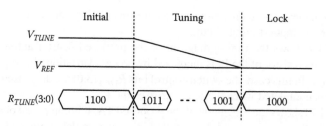

FIGURE 4.9 The timing diagram of the proposed resistor automatic-tuning method.

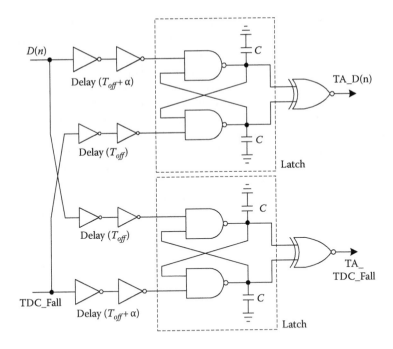

FIGURE 4.10   The schematic of the proposed TA.

The TA will amplify the time interval between the rising edge of PI(n) and TDC_ Fall. The small-signal gain of the TA is expressed by

$$\text{Small-signal gain of the TA} = \frac{2C}{g_m \times \alpha} \tag{4.4}$$

where
    g_m is the transconductance of the NAND
    C is the capacitance at its output [9]

It should be noted that both the gain and linear range can be controlled by the time offset α. The TA uses a low value of C and α, to let it cover the high gain and high frequency. The conventional TA is designed with two latches and delays with opposite inputs. Therefore, it has a problem in that it cannot implement a high gain due to having a delay time larger than the inverter delay time. The proposed TA is improved by using a fractional delay time (α) that is less than that of the inverter; this is the difference between two delay cells with separate delay times (T_{off}, $T_{off} + \alpha$).

4.2.3   CYCLIC TDC

In the two-step or three-step TDC architecture, the areas of the coarse and fine TDC stages are very large to be integrated into the ADPLL. Thus, coarse and fine TDC cores can be shared in the time domain, since they are not activated at the same time during the conversion. In this chapter, the cyclic TDC that has only one TDC core is proposed to reduce the area and also the current consumption [10].

The block diagram of the proposed cyclic TDC architecture is shown in Figure 4.11. It consists of the TDC core and TA. Two inputs, F_{DCO} and F_{REF}, are connected to INA and INB of TDC core through the MUX when Sel is *low* at the coarse conversion stage. In order to acquire a resolution less than that of the inverter, the delay cell is implemented with inverters and PI is composed of resistor arrays.

The ED in the TDC core determines which edge of PI(n) is closest to the F_{REF} and then, the MUX passes the residue to the fine TDC through the TA. Outputs of TA, TA_O_A and TA_O_B, are connected to INA and INB of TDC core through MUX, when Sel is high at the final conversion stage.

Figure 4.12 shows the block diagram of the TDC core in Figure 4.11. TDC core is composed of a delay cell, which consists of two cascaded inverters, PI, F/F, ED, and MUX. TDC core uses a delay cell chain. Each stage in the chain consists of two inverters and PIs for which the delay is 5 ps. Flip-flop is used as a comparator to determine the thermometer codes of the TDC. By locating the transition from 1 to 0, the ED logic identifies the critical residue which is to be sent to the TA in the TDC core during fine conversion.

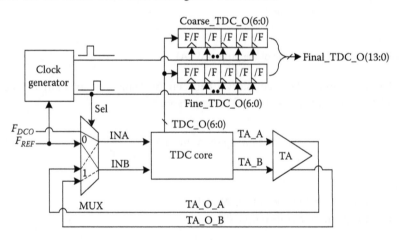

FIGURE 4.11 Block diagram of proposed cyclic TDC.

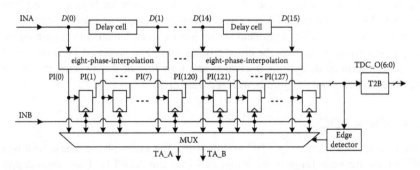

FIGURE 4.12 Block diagram of TDC core.

As shown in Figure 4.12, multiphase signals, PI(0)–PI(127), generated from INA by the delay cell and eight-phase-interpolation are sampled at the rising edge of INB. When INB is earlier than INA, the sampled values are all zero and the ED could not detect the transition. As a result, the fine TDC is not activated and the output of TDC is also all zero in this case, which is ignored by the following block after the TDC. The output of TDC is valid only when INA is earlier than INB.

Figure 4.13 shows the timing diagram of the proposed cyclic TDC. At the coarse TDC stage, F_{DCO} and F_{REF} are connected to INA and INB, respectively.

TDC core converts the time difference between INA and INB into 7-bits TDC code, Coarse_TDC_O(6:0). The residue is also amplified by the TA. At the fine TDC stage, outputs of TA, TA_O_A, and TA_O_B, are connected to INA and INB, respectively. TDC core converts the time difference between INA and INB into 7-bits TDC code, Fine_TDC_O(6:0). The outputs of the coarse TDC stage and fine TDC stage are Coarse_TDC_O(6:0) and Fine_TDC_O(6:0), respectively. They are merged for the generation of the final TDC output, TDC_O(13:0).

Figure 4.14a shows the schematic of the TA. It is composed of the latch with a large output capacitance and delay cells with separate delay times (T_{off}, T_{off} + α).

In Equation 4.2, the gain of the TA is limited by T_{off} and the delay of the inverter. In order to overcome this problem, the gain of TA is improved using the delay time difference (α) between two delay cells in this work.

TA will amplify the time interval between the rising edge of PI(n) and the rising edge of F_{REF}. The small signal gain of the TA in this work is same as Equation 4.4.

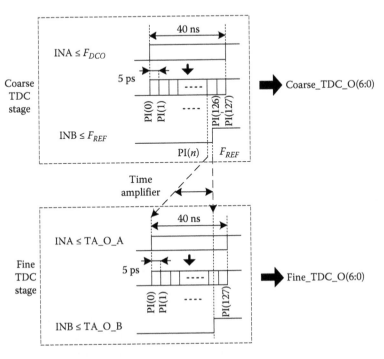

FIGURE 4.13 Timing diagram of the proposed cyclic TDC.

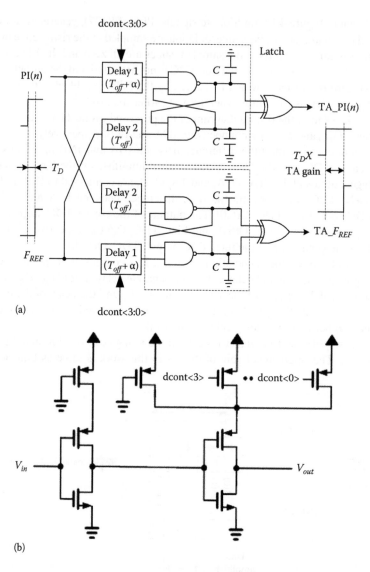

FIGURE 4.14 Schematic of (a) time amplifier and (b) delay cell (Delay 1).

It should be noted that both the gain and linear range can be controlled by the time off-set α. The TA uses the low value of C and α, to cover the high gain and high frequency.

TA exploits the variable delay of an SR latch subject to nearly coincident input edges. If rising edges are applied to S and to R at almost the same time, the latch will be metastable. After both inputs go to high, the initial voltage developed at the output of SR latch is proportional to the input initial time difference, and the positive feedback in the latch forces the output eventually to a binary level. Conventional TA is implemented in two latches and delays in opposite inputs. Thus, it has a problem that it cannot implement high gain due to a delay time larger than the inverter delay time. TA is improved by using a fractional delay time (α) less than that of the inverter; this is difference between two delay cells with separate delay times (T_{off}, $T_{off} + \alpha$). The resolution of TDC is determined by gain of the TA. The gain of TA is determined by a capacitor and the transconductance of a NAND gate, which are influenced by PVT variation.

Since the delay time difference (α) in Figure 4.14a is very small and more sensitive to the PVT variations compared with T_{off}, it is necessary to do layout very carefully and perform the calibration of the delay manually to adjust the gain of the TA. Also, it requires additional area to implement the small delay cell (Delay 1). Figure 4.14b shows the schematic of the delay cell, Delay 1, whose delay is controlled by dcont<3:0>. The additional delay time (α) of TA is manually controlled by dcont<3:0> in order to compensate for the PVT variation. Although T_{off} is changed for the PVT variation, gain of the TA has little effect on it, because α, difference between two delay cells, can be controlled manually.

Offset and delay mismatches in the TA are minimized by the careful layout and postlayout simulation. They are very sensitive to the parasitic capacitance and resistance of the interconnection wires. We finely adjusted them during the postlayout simulation stage. Also, the sizes of the transistors in TA are increased to improve the matching characteristics.

4.3 EXPERIMENTAL RESULTS

The three-step TDC with PI in Figure 4.5 was fabricated using the 0.13 μm technology CMOS process with a single polylayer, six layers of metal, MIM capacitors, and high sheet resistance poly resistors. Figure 4.15 shows the chip microphotograph of the ADPLL including TDC. The die areas of ADPLL and TDC are 0.8 and 0.18 mm², respectively.

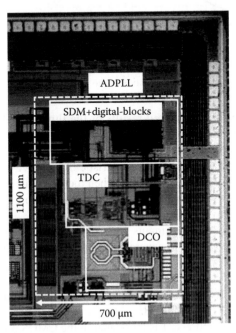

FIGURE 4.15   The chip microphotograph of three-step TDC with PI.

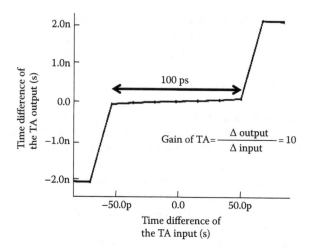

FIGURE 4.16   The measurement results of the TA difference characteristic.

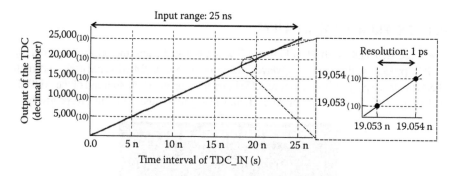

FIGURE 4.17   The measured characteristics of the three-step TDC.

Figure 4.16 shows the measured results of the three-step TDC in Figure 4.5. The relationship between the regeneration time and the initial time difference is a linear function. The gain and linear ranges of the TA are 5 and 20 ps, respectively.

Figure 4.17 shows the measured characteristic of the three-step TDC. The decimal number output of the three-step TDC when the time interval swept over TDC_IN is presented. The resolution and input range of the proposed three-step TDC are 1 ps and 25 ns, respectively.

To measure the linearity, two inputs with a 1 Hz difference at the reference frequency of 26 MHz are applied to generate a ramp input. The differential non linearity (DNL) and the integral nonlinearity (INL) of the three-step TDC are calculated, sweeping the value of the difference in the frequency and then performing a code density statistic; both are reported in Figure 4.18. The maximum DNL is ±0.6 LSB, while the maximum INL is ±1.9 LSB.

This cyclic TDC in Figure 4.11 is fabricated in the CMOS process with a feature size of 0.13 μm technology. Figure 4.19 shows the chip layout pattern. The chip has a

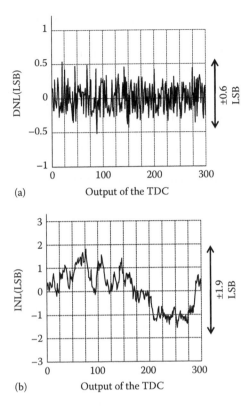

(a)   Output of the TDC

(b)   Output of the TDC

FIGURE 4.18 The measured (a) DNL and (b) INL of the three-step TDC in LSB.

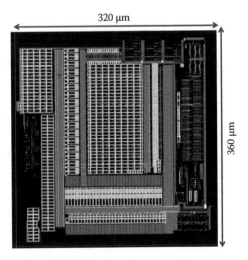

FIGURE 4.19 Chip layout pattern of Cyclic TDC.

single polylayer, six layers of metal and high sheet resistance poly resistors. The total die of the TDC with PI and TA is 0.12 mm².

Figure 4.20 shows the simulation result of the PI in the cyclic TDC when RAT is enabled. The phase error of PI with RAT is within ±5% including buffer mismatches.

The relationship between the input and output of the TDC is a linear function within the limited input range. In spite of the PVT variation, the gain of TA using the fixed delay time (α) is not changed in the input linear range of it. The linear input range of the TA is 1 ns, where the simulated gain of the TA is about 50 at the operating frequency of 25 MHz. Within this linear input range, the gain of TA is constant and can be controlled by adjusting the delay (α) of delay cell with dcont<3:0>. Since the residue from the coarse TDC is within the linear input range of TDC, the gain is kept almost constant.

Figure 4.21 depicts the measured characteristic of the two-step TDC. It shows the decimal number output of the cyclic two-step TDC when the time difference between F_{DCO} and F_{REF} is swept from 0 ps to 40 ns. Moreover, the resolution and the input range of the cyclic TDC are 5 ps and 40 ns, respectively.

To measure the linearity, two inputs with a 3 kHz difference at a reference frequency of 25 MHz are applied to generate a ramp input. The DNL and the INL of the cyclic TDC are calculated, sweeping the value of the difference in the frequency domain and then analyzing the code density statistic. The DNL and

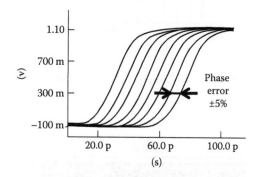

FIGURE 4.20 Simulation result of the PI in the two-step TDC when RAT is enabled.

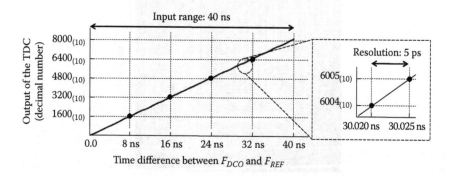

FIGURE 4.21 Measured characteristics of the proposed cyclic TDC.

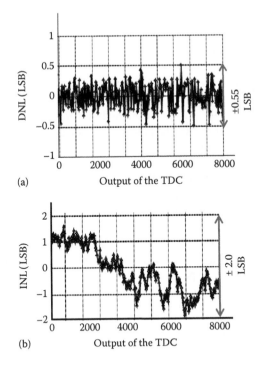

FIGURE 4.22 Measured (a) DNL and (b) INL of the proposed cyclic TDC.

INL are shown in Figure 4.22a and b, respectively. The maximum DNL is ±0.55 LSB, while the maximum INL is ±2.0 LSB.

Table 4.1 summarizes the performances of the three-step TDC with PI in Figure 4.10 and cyclic TDC. When the input frequency is 25 MHz and the frequency resolution is 5 ps, the power consumption and die area of the cyclic TDC are 2.4 mW and 0.12 mm², respectively. The power consumption and die area of the cyclic TDC is smallest compared with that mentioned in References 9,11,12, which are implemented in 90 nm and 0.13 µm process. The resolution of the cyclic TDC is larger than that mentioned in [9,12,13]. There exists a trade-off among the resolution of TDC, the die area, and power consumption. The design

TABLE 4.1

Summary of Measured Performance

Reference	[9]	[11]	[12]	[13]	Three-Step TDC with PI	Cyclic TDC
Process	90 nm	0.13 µm	90 nm	65 nm	0.13 µm	0.13 µm
Frequency	10 MHz	2 GHz	1.8 GHz	50 MHz	25 MHz	25 MHz
Resolution	1.25 ps	12 ps	0.75 ps	4.8 ps	1 ps	5 ps
Power consumption	3 mW	2.5 mW	70 mW	1.7 mW	3.5 mW	2.4 mW
Die area	0.6 mm²	N.A.	0.32 mm²	0.02 mm²	0.18 mm²	0.12 mm²

target of the cyclic TDC is to minimize the die area and power consumption of TDC. The resolution of the cyclic TDC can be further improved by increasing the die area and power consumption.

4.4   CONCLUSION

This chapter investigates the various TDC architectures for ADPLL. In order to improve the resolution and reduce the power consumption and die area of the TDC compared with two-step TDC, three-step TDC with PI and cyclic TDC are presented. The resolution of the three-step TDC with PI is improved by using a PI and a TA for the improvement of the in-band phase noise of the ADPLL. The input range of the TDC is 25 ns and the frequency resolution of the TDC is 1 ps.

In the cyclic TDC, the coarse and fine TDC stages in the conventional two-step TDC architecture are shared to reduce the area. It is implemented in the 0.13 μm CMOS process with a die area of 0.12 mm^2. The power consumption is 2.4 mW at 1.2 V supply voltage. Furthermore, the resolution and input frequency of the cyclic TDC are 5 ps and 25 MHz, respectively.

REFERENCES

1. Y. Arai and T. Baba, A CMOS time to digital converter VLSI for high energy physics, in *VLSI Circuits Symposium on Digest of Technical Papers*, San Diego, CA, August 1988, pp. 121–122.
2. R. B. Staszewski, D. Leipold, C.-M. Hung, and P. T. Balsara, TDC-based frequency synthesizer for wireless applications, in *Proceedings of IEEE Radio Frequency Integrated Circuits (RFIC) Symposium*, Fort Worth Convention Center, Fort Worth, TX, June 2004, pp. 215–218.
3. Y. G. Pu, A. S. Park, J.-S. Park, and K.-Y. Lee, Low-power, all digital phase-locked loop with a wide-range, high resolution TDC, *ETRI Journal*, 33(3), 366–373, June 2011.
4. R. B. Staszewski et al., 1.3 V 20 ps time-to-digital converters for frequency synthesis in 90-nm CMOS, *IEEE Transactions on Circuits and Systems II: Express Briefs*, 53(3), 220–224, March 2006.
5. P. Dudek, S. Szczepanski, and J. V. Hatfield, A high-resolution CMOS time-to-digital converter utilizing a Vernier delay line, *IEEE Journal of Solid-State Circuits*, 35(2), 240–247, February 2000.
6. K. Nose et al., A 1 ps-resolution jitter-measurement macro using interpolated jitter oversampling, in *IEEE ISSCC Digest of Technical Papers*, San Francisco, CA, February 2006, pp. 520–521.
7. S. Henzler, S. Koepp, D. Lorenz, W. Kamp, R. Kuenemund, and D. Schmitt-Landsiedel, A local passive time interpolation concept for variation-tolerant high-resolution time-to-digital conversion, *IEEE Journal of Solid-State Circuits*, 43(7), 1666–1676, July 2008.
8. M. Lee and A. A. Abidi, A 9b, 1.25 ps resolution coarse-fine time-to digital converter in 90 nm CMOS that amplifies time residue, in *VLSI Circuits Symposium on Digest of Technical Papers*, June 2007, pp. 168–169.
9. M. Lee and A. A. Abidi, A 9 b, 1.25 ps resolution coarse-fine time-to-digital converter in 90 nm CMOS that amplifies a time residue, *IEEE Journal of Solid-State Circuits*, 43(4), 769–777, April 2008.

10. H. Kim, S. Y. Kim, and K.-Y. Lee, A low power, small area cyclic time-to-digital converter in all-digital PLL for DVB-S2 application, *Journal of Semiconductor Technology and Science*, 13(2), 145–151, May 2013.
11. R. Tonietto, E. Zuffetti, R. Castello, and I. Bietti, A 3 MHz bandwidth low noise RF all digital PLL with 12 ps resolution time to digital converter, in *Proceedings of 32nd European Solid-State Circuits Conference (ESSCIRC)*, Montreux Convention & Exhibition Center, Montreux, Switzerland, 2006, pp. 150–153.
12. M. Lee, M. E. Heidari, and A. A. Abidi, A low-noise wideband digital phase-locked loop based on a coarse–fine time-to-digital converter with subpicosecond resolution, *IEEE Journal of Solid-State Circuits*, 44(10), 2808–2816, October 2009.
13. L. Vercesi, A. Liscidini, and R. Castello, Two-dimensions Vernier time-to-digital converter, *IEEE Journal of Solid-State Circuits*, 45(8), 1504–1512, August 2010.

5 Design Principles for Accurate, Long-Range Interpolating Time-to-Digital Converters

Pekka Keränen, Jussi-Pekka Jansson, Antti Mäntyniemi, and Juha Kostamovaara

CONTENTS

5.1   INTRODUCTION

The basic function of a time interval measurement device is to produce a quantitative measure of the length of a time interval at the input to the device. Typically, this time interval is represented by logic-level timing pulses, which are related to the beginning and end of a physical phenomenon whose time behavior is to be studied. The output of a time interval measurement device is typically given in the form of a digital number, which expresses the measured length of the time interval as a multiple of the smallest resolvable time interval of the device, see Figure 5.1. The smallest resolvable time interval characterizing the measurement system can be regarded as the value of its least significant bit (LSB) by analogy with ADCs. It is this that defines the *resolution* of the device. Accordingly, a time interval measurement device is typically referred to as a TDC.

Among its many applications, TDCs can serve as one of the critical components of a pulsed time-of-flight laser radar, in which the accuracy of the distance measurement is directly dependent on that of the time measurement. Pulsed time-of-flight laser radars are used in geodesy, space measurements and also in various industrial inspection applications, and it is foreseen that the measurement of 3D data, which could be based on laser scanning, will be quite essential for many control and navigation applications in the future. The development of an autonomous, or "driverless," car is one example of such an activity that obviously calls for high-speed environment-sensing techniques [1–4]. It is also widely accepted that low-cost 3D imaging, again potentially based on laser scanning, could open up a vast field of applications in surveying, map capturing, consumer electronics (games), robotics (man–machine interfaces), medical technology, and the controlling of machines (gesture control), for example [5].

Another, more "traditional" application area for TDCs is in nuclear science, where many types of time-of-flight investigations are being undertaken [6]. TDCs are also used in measurement instruments and in the calibration of test equipment for electronic circuits [7,8], and they have also found applications recently in digital frequency synthesizers, sensors, AD/DA converters, and other "time-mode" circuits [9,10]. The time interval range to be digitized in a time mode circuit is typically limited to a

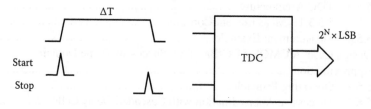

FIGURE 5.1   The time interval measurement circuit of a TDC.

few nanoseconds only, and the important parameter is the resolution of the measurement, which has an effect on the achievable phase noise in DPLLs, for example. In traditional applications such as laser radar, however, it is important not only to have high resolution and precision but also to cover a wide range of time intervals. For example, if a distance range of 100 m and a resolution of 1 cm are aimed at, the corresponding time scale parameters will be 670 ns and ~50 ps (LSB), respectively.

The most straightforward method for realizing a *wide-range TDC* is to have an electronic counter that counts the pulses of an accurate oscillator during the start–stop time interval. The resolution of this method is limited, however, by the period of the oscillator. This means that a resolution better than the nanosecond range is difficult to achieve due to problems in the realization of high-frequency, high-performance oscillators. The purpose of this chapter is to describe a time interval measurement principle (and a few of its technical variants) in which the fundamental resolution limitation on the counting technique is overcome by interpolating the position of the timing pulses (start and stop) within the clock period of the oscillator. By doing this, a wide dynamic measurement range limited basically by the number of bits in the counter and high resolution defined by the interpolator can be achieved *simultaneously*.

This chapter will be organized as follows. First, in Section 5.2, some important performance parameters of a time interval measurement circuit will be explained. Then, in Section 5.3, the basic counting method will be analyzed. After this, in Section 5.4, the interpolation method will be presented and analyzed in greater depth. Sections 5.5 and 5.6 will describe in detail CMOS TDC realizations based on delay line interpolation and successive approximation principles, respectively. Finally, the work will be summarized in Section 5.7.

5.2   PERFORMANCE PARAMETERS IN TIME
INTERVAL MEASUREMENT

The main performance parameters of a time interval measurement circuit are resolution, precision, and accuracy [11].

The term *resolution* is used for the smallest time interval, which theoretically can be resolved with the TDC in a single measurement, that is, the quantization step (LSB). The term (single-shot) precision is used for the standard deviation (s) of the distribution of measurement results around the mean value when a constant time interval is measured repeatedly, see Figure 5.2. In a practical situation, the single-shot precision is influenced not only by the quantization error but also by nonidealities such as jitter in the timing signals and power supply noise, and especially by nonlinearities in the clock period interpolators. Single-shot precision can be used to estimate the smallest real time interval that can be resolved in a single isolated measurement. Since precision is limited by the quantization error and statistical error sources, it can usually be improved by averaging.

While precision gives the statistical variation in the measurement result around a mean value, single-shot accuracy is affected both by this statistical variation and by any systematic errors in the mean value. The statistical variation can be reduced by

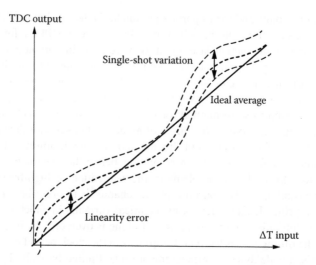

FIGURE 5.2 Performance parameters of a TDC circuit.

averaging, but the systematic errors remain, that is, the precision may be good, especially after averaging, but the accuracy may nonetheless be poor. Systematic errors are typically dominated by the nonlinearity of the measurement. Integral linearity error (INL) is the deviation of the input–output characteristics from the ideal, straight-line input–output relation (Figure 5.2), while the differential linearity error (DNL) is the deviation of each individual quantization step from the average value of the LSB. Consequently, DNL and INL are related, in that the INL in a particular TDC channel is the sum of the DNLs of all the preceding channels. Evidently, the statistical variation should be reduced to a negligible level by averaging to measure the systematic errors reliably. Another important performance parameter for a TDC is stability, which reflects the sensitivity of its characteristics to temperature, supply voltage, and time, for example.

Important performance parameters from the application point of view are the measurement range and conversion time of the TDC. The measurement range defines the maximum time interval that can be measured or digitized, while the conversion time is the time between the end mark of the input time interval and the moment when the measurement result is available at the TDC interface. In addition, the dead time, indicating the period during which the system is incapable of accepting a new start after a registered start signal (conversion time + possible recovery time), is sometimes also used, and this could be an important parameter, especially in applications where the mean time between the time intervals to be measured is random, which may happen in nuclear measurements, for example. For instance, if the start detector efficiency is high and the stop efficiency low, a long dead time may result in reduced measurement efficiency, since many of the start signals which trigger time interval measurement may not be followed by a valid stop signal. In this case, it is important for the TDC to recover from a useless start pulse as quickly as possible. In some cases, however, it is also possible to use the true stop signals as markers of the start of measurement, in order to increase the conversion efficiency.

5.3 TIME-TO-DIGITAL CONVERSION BASED ON COUNTING

In the "counting method," pulses of a stable clock are counted during the input time interval, see Figure 5.3. Provided that the reference clock is accurate, as in the case of a crystal oscillator, for example, this technique can achieve a wide linear range and good stability. In asynchronous measurements, that is, when measurement begins in a random phase with respect to the clock, the maximum error in one measurement is $\pm T_{clk}$, where T_{clk} is the clock period. For each input interval, the counter will measure either N or N + 1 counts, and for this binomial distribution with an input interval of $(Q + F)T_{clk}$, where Q is an integer and $0 \le F < 1$, the expected value of the measurement result will be Q + F (the counter reading divided by the total number of measurement results) and the standard deviation [12]

$$\sigma = \sqrt{F(1-F)} \tag{5.1}$$

Thus, the measurement precision varies with the length of the input time interval, the worst-case precision of $0.5T_{clk}$ is achieved when F = 0.5. With a 1 GHz clock, for example, the maximum single-shot error is ± 1 ns and the worst-case value for the single shot precision (σ value) is 500 ps.

Averaging can be used to improve the precision. Given N_{av} samples in asynchronous measurement, the precision is

$$\sigma = \frac{\sqrt{F(1-F)}}{\sqrt{N_{av}}} \tag{5.2}$$

as shown in Figure 5.4.

$$F, \Delta T = (Q+F) \cdot T_{clk}$$

In asynchronous measurement, the signal (time interval to be measured) repetition rate is not coherent with the counter clock, which is an essential requirement, as the time relationship between the signal and the counter clock must be such as to sweep through the full range of the N/N + 1 count ambiguity in a random manner to

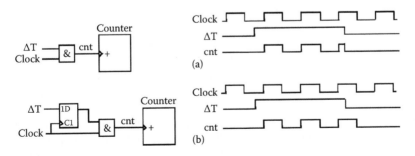

FIGURE 5.3 The counting method. (a) Clock gating and (b) synchronized clock gating.

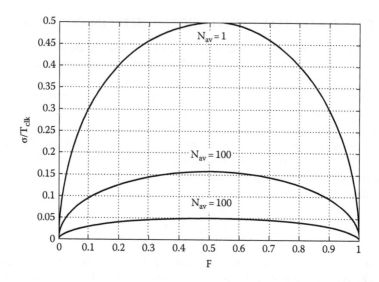

FIGURE 5.4 Precision in the counting method as a function of the fractional input time interval.

satisfy the statistical requirement of averaging. By contrast, the start timing signal in synchronous measurement is repeated at a constant rate and with a constant phase relationship to the TDC clock. In this case, averaging does not improve the precision, but if the position of the timing signal can be controlled within the clock period, it is possible to achieve a faster precision improvement rate than in asynchronous measurement. For example, if the phase of the start signal and oscillator has M discrete values evenly distributed within the clock period, the measurement precision can be improved by averaging at a rate proportional to $1/M$ (rather than to $1/\sqrt{M}$ as in synchronous measurement), but the improvement is limited by the factor $1/M$.

The basic counting method has good accuracy, since with a stable crystal oscillator reference, the linearity and both the short-term and long-term stabilities are excellent. Moreover, it is easy to increase the measurement range simply by adding more bits to the counter. With standard commercially available high-performance crystal oscillators, the maximum frequency is limited to a few hundred MHz, corresponding to ns-range resolution, although sub-ns-range resolution can be achieved with on-chip clock generation techniques.

The gating of the clock may have a powerful impact on the measurement accuracy of the counting method. If the input time interval is used to gate the counter, short clock pulses may occur at the input to the counter (Figure 5.3, upper circuit), assuming that the input time interval is asynchronous with respect to clock. These may or may not be long enough to increment the counter and may cause an unpredictable error in the averaged result. Therefore, a *synchronizer*, in which the input interval is synchronized with the clock, is needed. One possible realization example is shown in Figure 5.3 (lower circuit). Note that now only full-length clock pulses are allowed to pass to the counter. When using a synchronizer, the measurement result will be unbiased and time intervals shorter than the clock period can also be reliably measured. It should be noted, however, that even with the synchronizer shown in Figure 5.3, the edges of the time interval signal and the clock may occur simultaneously at the input to the flip-flop, possibly resulting in meta-stability in the flip-flop [13] and eventually in an erroneous result. The cure for this will be discussed in more detail in the following.

5.4 INTERPOLATION

5.4.1 COMBINING COUNTING AND INTERPOLATION

A powerful method for the accurate measurement of time intervals is to combine the counter with an analogue or digital interpolation circuit, as shown in Figure 5.5 [14–16]. The input time interval is digitized roughly by synchronously counting the clock periods of a reference clock during the time interval, the counter being enabled by the first clock pulse following the start signal, and disabled by the first clock pulse following the stop signal. The resulting time interval T_{12} is thus synchronized to the clock and can be accurately digitized with a counter. The time fractions T_1 and T_2 are digitized separately with interpolators to improve the single-shot resolution. For an n-bit interpolator, the LSB of the measurement is equal to $T_{clk}/2^n$ and the input interval Δt can be calculated by combining the results of the interpolators and the counter

$$t_x = T_{12} + T_1 - T_2 = N_c T_{clk} + N_1 T_{clk}/2^n - N_2 T_{clk}/2^n \qquad (5.3)$$

It is important to note that if the system clock is asynchronous with respect to the time interval to be measured, the lengths of the time fractions T_1 and T_2 change randomly in repeated measurements, even though their difference has only two discrete

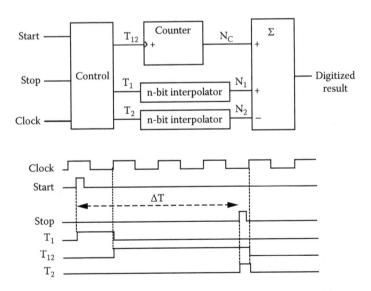

FIGURE 5.5 A TDC architecture combining counting and interpolation.

values, $T_1 - T_2$ if the main counter result is N_C or $1 - (T_1 - T_2)$ if the main counter result is $N_C + 1$. This means that the effect of the interpolator nonlinearities on the averaged results is strongly suppressed due to the averaging effect, and eventually, the accuracy of the system ceases to be limited by them. Interestingly, the drifts in the interpolators tend to cancel out, as it is the difference $T_1 - T_2$ that counts in the final result [16].

The interpolators may be based on analogue time-to-voltage conversion or digital delay lines [17–21], both structures that can achieve good single-shot resolution in a limited dynamic range, as in practice, they can be used for measuring time intervals from subnanoseconds to a few nanoseconds with a resolution of <1 ps, …, 10 ps depending on the measurement range. In conclusion, the method described earlier, generally known as the Nutt method, combines the inherently good single-shot resolution of an analogue time interval measurement method such as TAC or a Vernier digital delay line, for example, with the excellent accuracy and wide linear range of the counting method.

5.4.2 SYNCHRONIZATION ISSUE

In most cases, the timing signals and the clock are asynchronous, that is, they do not have a fixed phase relationship. As a result, there will be a synchronization issue in the generation of the time intervals T_1, T_{12}, and T_2. In the simplified scheme for one possible interpolation-based control block for a TDC shown in Figure 5.6a, the flip-flops D1a and b are used only to capture the start and stop signals, which are assumed to be short pulses. Note that the inputs to flip-flops D2a and b (data and clock) may now occur simultaneously, in which case the corresponding timing pulse may or may not be detected by the simultaneous clock edge. Assuming ideal flip-flops, the logic nevertheless ensures that the final result of the measurement is the same in both cases. For example, if the flip-flop D2a does not detect the start timing pulse edge, the time fraction T_1 will be one clock cycle longer, since the timing pulse will only be detected by the following clock pulse. In the opposite case, the counter will record one more clock cycle, and thus, the result will be unambiguous.

The situation is more complicated with real-life flips-flops, however, the difference arising from the fact that in reality, the propagation delay of the flip-flop D2a or b will depend on the relative time positions of the data and clock edges. In other words, when the setup time requirement of either the flip-flop D2a or D2b is not fulfilled, the propagation delay of that flip-flop may increase (Figure 5.6b) and in the extreme case, the flip-flop may enter a metastable state. Note that a change in the propagation delay of the flip-flop will manifest itself in the length of T_1 (or T_2). Thus, if the start (stop) pulse occurs near the rising clock edge, an erroneous measurement result is possible.

The probability of a synchronization error can be reduced, for example, by delaying the sampling of the output of flip-flop D2a (D2b) to the next clock edge. Note that, given the synchronizer in Figure 5.7, the measurement will be accurate if the flip-flop D2a (D2b) settles completely in less time than T_{clk} [16]. This results from the fact that even though the delay in the flip-flop D2a (D2b) might increase due to concurrent clock and data edges, this will have no effect on the length of $T_1(T_2)$, since

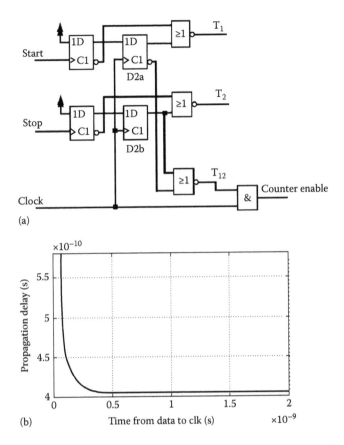

(a)

(b)

FIGURE 5.6 (a) Control logic producing T_1, T_{12}, and T_2 for the TDC but suffering from a nonconstant flip-flop delay problem; (b) dependence of a flip-flop propagation delay on the time interval between the data and clock edges.

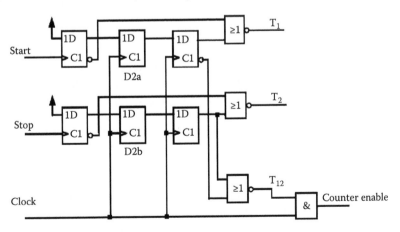

FIGURE 5.7 The logic configuration producing T_1, T_{12}, and T_2 for the TDC. Possible changes in the propagation delay of D2a and D2b will not affect the final result.

the back edge of $T_1(T_2)$ is defined by the next clock edge only. Note also that, even in the case of a timing pulse being completely missed with a particular clock edge, the measurement will still be correct, since this will simply mean that $T_1(T_2)$ will be one T_{clk} longer while T_{12} will correspondingly be one T_{clk} shorter. The sampling delay can obviously be increased by more clock periods if needed (adding flip-flops accordingly).

The scheme shown in Figure 5.7 increases the measurement range required of the interpolators, which can be considered a disadvantage in digital interpolators, where the integral nonlinearity due to delay element mismatch tends to increase with any increase in the length of the delay line. In time-to-voltage conversion, however, a time offset may be advantageous, since the nonlinear part of the characteristics due to switching effects is then similar in all the measurements. It is clear that the detailed construction of the synchronization mechanism and circuitry used depends largely on the particular interpolation method and architecture. The aforementioned examples are intended simply to demonstrate the typical challenges met with in designing synchronization circuits for TDCs utilizing interpolation.

5.4.3 INTERPOLATOR CIRCUIT TECHNIQUES

A straightforward method for digitizing the start and stop time fractions T_1 and T_2 is to apply time-to-amplitude conversion (TAC) followed by an A/D converter. In TAC, a capacitor is charged or discharged with a constant current during the input time interval. Any change in the capacitor voltage is thus proportional to the input time interval and can be converted to digital form with an A/D converter. After the conversion, the capacitance is again charged to the reference voltage and the cycle can be repeated. Recent TDCs based on this technique are those presented in [22,23], for example, which have achieved a single-shot precision of the order of 1, ..., 10 ps (sigma).

Another, more popular technique for realizing interpolation is based on a digital delay line (or variants of this approach) where the LSB is based on the propagation delay in a logic gate, usually that of an inverter, or to overcome the technological limit for the unit delay, on the difference between two delay elements [24]. In the basic configuration, the start mark of the input interval travels along the delay line, and when the stop mark (e.g., clock edge) arrives, it stores the status of the delay line in the flip-flops. The time interval between the start and stop can then be coded from these data as a multiple of one gate delay. The delay in the element is usually made controllable in order to compensate for the effects of process variations and temperature and supply changes. Alternatively, digital calibration methods can also be used.

Delay line interpolators based on inverter gates are easily implemented in CMOS technologies, have low power consumption, and can relatively easily be transferred to a low supply voltage environment. Also, the conversion time of a TDC based on delay line interpolators is short, since it depends on the propagation delay of the delay element chain and the coding logic. The main factor limiting the performance is the nonlinearity caused by random mismatch in the individual delay elements, as it is this that determines the achievable precision. Examples of CMOS TDCs operating on these principles are given in Sections 5.5 and 5.6.

5.4.4   PERFORMANCE LIMITATIONS

Assuming a general form $e_{1,2}(x)$ for the interpolator error (nonlinearity and/or gain error), using the subscripts 1 and 2 for the start and stop interpolators, respectively, and normalizing T_{clk} to 1, the TDC measurement error for a input time interval $Q + F$ (where Q is an integer and $0 \leq F < 1$) in a single measurement can be calculated as

$$e(x,F) = \begin{cases} e_1(x) - e_2(x - F + 1), & x < F \\ e_1(x) - e_2(x - F), & x \geq F \end{cases} \tag{5.4}$$

The maximum single-shot measurement error is now equal to the maximum difference $e_1(x) - e_2(x)$ and the distribution of measurement results is no longer binomial, as more than two results are possible. Since in asynchronous measurement, x varies randomly and with equal probability between $0 \leq x < 1$, the mean value of the TDC measurement error is

$$m(F) = \int_0^1 e(x,F)dx$$

$$= \int_0^1 e_1(x)dx - \int_0^F e_2(x - F + 1)dx - \int_F^1 e_2(x - F)dx$$

$$= \int_0^1 e_1(x)dx - \int_0^1 e_2(x)dx = m_1 - m_2 \tag{5.5}$$

where m_1 and m_2 are the mean values of the start and stop interpolator errors. Thus, the gain error or the INL of the interpolators is reduced to a constant bias error that is independent of the time to be measured [16,25]. This is an extremely important result, as it indicates that the error due to the interpolator nonlinearities is effectively averaged out in a series of repeated measurements due to the randomization process present in the asynchronous technique. This is the source of the excellent linearity of the method.

The nonlinearities of the interpolators do have an effect on the precision of the system, however. The variance (σ^2) of the TDC measurement result is

$$\sigma^2(F) = \int_0^1 e^2(x,F)dx - m^2(F)$$

$$= e_{1RMS}^2 + e_{2RMS}^2 - (m_1 - m_2)^2$$

$$-2\left(\int_0^F e_1(x)e_2(x - F + 1)dx + \int_F^1 e_1(x)e_2(x - F)dx \right) \tag{5.6}$$

The value of the last two terms in the Equation 5.6 depends on F, the fractional part of the time interval to be measured, and thus, the standard deviation of the result is a

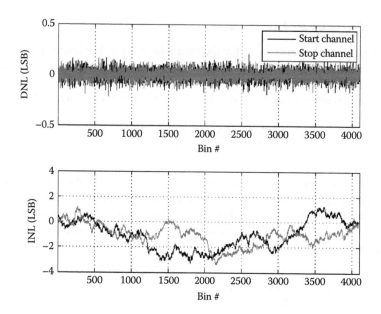

FIGURE 5.8   DNL and INL when DNL was generated with 0.05LSB random deviation.

function of the fractional part (F) of the input time interval (Q + F). Since $0 \leq F < 1$, the measurement precision due to interpolator nonlinearity or gain error now has a period of T_{clk} instead of LSB. It is also possible to correct the nonlinearity of the interpolators by measuring the linearity error and using this to correct the measurement result [26]. However, since this is technically more complicated than in the case of gain error, minimization of the interpolator nonlinearity is important.

As an example, Figure 5.8 shows simulated DNL and INL results for a TDC with 12-bit interpolators. The bin width deviation (DNL) was generated from a normal distribution with a σ value of 0.05LSB. The resulting single-shot precision is shown in Figure 5.9. It should be noted that the single-shot precision shown here does not include the variation due to quantization error. A more complete view of the simulated TDC error as a function of the fractional part F and the start event arrival time x, that is, e(x,F), is shown in Figure 5.10.

5.4.5   Effect of Reference Clock Phase Noise

When measuring short time intervals, the measurement precision is typically dominated by quantization noise and other noise sources that can be modeled as additive white noise sources with limited bandwidth. However, when long time intervals are measured, the time base provided by the reference oscillator may have a significant impact on the measurement precision [27].

Oscillator instabilities are usually defined in terms of phase noise, which is typically measured in the frequency domain as a power spectral density (PSD) and is expressed in units of dBc/Hz. Phase noise consists of power law noise processes having a PSD proportional to $1/f^{\beta}$ ($\beta = 0, 1, 2, 3, ...$) (Table 5.1) [28].

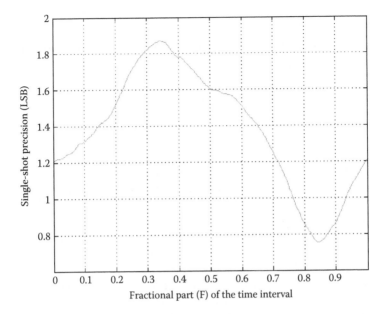

FIGURE 5.9 Simulated single-shot precision when DNL was generated with 0.05LSB random deviation.

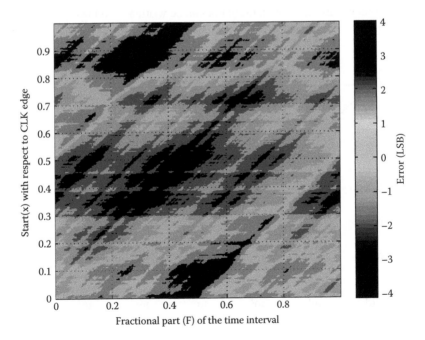

FIGURE 5.10 Time interval measurement error due to INL at different values of x and F.

TABLE 5.1

Power-Law Noise Processes Usually Observed in a Phase Noise PSD

β	S(f)	Noise Type
0	h_0/f^0	White phase noise
1	h_1/f^1	Flicker phase noise
2	h_2/f^2	White frequency noise/random walk phase noise
3	h_3/f^3	Flicker frequency noise

The phase noise PSD of a free running oscillator typically consists of a white phase noise floor, white frequency noise, and flicker frequency noise at small offset frequencies. The $1/f^2$ and $1/f^3$ slopes seen in a phase noise PSD arise from the white noise and flicker noise sources intrinsic to the oscillator, where they are upconverted to $1/f^2$ and $1/f^3$ noise [29]. On the other hand, white phase noise and flicker phase noise is created by clock buffers and other noise sources that are not part of the oscillator's intrinsic feedback loop. These phase noise processes are illustrated in Figure 5.11.

A major problem related to analyzing the higher-order noise processes is that they are divergent. The usual statistical tools such as a variance estimator do not converge for noise processes other than band-limited white noise. For example, the variance of a random walk noise process increases linearly with respect to time. This means that the time interval jitter is also a function of the time interval, and in some cases, even the observation time needs to be taken into account in order to estimate the jitter variance due to phase noise.

The instantaneous phase of an oscillator can be written as follows:

$$\varphi(t) = 2\pi f_{clk} t + \varphi_e(t) \tag{5.7}$$

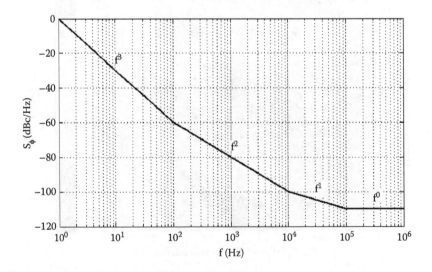

FIGURE 5.11   Phase noise processes in a free-running oscillator.

where
   f_{clk} is the nominal oscillation frequency
   $\varphi_e(t)$ is the instantaneous phase error

The resulting time error is roughly

$$t_e(t) = \frac{\varphi_e(t)}{2\pi f_{clk}} \tag{5.8}$$

and the time interval error for an interval of ΔT is then given by

$$\Delta T_e = t_e(t + \Delta T) - t_e(t) = \frac{\varphi_e(t + \Delta T) - \varphi_e(t)}{2\pi f_{clk}} \tag{5.9}$$

Consider the sample variance of the time interval error when M consecutive measurements are performed with no dead time in between:

$$
\begin{aligned}
\sigma_{\Delta T_e}^2 &= \frac{1}{M-1} \sum_{i=1}^{M} \left(\Delta T_e(i) - \frac{1}{M} \sum_{i=1}^{M} \Delta T_e(i) \right)^2 \\
&= \frac{1}{M-1} \left(\sum_{i=1}^{M} \Delta T_e(i) - \frac{1}{M} \left(\sum_{i=1}^{M} \Delta T_e(i) \right)^2 \right) \\
&= \frac{1}{(2\pi f_{clk})^2 (M-1)} \left(\begin{array}{c} \sum_{i=1}^{M} \left(\varphi_e(t + i\Delta T) - \varphi_e(t + (i-1)\Delta T) \right)^2 \\ - \frac{\left(\varphi_e(t + M\Delta T) - \varphi_e(t) \right)^2}{M} \end{array} \right)
\end{aligned}
\tag{5.10}
$$

Taking the expected value of the sample variance yields

$$
\begin{aligned}
E\left[\sigma_{\Delta T_e}^2 \right] &= \frac{2}{(2\pi f_{clk})^2 (M-1)} \left(\sum_{i=1}^{M} \left(R_{\varphi e}(0) - R_{\varphi e}(\Delta T) \right) - \frac{R_{\varphi e}(0) - R_{\varphi e}(M\Delta T)}{M} \right) \\
&= \frac{2M}{(2\pi f_{clk})^2 (M-1)} \left(\left(R_{\varphi e}(0) - R_{\varphi e}(\Delta T) \right) - \frac{R_{\varphi e}(0) - R_{\varphi e}(M\Delta T)}{M^2} \right)
\end{aligned}
\tag{5.11}
$$

where $R_{\varphi e}(\tau)$ denotes the autocorrelation of the phase noise process with a lag τ. The Wiener–Khinchin theorem can be used to express the autocorrelation function in terms of the phase noise PSD:

$$R(\tau) = \int_{-\infty}^{\infty} S(f) e^{j2\pi f \tau} df \tag{5.12}$$

where $S(f)$ is the PSD at an offset frequency f. We now obtain the following equation for the time interval variance:

$$E\left[\sigma_{\Delta T_e}^2\right] = \frac{2M}{\left(2\pi f_{clk}\right)^2 (M-1)} \int_{-\infty}^{\infty} S_{\varphi e}(f) \left(\left(1 - e^{j2\pi f\Delta T}\right) - \frac{\left(1 - e^{j2\pi fM\Delta T}\right)}{M^2}\right) df \quad (5.13)$$

Since the PSD is symmetrical around zero, we can write

$$E\left[\sigma_{\Delta T_e}^2\right] = \frac{2M}{\left(\pi f_{clk}\right)^2 (M-1)} \int_{0}^{\infty} S_{\varphi e}(f) \left(\sin^2(\pi f\Delta T) - \frac{\sin^2(\pi fM\Delta T)}{M^2}\right) df \quad (5.14)$$

This equation gives the expected sample variance if M successive ΔT time intervals were to be measured. In practice, M is usually taken to infinity in order to make the uncertainty in the variance estimate as small as possible. The variance is then given by

$$E\left[\sigma_{t_e}^2\right] = \frac{2}{\left(\pi f_{clk}\right)^2} \int_{0}^{\infty} S_{\varphi e}(f) \sin^2(\pi\Delta fT) df \quad (5.15)$$

The problem with Equation 5.15, however, is that M also accounts for the observation time $M\Delta T$. If the observation time is taken to infinity, the time interval jitter for the $1/f^3$ noise process does not converge but goes to infinity as well. The resulting time interval variances for different phase noise processes are summarized in Table 5.2, where γ is the Euler–Mascheroni constant and f_{BW} is the noise bandwidth.

TABLE 5.2
Time Interval Variances due to $1/f^2$ and $1/f^3$ Phase Noise Processes

Noise Type	$S_{\varphi e}(f)$	Time Interval Variance $\left(\sigma_{t_e}^2\right)$ (M Sample)	Time Interval Variance $\left(\sigma_{t_e}^2\right)$ (M → ∞)
White phase noise	h_0/f^0	$\dfrac{h_0 f_{BW}}{\left(\pi f_{clk}\right)^2} \dfrac{M+1}{M}$	$\dfrac{h_0 f_{BW}}{\left(\pi f_{clk}\right)^2}$
Flicker phase noise	h_1/f^1	$\dfrac{h_1}{\left(\pi f_{clk}\right)^2} \dfrac{M+1}{M}\left(\ln(2\pi f_{BW}\Delta T) + \gamma - \dfrac{\ln(M)}{M^2-1}\right)$	$\dfrac{h_1}{\left(\pi f_{clk}\right)^2}\left(\ln(2\pi f_{BW}\Delta T) + \gamma\right)$
White frequency noise	h_2/f^2	$\dfrac{h_2}{f_{clk}^2}\Delta T$	$\dfrac{h_2}{f_{clk}^2}\Delta T$
Flicker frequency noise	h_3/f^3	$\dfrac{2h_3}{f_{clk}^2} \dfrac{\ln(M)M}{M-1}\Delta T^2$	∞

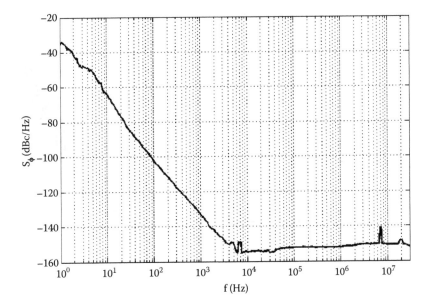

FIGURE 5.12   Measured phase noise of the 200 MHz reference oscillator used in [23].

It should be noted that although the variance goes to infinity in the case of $1/f^3$ noise, it grows logarithmically with respect to M, that is, very slowly, and is usually masked by other noise sources unless very long time intervals are measured.

An example of a measured time interval deviation up to 100 ms for the TDC presented in [23] is shown in Figure 5.13. The TDC is based on time-to-voltage interpolators, has a single-shot precision of about 2 ps due to nonlinearities, and uses a 200 MHz reference oscillator. The measured phase noise PSD of the reference oscillator is shown in Figure 5.12.

The time interval deviation was measured by using a similar 200 MHz clock to generate the input time intervals. The time deviation includes the combined jitter of both the input generating clock and the reference clock. Thus, the measured time deviation is $\sqrt{2}$ larger than that due to the reference clock only.

As can be seen in Figure 5.13, the time deviation for short time intervals is dominated by the single-shot precision of the TDC, while the white noise floor of the oscillator has little impact. After about 1 ms, however, the $1/f^3$ noise of the oscillator starts to dominate and the single-shot precision deteriorates.

5.5 DESIGN CASE: A MULTICHANNEL CMOS TDC BASED ON A LOW-FREQUENCY REFERENCE

This section presents the operation and architecture of a long-range, multichannel CMOS TDC. The TDC architecture presented here has been developed especially for the needs of a pulsed laser radar, where high precision is needed over a long dynamic range [30]. The architecture supports several parallel measurement

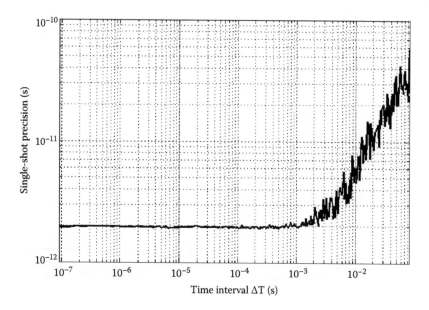

FIGURE 5.13   Measured single-shot precision (M-sample deviation) of the TDC in [23]. The input time intervals were generated by a similar 200 MHz clock oscillator.

channels, which can operate simultaneously, and the design includes automatic stabilization against process, voltage, and temperature variations. The flash mode of operation supports a high measurement rate and the reference recycling method presented here makes it possible to use a low-frequency crystal as the reference source.

5.5.1   OPERATION

5.5.1.1   Measurement Modes

Multiple parallel measurement channels offer versatility in time interval measurements and improve the performance in some applications. Here the multiple channels make it possible to estimate the shape of the analogue input pulses, which helps to compensate for the error due to the varying amplitude of the reflected pulse in pulsed laser distance measurement [30,31].

The multichannel TDC presented here has seven parallel measurement channels, CH_{start}, CH_{stop1}, ..., CH_{stop6}, which can detect and register seven time instants per measurement. It has three input paths (input pins) for the timing signals, one for the start signal and two for stop signals (stop1 and stop2) and can operate in two modes, as presented in Figure 5.14. Mode 1 is suitable for pulse width measurements and can resolve time intervals between the start signal and three successive stop pulses in the input path stop1, $T_{SP1} - T_{SP3}$ in Figure 5.14. In addition, it can measure the widths of the stop pulses, $T_{w1} - T_{w3}$ in Figure 5.14. The channels CH_{start}, CH_{stop1}, CH_{stop3}, and CH_{stop5} are active at the rising edge and CH_{stop2}, CH_{stop4}, and CH_{stop6} at the falling edge, and the timing signals registered

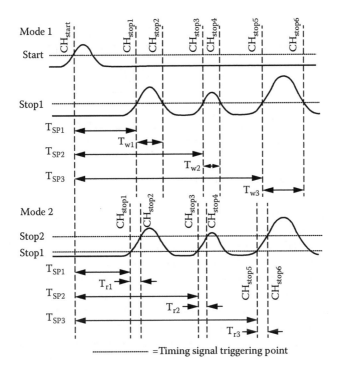

FIGURE 5.14   Measurement modes.

in CH_{stop1} enable CH_{stop3}, which in turn enables CH_{stop5}, with the same sequence operating between CH_{stop2}, CH_{stop4}, and CH_{stop6}. Enabling a channel takes time, of course, and short pulse widths could not be measured if CH_{stop1} were to enable CH_{stop2}, for example.

The second operation mode (Mode 2) is indicated for slew rate measurements. This measures the time intervals between the start and stop pulses T_{SP1} – T_{SP3} in Figure 5.14, but it also uses the third timing signal input path stop2. The previous stage before the TDC (e.g., the receiver channel) can detect pulse echoes at two threshold levels. The first path, with a lower detection threshold, is connected to stop1 inputs into the TDC, and the second, with a higher detection threshold, to stop2. Now the slew rates of three successive stop pulses can be measured with the TDC, T_{r1} – T_{r3} in Figure 5.14. All the channels are now active at rising edges. The channels are enabled in the same sequence as in Mode 1. Again a timing signal in CH_{stop1} can be so close to one in CH_{stop2} that channel enabling in a different order could cause a loss of results.

5.5.1.2   Measurement Method

The TDC digitizes the time intervals by means of the Nutt method (counter and interpolators), as presented earlier. The counter counts the clock edges between the timing signals (start and stop) and the interpolators resolve the time intervals between the timing signals and clock edges, as presented in Figure 5.5. The time

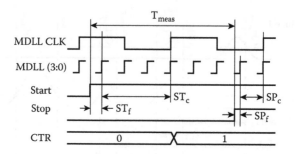

FIGURE 5.15 Timing diagram for the TDC.

reference source in this design is an external low frequency crystal, however, and the reference frequency is multiplied inside the circuit with a multiplying delay-locked loop structure (MDLL), to be presented later, whereupon the counter, CTR, counts the rising edges of this MDLL clock, as shown in Figure 5.15.

The interpolation of the timing signal location within the MDLL clock cycle is made at two resolution levels, coarse and fine interpolations. Four time samples with an even delay within every MDLL clock cycle, MDLL(3:0) in Figure 5.15, act as the coarse interpolation measurement phases, and the timing signal registers the prevailing phase, which reveals the coarse time difference between the timing signal and the reference edge, ST_c and SP_c in Figure 5.15. The fine interpolator resolves the time difference between the timing signal and the phase registered in the coarse interpolation, ST_f and SP_f in Figure 5.15.

The two-level interpolation architecture entails several benefits and is used widely in high-resolution TDCs [32–35]. The number of delay elements and registers is reduced, because the second, high-resolution measurement is performed only within a small range, as in Figure. 5.15. A 100 ns reference clock cycle, for example, can be interpolated with 10 ps resolution by means of 10,000 delay elements and registers on one interpolation level, but the same task requires only 100 delay elements and registers for both the coarse and fine interpolations in a two-level setup. The fact that less delay elements participate in the measurement also means that the INL and power consumption can be lower.

5.5.2 TDC Structures and Design Principles

The TDC presented here is composed of several blocks and structures. Measurement blocks, counters, and interpolators digitize the time interval, and synchronizing structures synchronize the asynchronous timing signals to a synchronous measurement logic, which provides compatible operations in the measurement blocks. Stabilization structures lock the measurement resolutions to the cycle time of the reference source, which provides stable operation with PVT variation, and the arithmetical logic unit decodes the raw measurement data to binary words and calculates the total measurement result. This section will present the main individual measurement structures, which are then combined in Section 5.5.3.

5.5.2.1 MDLL Frequency Multiplier and Coarse Interpolator

The MDLL was originally designed to prevent excessive jitter accumulation in frequency synthesis [36], but frequency multiplication with a short delay line without excessive jitter is also desirable in TDCs [32]. A suitable MDLL delay line structure for coarse timing signal interpolation is shown in Figure 5.16. The delay elements, as in the transistor-level diagram in Figure 5.17, are differential delay-adjustable multiplexers with two input channels. Identical delay elements provide time samples with an even time delay at their outputs when the signal propagates in the delay line. Note that the end of the delay line in Figure 5.16 is cross-coupled back to the second channel of the first element. This is necessary in order to achieve pulse propagation.

The delay line operates as follows. The rising edge of the external reference signal is first converted to a differential and then passed to the delay line. After this, the input channel of the first element is changed and the reference begins to circulate in the closed loop. After a certain number of rounds (counted by the recycling counter), the first element lets the next jitter-free reference edge enter the loop. A phase detector (PD) and charge pump (CP) control the element delays so that the delay in the recycled delay line becomes equal to the reference clock cycle time.

The recycling delay line structure offers several benefits. The frequency of the delay line is multiplied, and thus, a low-frequency reference source can be used (crystal), the delay line can be short, which reduces the size of the device and improves the INL, and the inherent accumulating delay line jitter is nulled when the recycling is refreshed with a new reference edge.

5.5.2.2 Counter Synchronization

The synchronous counter counts the rising edges of the MDLL clock between the asynchronous timing signals start and stop. Counter enabling and disabling needs to be synchronized to the same clock periods that were registered at the coarse

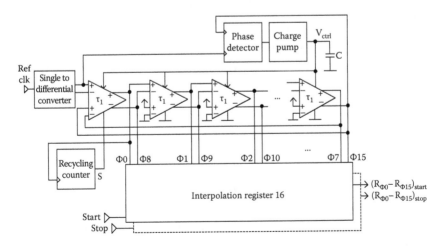

FIGURE 5.16   The MDLL-based coarse interpolator.

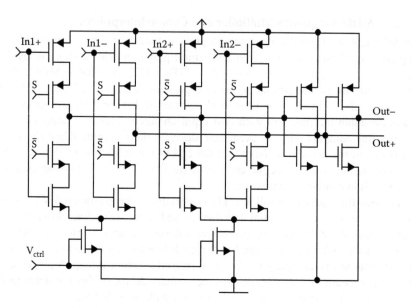

FIGURE 5.17 The delay-adjustable differential multiplexer.

interpolation level, so that the counter and interpolation results are compatible and the total measurement result correct [37].

Counter synchronization can be handled with a dual-edge synchronization method [38], as presented in Figure 5.18, where counting is delayed by one reference clock cycle and the counter control (enabling/disabling) is steered into the correct, safe position by the settled register result from the coarse interpolation level, INT1. In the counter synchronization structure, two parallel flip-flops, A and B in Figure 5.18, in the start and stop channels record the arrival of the timing signal together with the falling and rising edges of the reference signal (the rising edge also changes the state of the counter). This dual-edge method ensures reliable recording of the timing signal, because one of the two flip-flops always has enough delay between the data change and the clock edge. The selection of which flip-flop is used for counter control depends on the register results at the coarse interpolation level.

The time sample $\Phi 0$ (see Figure 5.16) always begins a new cycle of interpolation and counting by the counter, so its state $R_{\Phi 0}$ can be used to control the counting. If $\Phi 0$ was high when the timing signal arrived (to the right of the dashed line in Figure 5.18), the result $R_{\Phi 0}$ will also be high, but if $\Phi 0$ was low (to the left of the dashed line in Figure 5.18), $R_{\Phi 0}$ will also be low. The register result $R_{\Phi 0}$ controls a multiplexer, which chooses which of the two flip-flops is used to enable/disable the counter. If $R_{\Phi 0}$ was low (to the left in Figure 5.18), the next falling edge of the reference after the timing signal (flip-flop A) will enable/disable the counter, but if $R_{\Phi 0}$ was high (to the right in Figure 5.18), the next rising edge of the reference after the timing signal (flip-flop B) will enable/disable the counter. As shown in the timing diagram, the counter is enabled/disabled one clock cycle later in the latter case than in the first, because the counter is also clocked in conjunction with the rising edge.

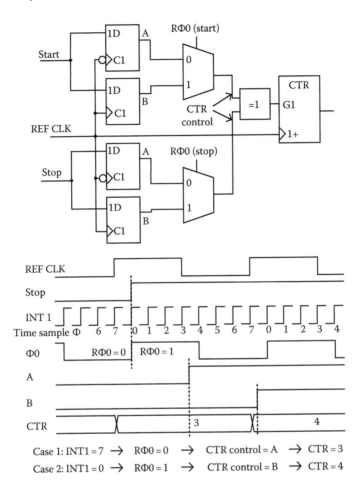

FIGURE 5.18   Counter synchronization.

5.5.2.3   Synchronization for the Second Interpolation Level

The second interpolation level (fine interpolator) resolves the time interval between the asynchronous timing signal and the synchronous MDLL clock edge, which is registered in the coarse interpolation, ST_f and SP_f in Figure 5.15. These can be referred to as the async and sync signals, respectively. Metastable registration (variation in the decision time of the register) is to be expected in the coarse interpolation, because the signals are very close to each other (less than or equal to the delay in one delay element). A proper synchronizer between the interpolation levels is needed to deal with two issues. Since the total result of the two levels must be unambiguous, the operation of the second level must be steered by the final, settled result of the first level. The varying decision time of the register on one level due to metastable operation cannot be allowed to influence (shift) the measurement signals on the second level.

Unambiguous operation of the measurement registers can be resolved with the synchronization structure presented in Figure 5.19 [38]. The measurement signals on the second interpolation level, sync and async, are delayed by a synchronization delay $\Delta t_S = \Delta t_A = 4\tau_1$, which provides compatible and correct results in the event of metastable register operation at the first interpolation level. The timing diagram and the numbers (0 or 1) next to the flip-flops describe the signal states in one possible situation after the timing signal. The upper flip-flops in Figure 5.19 describe the register bank of the first interpolation level, INT1, from where the result can then be decoded to a binary word. It can be seen from the recorded values and the timing diagram that the timing signal arrived after the time sample $\Phi 1$. The complementary outputs of these flip-flops are connected to data inputs into the flip-flops in the second row in Figure 5.19, the second row being clocked with first interpolation level time samples having a phase shift of four samples. In other words, the first interpolation level registers have a settling time of $4\tau_1$ before the data are used for the generation of a synchronous measurement signal for the second level. The OR gate with eight input lines in Figure 5.19 finds the first settled flip-flop after the synchronization delay and generates the sync signal. The total delay due to synchronization, Δt_S, must be compensated for in the asynchronous timing signal path by delaying it correspondingly by Δt_A.

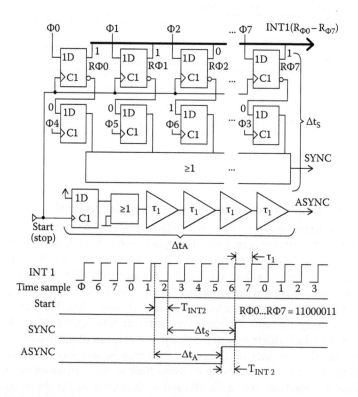

FIGURE 5.19 Synchronization for fine interpolation.

5.5.2.4 Parallel Capacitor-Scaled Delay Line Interpolation

Since the second interpolation level needs to resolve the time interval with very high resolution, structures reaching less than the gate delay need to be used. Here the fine interpolation is based on load capacitor scaling in delay elements connected in parallel [39]. The propagation delay of a buffer-type delay element, as presented in Figure 5.20, can be increased by adding extra capacitance inside the element. When a rising edge propagates through the element, the extra charge needs to be discharged, which creates an additional delay. Changes in the propagation delay resulting from the placement of different numbers of unit capacitors of the same size within the elements are shown in Figure 5.20. The total propagation delay τ_{tot} consists of the basic delay τ_e in a delay element and the delay from a single unit capacitor τ_2 multiplied by the number of unit loads n.

Actual high-resolution interpolation can be performed using two capacitor-scaled delay lines and a register bank, which form a two-dimensional array as shown in Figure 5.21, a setup in which the structure digitizes the time interval between the synchronous reference edge (sync signal in Figure 5.21) and an asynchronous timing signal (async). The async signal creates four time samples, A0–A3 in Figure 5.21, with an $8\tau_2$ time difference, while the sync creates eight time samples, S0–S7, with a $1\tau_2$ time difference. The delay lines are connected to an arbiter register bank where each arbiter records which out of the two time samples arrived earlier. The register data change as the timing signal moves in the interpolation region, and the interpolation result can be decoded from the register result. Note that the effect of the basic delay in the delay element, τ_e, is cancelled out within the limits of mismatch, because the same delay is used in all the elements of both delay lines.

The interpolation range of the fine interpolator is $\tau_1 = 32\tau_2$, as presented in Figure 5.21. The delay τ_1 was stabilized with a DLL structure on the first interpolation level. The delay τ_2 in the fine interpolator delay element can be stabilized to be equal to $\tau_1/32$ in all PVT circumstances with the DLL structure presented in Figure 5.22.

The time difference τ_1 (between the coarse interpolator time samples) is fed to two parallel fine interpolator delay elements with the unit capacitor difference of 32, equaling the delay in the fine interpolator measurement range. The PD and CP adjust the control voltage V_{ctrl2} until the two input signals to the PD arrive at the same time.

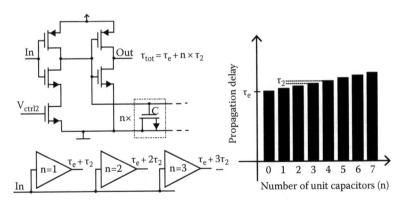

FIGURE 5.20 A load capacitor-scaled delay element.

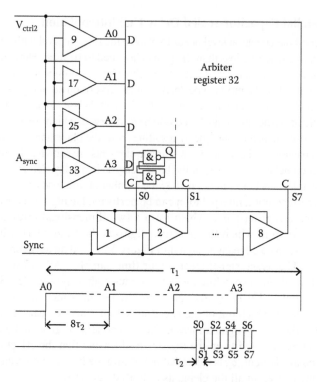

FIGURE 5.21 Fine interpolation with load capacitor-scaled delay elements.

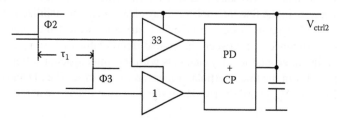

FIGURE 5.22 A DLL structure for capacitor-scaled delay element stabilization.

The two delay elements with a unit capacitor difference of 32 compensate for the delay of τ_1 at a certain control voltage V_{ctrl2}, which leads to a situation in which one unit capacitor is equal to a delay of $\tau_1/32 = \tau_2$.

5.5.3 TDC Architecture

5.5.3.1 Measurement Core

The measurement core, in which the individual structures presented here have been connected together, is shown in Figure 5.23. The system includes seven parallel measurement channels, and thus, identical counter registers, coarse interpolator registers, and fine interpolators are repeated seven times in the circuit.

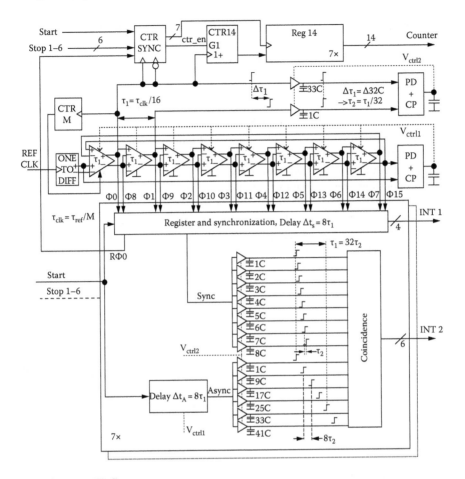

FIGURE 5.23 TDC measurement core.

The reference signal from the reference signal clock (REFCLK) is first converted to a differential and then passed to the MDLL delay line, which is composed of 8 differential delay elements, creating a total of 16 time samples $\Phi0$, ..., $\Phi15$ as the reference signal first propagates through the delay line and then does so again in the opposite phase due to cross-coupling in the first element. The 0.35 µm CMOS technology provides a ~280 ps propagation delay τ_1 for the adjustable delay elements over a large temperature range. With a 20 MHz reference crystal and a recycling factor $M = 11$, the internal delay line frequency $1/\tau_{clk}$ becomes 220 MHz.

The second delay-locked loop above the recycling delay line adjusts the control voltage V_{ctrl2} and stabilizes the second interpolation level delays against PVT variations. The 14-bit counter with counter synchronization at the top of Figure 5.23 counts the rounds of the MDLL delay line between the timing signals and provides a ±74 µs measurement range between the timing signals. The negative measurement results describe a situation in which the stop signals arrive before the start, which is a possible feature in some applications. The first timing signal (start or stop) enables the counter, and all the arriving timing signals register its current state.

The register and synchronization block composes the first interpolation level result INT1 and also provides the synchronous reference signal, sync, for higher resolution interpolation at the second level. Each asynchronous timing signal is delayed by the synchronization delay Δt_A, after which the signal async creates six parallel time samples with a time difference of $8\tau_2$ by means of the parallel capacitor-scaled delay line, also presented in Figure 5.21. Actually, four high-resolution delay elements are needed to cover the interpolation region of $32\tau_2$, but one extra element is inserted on each side of the targeted interpolation region to allow for a mismatch between the synchronization delays Δt_A and Δt_S. The sync signal creates eight parallel time samples with a resolution of $\tau_2 = 8.8$ ps. The synchronous and asynchronous high-resolution time samples confront each other in the register bank presented in Figure 5.21, and the second interpolation level result, INT2, can be determined from their degree of coincidence. Each timing signal registers the counter state, CTR, the first interpolation level result, INT1, and the second interpolation level result, INT2. The time intervals between the start and stop signals, stop pulse widths, or slew rates can be counted by subtracting the corresponding timing signal results. The subtraction for the time interval between the start and stop pulses, for example, is of the form:

$$T_m = \left(CTR_{stop} - CTR_{start}\right) \times \tau_{clk} + \left(INT1_{stop} - INT1_{start}\right) \times \tau_1 + \left(INT2_{stop} - INT2_{start}\right) \times \tau_2$$

$$(5.16)$$

The subtraction (stop − start), which is also applied to the counter result, makes it possible to express a negative overall result when the stop pulse arrives before the start pulse.

5.5.4 MEASUREMENT EXAMPLE

One start and three successive stop pulses with different widths and time intervals over several microseconds were generated with a signal generator. The measurements were repeated 100,000 times in both measurement modes, yielding distributions of the results with average values (μ) and standard deviations (σ) as presented in Figure 5.24.

The diagram at the top of Figure 5.24 shows the measured time intervals between the start and three successive pulses in the stop1 input channel, the results are similar in both measurement modes. The data for the next diagram were collected in Mode 1 and show the stop pulse widths, while the last diagram describes operation in Mode 2 and shows the propagation time difference between the two stop input channels with coaxial cables of different lengths.

As noted earlier, the precision of time interval measurement is affected by random jitter, quantization noise, and interpolation nonlinearities. More comprehensive measurements show that the interpolation nonlinearities, especially in the recycling delay line, dominate here and limit the attainable precision [30].

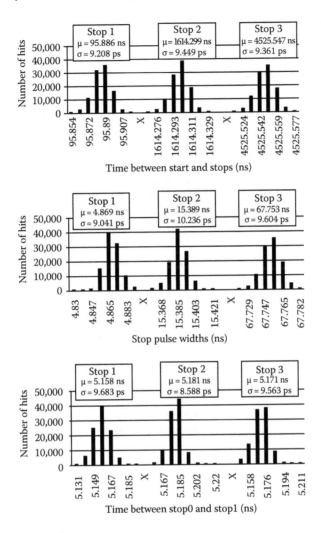

FIGURE 5.24 Measurements made with three successive stop-pulses.

5.6 DESIGN CASE: A CMOS TDC-BASED ON SUCCESSIVE TIME DOMAIN APPROXIMATION

Successive approximation register (SAR) analogue-to-digital converters (ADCs) using binary search as the conversion algorithm are frequently used when a medium to high resolution is needed simultaneously with moderate conversion speed and low power consumption. Successive approximation method can be applied to time domain by replacing the analogue voltage comparator with a phase detector (PD), and the digital-to-analogue converter (DAC) with a digital-to-time converter (DTC). The conversion result is then the final content of the SAR after the conversion is completed. Ps-level resolution is feasible with successive approximation even when using low-cost, moderate-linewidth CMOS technologies.

5.6.1 OPERATING PRINCIPLE

The operating principle of time domain successive approximation is illustrated in Figure 5.25. The arrival of the rising edges of the signals t1 and t2, the delay difference of which represents the time interval to be converted, is compared in the time domain with a PD. If the signal t1 arrives before t2, the bit decision is 1; otherwise, it is 0. After each bit decision, the delay difference between the signals t1 and t2 is adjusted in the direction, which will reduce it, and the amount of adjustment is halved in each following step from $2^{n-1} \cdot t$ to $2^0 \cdot t$, where n is the bit to be converted and t is the resolution of the TDC. At the end of the conversion cycle, the delay difference in the input to the PD is less than the LSB resolution t of the converter, as shown in Figure 5.25.

5.6.2 SUCCESSIVE APPROXIMATION WITH CASCADED DELAY CELLS

The concept of a TDC based on successive approximation was presented in [40,41] and implemented in [42] with a dynamic range of 10 ns and a resolution of 9.77 ps. The proposed architecture (Figure 5.26) consists of a cascade of blocks in which either one of the input signals can be delayed with respect to the other depending on the order of arrival of their rising edges, as shown in Figure 5.27. By scaling the delay difference in powers of two from the first block, converting the most significant bit (MSB), to the last, converting the LSB, a binary search can be conducted to resolve the initial phase difference between the input signals. In order to operate reliably, the minimum propagation delay T of the delay blocks must be longer than the recovery time of the PD from a possible metastable state, as shown in Figure 5.27.

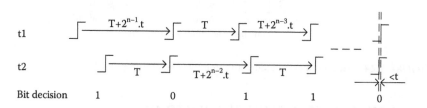

FIGURE 5.25   Conceptual timing diagram for time domain successive approximation.

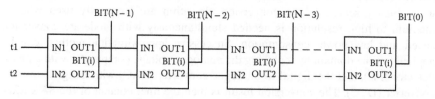

FIGURE 5.26   Successive approximation with cascaded binary-scaled selectable delays.

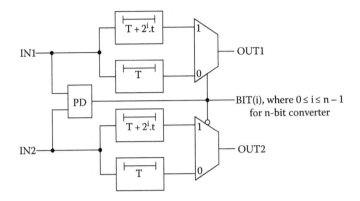

FIGURE 5.27 Binary-scaled selectable delays.

5.6.3 PHASE DETECTOR

The PD compares the arrival of the rising edges of the input signals, that is, it operates as an arbiter and indicates which signal arrived first. Two examples of symmetric PDs with zero setup time are shown in Figure 5.28a [40] and b [43,44]. The PD retains the result for as long as the input signals are logically high, but as soon as the input signals are cleared [44], the PD outputs are also cleared. This is not a major problem with the cascaded architecture presented here, in which rising edges propagate in the delay cells and the conversion result is stored before a new measurement, but the cyclic architectures presented in the following sections use pulse trains as input signals for the PD, so that the result must be stored in an additional flip-flop [35,44]. The clock signal for this flip-flop can be coded from the

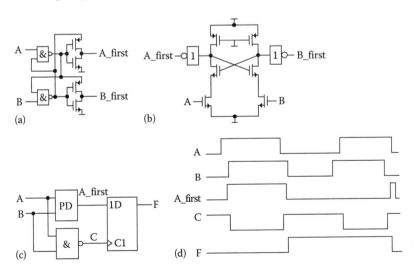

FIGURE 5.28 (a, b) Phase detectors and (c) phase detector with flip-flop, (d) timing diagram of the phase detector with a flip-flop.

inputs of the PD with a NAND gate, as shown in Figure 5.28c, making sure that the propagation delay of the gate is shorter than that of the PD and that the input signals are overlapping. Now, the first falling edge in the inputs to the PD will generate a rising edge in the NAND gate output just before the output from the PD disappears, as shown in Figure 5.28d [35].

5.6.4   DIGITAL-TO-TIME CONVERTER

A digitally controllable delay, or DTC, can be implemented with the circuits shown in Figure 5.29, for example. A simple current-starved inverter with a digitally controlled capacitor matrix as a load, implemented with moscaps, for example, and a simple CMOS inverter as a comparator can provide for adequate linearity over a dynamic range of a few hundred picoseconds [35]. For a longer dynamic range a circuit topology combining a differential switch with an analogue comparator can be used [45]. By making the delay cells adjustable with a bias signal, the operating point of the delay cell can be made tolerant to process, voltage, and temperature (PVT) changes, and the resolution of the TDC can be automatically calibrated and stabilized.

5.6.5   CYCLIC TIME-DOMAIN SUCCESSIVE APPROXIMATION

A more area-efficient architecture employing a cyclic conversion principle is presented in Figure 5.30. Instead of cascading the delay adjustment cells, a pair of DTCs is used to make the necessary delay adjustments for the binary search. The architecture contains two identical loops composed of a multiplexer, monostable and a DTC, in which the signals are recycled and the outputs of the DTCs are compared with the PD after each cycle to make the new bit decision, which is then used as a control signal to adjust the propagation delays. At the beginning of a new measurement, the timing signals t1 and t2 propagate through the multiplexers, after which each multiplexer selects the feedback signal as its input. Only the latest bit (BIT(0)) of the shift register SREGn operating as the SAR) is needed to select the new delay for the conversion of the next bit, because the necessary adjustments are integrated into the phases of the signals recycling in the two loops. The adjustments are scaled in powers of two toward the conversion of the LSB. In this architecture, the width of the pulses recycling in the loops has to be long enough for the PD to recover from

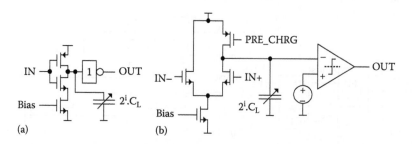

FIGURE 5.29   (a) A DTC with inverters and (b) a DTC with differential switch and comparator.

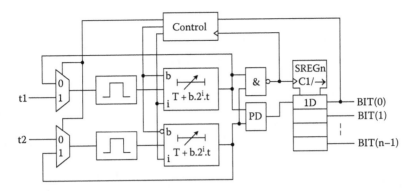

FIGURE 5.30 Cyclic successive approximation with recycling loops.

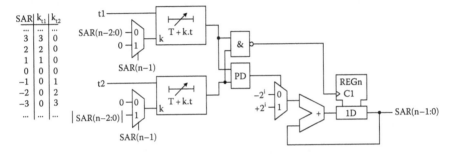

FIGURE 5.31 Cyclic successive approximation with clock signal inputs.

a possible metastable state. The rising edges of the pulses are compared with the PD, and after the first falling edge at the outputs of the delay cells, the NAND gate produces a rising edge to store the result in the shift register SREGn just before the output of the PD is cleared. The cycle time of the signal recycling in the loops has to be long enough to make the delay adjustments before the next rising edge arrives at the input to the delay cell. A TDC with 1.2 ps resolution and 328 μs dynamic range is reported in [35].

If the signals t1 and t2, the phase difference of which is to be measured, are clock signals, the recycling loops can be removed [46], as shown in Figure 5.31. The input signal of the TDC could enable a startable ring oscillator (SRO), which then retains the phase of the input signal in the signal t1 with respect to the phase of the reference clock signal t2. The SRO operates as a time domain sample and hold [46]. In this architecture, the required adjustments accumulate into the SAR REGn in Figure 5.28, rather than into the phases recycling in the delay cells as in the previous architectures, and the delay adjustments required to align the phases of the outputs of the adjustable delay cells are decoded from the content of the SAR. If the content of the SAR-REG is positive, that is, the MSB of SAR is 0, the signal t1 is adjusted by the value of the SAR and the phase of the signal t2 is kept intact. Otherwise, the signal t2 is adjusted by the absolute value of the SAR, as shown in Figure 5.31. This architecture can also be used for clock deskewing [47,48], because after the conversion, the outputs of the delay cells are aligned within one LSB.

5.7 CONCLUSIONS

We have shown earlier that a combination of the counter method with the interpolation of timing pulse positions within the clock period is a very efficient method for realizing a high-precision, accurate TDC with a wide operation range. This approach combines the inherently good single-shot resolution of a short-range interpolator based on digital delay line techniques, for example, with the excellent accuracy and wide linear range of the counting method. It was also shown that in a general measurement situation where the timing pulses are asynchronous with respect to the system clock, the effect of interpolator nonlinearities on the final averaged output is strongly suppressed due to the inherent averaging effect of the interpolation method. On the other hand, these nonlinearities widen the distribution of the measured single-shot results and, in many cases, limit the single-shot precision of the TDC. It was also pointed out that careful synchronization of the timing signals is needed in order to get unambiguous measurement results that are free from systematic errors. Finally, two case studies showed that, with the approaches presented here, a TDC realized in standard nonaggressive CMOS technologies can achieve ps-level resolution and a single-shot precision better than 10 ps (sigma value) over a wide operation range of hundreds of microseconds.

REFERENCES

1. B. Schwarz, LIDAR: Mapping the world in 3D, *Nature Photonics*, 4, 429–430, 2010.
2. K. Fuerstenberg, F. Ahlers, Development of a low cost laser scanner—The EC project MiniFaros, in: G. Meyer, J. Valldorf (Eds.), *Advanced Microsystems for Automotive Applications 2011, Smart Systems for Electric, Safe and Networked Mobility*, Springer, Berlin, Germany, 2011.
3. C. Niclass, K. Ito, M. Soga, H. Matsubara, I. Aoyagi, S. Kato, M. Kagami, Design and characterization of a 256 × 64-pixel single-photon imager in CMOS for a MEMS-based laser scanning time-of-flight sensor, *Optics Express*, 20(11), 11863–11881, 2012.
4. K. Ito, C. Niclass, I. Aoyagi, H. Matsubara, M. Soga, S. Kato, M. Maeda, M. Kagami, System design and performance characterization of a MEMS-based laser scanning time-of-flight sensor based on a 256\times 64-pixel single-photon imager, *IEEE Photonics Journal*, 5(2), Article: 6800114, 2013.
5. V.C. Coffey, Imaging in 3-D: Killer Apps coming soon to a device near you! *Optics and Photonics News*, 25(6), 36–43, 2014.
6. J. Christiansen, Picosecond stopwatches: The evolution of time-to-digital converters, *IEEE Solid-State Circuits Magazine*, 4(3), 55–59, 2012.
7. C. Veerappan, J. Richardson, R. Walker, D.-U. Li, M.W. Fishburn, Y. Maruyama, D. Stoppa et al., A 160 × 128 single-photon image sensor with on-pixel 55 ps 10b time-to-digital converter, *IEEE ISSCC Digest of Technical Papers*, 2011, pp. 312–314.
8. J.B. Rettig, L. Dobos, Picosecond time interval measurements, *IEEE Transactions on Instrumentation and Measurement*, 44(2), 284–287, 1995.
9. G.W. Roberts, Reducing the analog-digital productivity gap using time-mode signal processing, *2014 IEEE International Symposium on Circuits and Systems (ISCAS)*, Melbourne, Australia, June 1–5, 2014, pp. 782–785.
10. R.B. Staszewski et al., 1.3 V 20 ps time-to-digital converter for frequency synthesis in 90-nm CMOS, *Proceedings of IEEE Transactions on Circuits and Systems-II: Express Briefs*, 52(3), 220–224, 2006.

11. Institute of Electrical and Electronics Engineers Inc., IEEE Standard Dictionary of Electrical and Electronics Terms, 6th Edition, 1984.
12. Hewlett-Packard Inc., Time Interval Averaging. Application Note 162-1.
13. J. Rabaey, *Digital Integrated Circuits*, Prentice Hall, Englewood Cliffs, NJ, 1996.
14. R. Nutt, Digital time intervalometer, *The Review of Scientific Instruments*, 39(9), 1342–1345, 1968.
15. B. Turko, A picosecond resolution time digitizer for laser ranging, *IEEE Transactions on Nuclear Science*, 25(1), 75–80, 1978.
16. J. Kostamovaara, R. Myllylä, Time-to-digital converter with an analog interpolation circuit, *Review of Scientific Instruments*, 57(11), 2880–2885, 1986.
17. E. Raisanen-Ruotsalainen, T. Rahkonen, J. Kostamovaara, A low-power CMOS time-to-digital converter, *IEEE Journal of Solid-State Circuits*, 30(9), 984–990, 1995.
18. K. Karadamoglou et al., An 11-bit high-resolution and adjustable-range CMOS time-to-digital converter for space science instruments, *IEEE Journal of Solid-State Circuits*, 39(1), 214–222, 2004.
19. E. Raisanen-Ruotsalainen, T. Rahkonen, J. Kostamovaara, An integrated time-to-digital converter with 30-ps single-shot precision, *IEEE Journal of Solid-State Circuits*, 35(10), 1507–1510, 2000.
20. M. Mota, J. Christiansen, A high-resolution time interpolator based on a delay locked loop and an RC delay line, *IEEE Journal of Solid-State Circuits*, 34(10), 1360–1366, 1999.
21. J. Kalisz, R. Szplet, J. Pasierbinski, A. Poniecki, Field-programmable-gate-array-based time-to-digital converter with 200-ps resolution, *IEEE Transactions on Instrumentation and Measurement*, 46(1), 51–55, 1997.
22. K. Maatta, J. Kostamovaara, A high-precision time-to-digital converter for pulsed time-of-flight laser radar applications, *IEEE Transactions on Instrumentation and Measurement*, 47(2), 521–536, 1998.
23. P. Keranen, K. Maatta, J. Kostamovaara, Wide-range time-to-digital converter with 1-ps single-shot precision, *IEEE Transactions on Instrumentation and Measurement*, 60(9), 3162–3172, 2011.
24. T. Rahkonen, J. Kostamovaara, The use of stabilized CMOS delay lines for the digitization of short time intervals, *IEEE Journal of Solid-State Circuits*, 28, 887–894, 1993.
25. J. Kalisz, M. Pawlowski, R. Pelka, Error analysis and design of the Nutt time interval digitiser with picosecond resolution, *Journal of Physics E: Scientific Instruments*, 20, 1330–1341, 1987.
26. R. Pelka, J. Kalisz, R. Szplet, Nonlinearity correction of the integrated time-to-digital converter with direct coding, *IEEE Transactions on Instrumentation and Measurement*, 46(2), 449–453, 1997.
27. P. Keranen, J. Kostamovaara, Oscillator instability effects in time interval measurement, *IEEE Transactions on Circuits and Systems I: Regular Papers*, 60(7), 1776–1786, 2013.
28. IEEE Standard Definitions of Physical Quantities for Fundamental Frequency and Time Metrology—Random Instabilities, IEEE STD 1139-2008.
29. A. Hajimiri, T.H. Lee, A general theory of phase noise in electrical oscillators, *IEEE Journal of Solid-State Circuits*, 33(2), 179–194, February 1998.
30. J.-P. Jansson, V. Koskinen, A. Mäntyniemi, J. Kostamovaara, A multi-channel high precision CMOS time-to-digital converter for laserscanner based perception systems, *IEEE Transactions on Instrumentation and Measurement (TIM)*, 81, 2581–2590, September 2012.
31. S. Kurtti, J. Kostamovaara, An integrated laser radar receiver channel utilizing a time-domain walk error compensation scheme, *IEEE Transactions on Instrumentation and Measurement*, 60, 146–157, January 2011.

32. J.-P. Jansson, A. Mäntyniemi, J. Kostamovaara, A CMOS time-to-digital converter with better than 10 ps single-shot precision, *IEEE Journal of Solid-State Circuits*, 41, 1286–1296, June 2006.

33. S. Henzler, S. Koeppe, W. Kamp, H. Mulatz, D. Schmitt-Landsiedel, 90 nm 4.7 ps-resolution 0.7-LSB single-shot precision and 19 pJ-per-shot local passive interpolation time-to-digital converter with on-chip characterization, *IEEE International Solid-State Circuits Digest of Technical Papers*, San Francisco, CA, February 2008, pp. 548–635.

34. M. Lee, A.A. Abidi, A 9b, 1.25 ps resolution coarse-fine time-to-digital converter in 90 nm CMOS that amplifies a time residue, *IEEE Journal of Solid-State Circuits*, 43, 769–777, April 2008.

35. A. Mäntyniemi, T. Rahkonen, J. Kostamovaara, A CMOS time-to-digital converter (TDC) based on a cyclic time domain successive approximation interpolation method, *IEEE Journal of Solid-State Circuits*, 44, 3067–3078, November 2009.

36. A. Waizman, A delay line loop for frequency synthesis of de-skewed clock, *IEEE International Solid-State Circuits Digest of Technical Papers*, San Francisco, CA, February 1994, pp. 289–299.

37. J.-P. Jansson, A. Mäntyniemi, J. Kostamovaara, Synchronization in a multi-level CMOS time-to-digital converter, *IEEE Transactions on Circuits System I*, 56, 1622–1634, August 2009.

38. A. Mäntyniemi, T. Rahkonen, J. Kostamovaara, A 9-channel integrated time-to-digital converter with sub-nanosecond resolution, *Proceedings of IEEE MWSCAS'97*, 1, 189–192, August 1997.

39. A. Mäntyniemi, T. Rahkonen, J. Kostamovaara, A nonlinearity-corrected CMOS time digitizer IC with 20 ps single-shot precision, *Proceedings of the IEEE International Symposium on Circuits and Systems*, 1, 513–516, May 2002.

40. D.J. Kinniment, O.V. Maevsky, A. Bystrov, G. Russell, A.V. Yakovlev, On-chip structures for timing measurement and test, *Proceedings of Eighth International Symposium on Asynchronous Circuits and Systems*, Manchester, U.K., 2002, pp. 190–197.

41. M.A. Abas, G. Russell, D.J. Kinniment, Built-in time measurement circuits—A comparative design study, *Computers & Digital Techniques, IET*, 1, 87–97, 2007.

42. H. Chung, H. Ishikuro, T. Kuroda, A 10-bit 80-MS/s decision-select successive approximation TDC in 65-nm CMOS, *IEEE Journal of Solid-State Circuits*, 47, 1232–1241, 2012.

43. V. Gutnik, A. Chandrakasan, On-chip picosecond time measurement, *Symposium on VLSI Circuits Digest of Technical Papers*, Honolulu, HI, 2000, pp. 52–53.

44. P.M. Levine, G.W. Roberts, A high-resolution flash time-to-digital converter and calibration scheme, *Proceedings of ITC International Test Conference 2004*, Charlotte, NC, 2004, pp. 1148–1157.

45. S. Alahdab, A. Mantyniemi, J. Kostamovaara, A time-to-digital converter (TDC) with a 13-bit cyclic time domain successive approximation interpolator with sub-ps-level resolution using current DAC and differential switch, *2013 IEEE 56th International Midwest Symposium on Circuits and Systems (MWSCAS)*, Columbus, OH, 2013, pp. 828–831.

46. A. Mantyniemi, J. Kostamovaara, Time-to-digital converter (TDC) based on startable ring oscillators and successive approximation, *2014 Norchip*, IEEE, October 27–28, 2014, Tampere, Finland, pp. 1–4.

47. G.-K. Dehng, J.-M. Hsu, C.-Y. Yang, S.-I. Liu, Clock-deskew buffer using a SAR-controlled delay-locked loop, *IEEE Journal of Solid-State Circuits*, 35, 1128–1136, 2000.

48. H.-H. Chang, S.-I. Liu, A wide-range and fast-locking all-digital cycle-controlled delay-locked loop, *IEEE Journal of Solid-State Circuits*, 40, 661–670, 2005.

6 Fundamentals of Time-Mode Analog-to-Digital Converters

Fei Yuan

CONTENTS

Analog-to-digital converters (ADCs) are the key components of mixed-mode systems. High-resolution ADCs are often realized using switched-capacitor $\Delta\Sigma$ modulators to tap the following advantages: the performance of switched-capacitor networks is only determined by capacitance ratios (provided that operational amplifiers are ideal), the accuracy of capacitance ratios provided by standard CMOS technologies is typically 0.1%, and noise-shaping and nonlinearity suppression is provided by $\Delta\Sigma$ operation to achieve superior performance with the use of only standard CMOS technologies. A key component of voltage-mode $\Delta\Sigma$ modulators is the quantizer implemented using voltage comparators. Multibit quantization for better quantization noise and stability requires that the number of comparators be $2^N - 1$ where N is the number of quantization bits. As a result, not only the power and silicon consumption of voltage-mode multibit quantizers become prohibitively high especially when N is large, mismatch between comparators also greatly degrades the performance. In addition to the mismatch of comparators, mismatches present in the digital-to-analog converter (DAC) that converts the output of the quantizer to a voltage so as to perform the Δ-operation of $\Delta\Sigma$ modulators also severely deteriorate the performance of the modulators. Their effect needs to be suppressed using dynamic element matching (DEM).

This chapter explores time-mode techniques that overcome the difficulties encountered in realization of multibit voltage-mode ADCs. We show that voltage-controlled ring oscillators provide technology-friendly multibit quantization of analog signals with built-in DEM. We further show that VCO phase quantizers offer a number of unique and desirable characteristics such as first-order noise-shaping, a large oversampling ratio (OSR), and low power consumption, to name a few.

The chapter is organized into the following: Section 6.1 examines the key parameters and figure-of-merits (FOMs) that quantify the performance of ADCs.

Section 6.2 studies the principles and properties of multibit quantizers realized using voltage-controlled ring oscillators. Section 6.3 studies VCO-based frequency quantizers. Section 6.4 investigates open-loop ADCs utilizing VCO phase quantizers and VCO frequency quantizers. Section 6.5 reviews the fundamentals of $\Delta\Sigma$ modulators. Section 6.6 introduces time-mode $\Delta\Sigma$ modulators. The characteristics of time-mode $\Delta\Sigma$ modulators are investigated in Section 6.7. Time-mode $\Delta\Sigma$ modulators with VCO phase quantizers are investigated in Section 6.8, while those with VCO frequency quantizers are explored in Section 6.9. $\Delta\Sigma$ modulators with phase feedback are examined in Section 6.10. In Section 6.11, $\Delta\Sigma$ modulators with pulse-width-modulation (PWM) for linearity improvement are studied. Section 6.12 is devoted to multistage also known as MASH time-mode $\Delta\Sigma$ modulators, both single-rate and multirate time-mode MASH $\Delta\Sigma$ modulators are examined. DEM, an effective technique to minimize the effect of the mismatch of DACs, is briefly studied in Section 6.13. Section 6.14 compares the performance of some recently reported time-mode $\Delta\Sigma$ modulators. The chapter is summarized in Section 6.15.

6.1 CHARACTERIZATION OF ANALOG-TO-DIGITAL CONVERTERS

6.1.1 FUNDAMENTALS OF ANALOG-TO-DIGITAL CONVERTERS

An ADC accepts an analog voltage v_{in} and generates a N-bit binary output $D_N...D_1$ that satisfies

$$v_{in} \approx \left(\frac{D_N}{2^1} + \frac{D_{N-1}}{2^2} + \cdots + \frac{D_2}{2^{N-1}} + \frac{D_1}{2^N} \right) V_{ref}, \tag{6.1}$$

where
  V_{ref} is the reference voltage, the full-scale range (FSR) value or the maximum value of the input voltage
  D_k, k = 1,2,...,N, is the kth digit of the binary number with D_1 the least significant bit (LSB)
  D_N is the most significant bit (MSB)

It is seen from Equation 6.1 that the contribution of the MSB is FSR/2 while that of the LSB is

$$\Delta = \frac{FSR}{2^N}. \tag{6.2}$$

The LSB or Δ is the minimum input that the ADC can sense and is therefore the resolution of the ADC.

The digital representation of v_{in} given in Equation 6.1 only provides an approximation of v_{in}. The difference between the actual and approximated values of v_{in} is the error of the approximation and is termed quantization error or quantization noise, as shown graphically in Figure 6.1a. The quantization error in Figure 6.1a is LSB. Since the quantization error is not centered at 0 V, an offset error exists. The quantization error can be

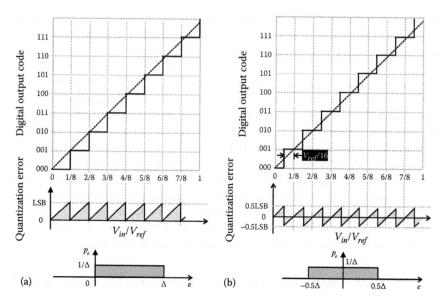

FIGURE 6.1   Quantization error. (a) Quantization error without input offset and (b) quantization error with input offset.

reduced from 1 LSB or Δ to 0.5 LSB or 0.5Δ by introducing an offset $V_{ref}/16$ to the input to eliminate the offset error so as to move the center of the quantization error to 0 V, as shown in Figure 6.1b. If the input varies sufficiently fast with time, the quantization error can be considered to be random with its value distributed uniformly over $[-0.5\Delta, 0.5\Delta]$. The probability density function p_e of the quantization error is given by

$$p_e = \begin{cases} 1/\Delta, & \text{if } -0.5\Delta \le e \le 0.5\Delta \\ 0, & \text{otherwise.} \end{cases} \tag{6.3}$$

Equation 6.3 ensures that

$$\int_{-\infty}^{\infty} p_e de = \int_{-0.5\Delta}^{0.5\Delta} \frac{1}{\Delta} de = 1. \tag{6.4}$$

The power of the quantization error, denoted by P_e, is obtained from

$$P_e = \int_{-0.5\Delta}^{0.5\Delta} e^2 p_e de = \frac{\Delta^2}{12} = \left(\frac{\text{FSR}}{2^N \sqrt{12}} \right)^2. \tag{6.5}$$

When the mismatch of the devices of an ADC exists, an offset error is introduced, as shown in Figure 6.2. The offset error of the ADC increases the quantization error. Since the offset error of the ADC is a systematic error, it can be eliminated by employing mismatch compensation techniques.

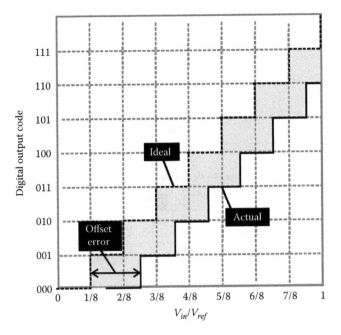

FIGURE 6.2 Offset error.

Unlike offset error, the gain error of an ADC is due to the variation of the conversion gain of the ADC when the amplitude of the input changes, as shown in Figure 6.3. It is defined as the difference between the nominal gain and actual gain of the ADC after the offset error of the ADC is removed. The gain error represents the difference in the slope of the actual and ideal transfer functions of the ADC. It is evident that the gain error of the ADC increases the quantization error of the ADC.

6.1.2 DYNAMIC RANGE

The dynamic range (DR) of an ADC is defined as the ratio of the full-scale range value of the input to the smallest input that the ADC can detect. Since for an ideal ADC, the smallest input that the ADC can detect is LSB: $\Delta = \text{FSR}/2^N$ whereas the largest input that the ADC permits is FSR, the DR of the ADC is obtained from

$$DR = \frac{\text{FSR}}{\text{LSB}} = 2^N. \tag{6.6}$$

In units of decibel (dB), it is given by

$$DR = 20\log\left(2^N\right) = 6.02N \quad (dB). \tag{6.7}$$

When harmonic tones (spurs) are present in the spectrum of an ADC with a sinusoidal input, since the amplitude of the spurs is typically much larger as compared with the quantization error of the ADC, spurious-free dynamic range (SFDR) defined as the

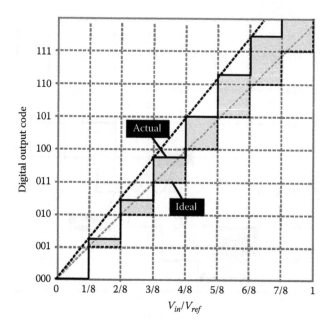

FIGURE 6.3   Gain error.

ratio of the full-scale range value of the input signal to the largest amplitude of the spurs provides a better quantification of the DR of the ADC. Clearly SRDR ≤ DR.

6.1.3   SIGNAL-TO-NOISE RATIO

The signal-to-noise ratio (SNR) of an ADC is defined as the ratio of the power of the full-scale input signal to that of the quantization noise of the ADC. If the input voltage of the ADC is a sinusoid with amplitude V_m = FSR/2, its power is given by

$$P_s = \left(\frac{V_m}{\sqrt{2}} \right)^2 = \left(\frac{\text{FSR}}{2\sqrt{2}} \right)^2. \tag{6.8}$$

Since the power of the quantization noise is given by Equation 6.5, we arrive at

$$\text{SNR} = 10\log\left(\frac{P_s}{P_e} \right) = 20\log\left(\frac{\dfrac{\text{FSR}}{2\sqrt{2}}}{\dfrac{\text{FSR}}{2^N\sqrt{12}}} \right) = 6.02N + 1.76 \quad \text{(dB)}. \tag{6.9}$$

If the input voltage is band-limited with bandwidth $2f_B$, as shown in Figure 6.4a, the input can be digitized by a sample-and-hold (S/H) block with sampling frequency f_s set to the Nyquist frequency of the input $f_N = 2f_B$ without aliasing, as shown in Figure 6.4b. These ADCs are termed Nyquist-rate ADCs. The band-limited input can also be

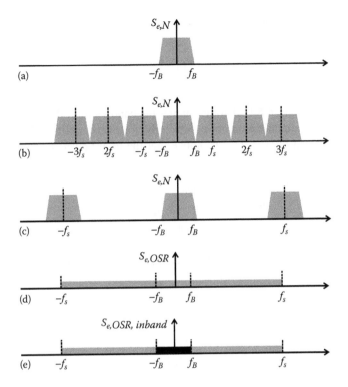

FIGURE 6.4 Spectrum of quantization noise of Nyquist and oversampling ADCs. (a) Band-limited quantization noise. (b) Quantization noise spectrum of Nyquist-rate ADCs. (c) Quantization noise spectrum of oversampling ADCs. (d) Equivalent quantization noise spectrum of oversampling ADCs. (e) In-band quantization noise of oversampling ADCs.

digitized at a higher frequency $f_s = \text{OSR} \times f_N$, where OSR is given by $\text{OSR} = f_s/(2f_B)$ and f_s is the sampling frequency. These ADCs are termed oversampling ADCs. If only quantization noise within the signal band is of interest and quantization noise outside the signal band is assumed to be zero, oversampling produces the replicas of the in-band quantization noise with a large spacing, as shown in Figure 6.4c. Consider frequency band $[-f_B, f_B]$. Since no quantization noise exists in frequency bands $[-f_B, -(f_s - f_B)]$ and $[f_B, (f_s - f_B)]$, oversampling is equivalent to spreading the quantization noise in $[-f_B, f_B]$ over the broader frequency range $[-f_s, f_s]$, as shown in Figure 6.4d.

To obtain the power spectral density (PSD) of the quantization noise $S_e(f)$, we notice that $S_e(f)$ is constant over $[-f_B, f_B]$ for Nyquist-rate ADCs and $[-f_s, f_s]$ for oversampling ADCs. For Nyquist-rate ADCs, the power of the quantization noise, denoted by $P_{e,Nyquist}$, is the total power of the quantization noise over the signal band $[-f_B, f_B]$ and is given by Equation 6.5. We therefore arrive at

$$P_{e,Nyquist} = \int_{-f_B}^{f_B} S_{e,Nyquist}(f)df = \frac{\Delta^2}{12}, \tag{6.10}$$

from which we obtain the PSD of Nyquist-rate ADCs, denoted by $S_{e,Nyquist}(f)$

$$S_{e,Nyquist}(f) = \frac{\Delta^2}{12} \frac{1}{2f_B}. \tag{6.11}$$

It is seen from Equation 6.11 that the smaller the bandwidth of the input, the higher is the PSD of the quantization noise. Similarly, for oversampling ADCs, to derive their PSD, denoted by $S_{e,OSR}(f)$, we notice that the total power of the quantization noise is given by

$$P_{e,OSR,total} = \int_{-f_B \times OSR}^{f_B \times OSR} S_{e,OSR}(f)df = \frac{\Delta^2}{12}. \tag{6.12}$$

Equation 6.12 enables us to derive the PSD of oversampling ADCs

$$S_{e,OSR} = \frac{\Delta^2}{12} \frac{1}{2f_B} \frac{1}{OSR}$$

$$= \frac{S_{e,Nyquist}}{OSR}. \tag{6.13}$$

It is seen from Equation 6.13 that oversampling lowers the PSD. Since the quantization noise is spread from $[-f_B, f_B]$ to $[-f_s, f_s]$, the total power of the quantization noise of oversampling ADCs in the signal band $[-f_B, f_B]$, as shown in Figure 6.4e, is obtained from

$$P_{e,OSR,inband} = \int_{-f_B}^{f_B} S_{e,OSR}(f)df$$

$$= \frac{\Delta^2}{12} \frac{1}{OSR}$$

$$= \frac{P_{e,OSR,total}}{OSR}. \tag{6.14}$$

It is evident from Equation 6.14 that increasing the OSR reduces quantization noise. Also, doubling OSR lowers the quantization noise by 3 dB or equivalently improves the resolution by 0.5 bit.

Let us now calculate the SNR of oversampling ADCs. Let the input voltage of an oversampling ADC be a sinusoid with amplitude $V_m = FSR/2$. Its power is given by

$$P_s = \left(\frac{V_m}{\sqrt{2}}\right)^2 = \left(\frac{FSR}{2\sqrt{2}}\right)^2 = \left(\frac{2^N \Delta}{2\sqrt{2}}\right)^2. \tag{6.15}$$

Since the power of the in-band quantization noise of the oversampling ADC is given by Equation 6.14, we have

$$\text{SNR} = 10 \log \left(\frac{P_s}{P_{e,OSR,inband}} \right)$$

$$= 10 \log \left(\frac{2^{2N} \Delta^2}{8} \frac{12 \times OSR}{\Delta^2} \right)$$

$$= 6.02N + 1.76 + 10 \log(OSR). \tag{6.16}$$

A comparison of Equations 6.9 and 6.16 reveals that oversampling improves SNR by $10 \log(OSR)$. Doubling OSR improves SNR by 3 dB.

6.1.4 SIGNAL-TO-NOISE-PLUS-DISTORTION RATIO

When the effect of the nonlinearity of an ADC with a sinusoidal input is considered, the nonlinearity manifests itself as harmonic tones in the spectrum of the ADC and a rising noise floor. These harmonic tones are typically much larger as compared with the quantization noise of the ADC. As a result, they set the lower bound of the input signal that the ADC can detect. The signal-to-noise-plus-distortion ratio (SNDR) of the ADC, defined as the ratio of the power of the full-scale range value of the input to the sum of the power of the noise and that of the harmonics of the ADC

$$\text{SNDR} = 10 \log \left(\frac{P_s}{P_n + P_d} \right), \tag{6.17}$$

where P_s, P_n, and P_d are the power of the signal, noise, and harmonics of the ADC over the Nyquist frequency range of the ADC, respectively, provides a better quantification of the DR of the ADC.

To obtain the spectrum of the output of an ADC with a sinusoidal input, the time-domain response of the ADC is recorded. A fast Fourier transform (FFT) analysis is then performed on the recorded time-domain response of the ADC in a postprocessing step. To ensure that FFT analysis is carried out properly, the key parameters of FFT analysis such as frequency resolution, the number of time-domain data samples used in FFT analysis, and windows for spectral leakage reduction must be chosen properly. As an example, let us assume that the input of an ADC is a sinusoidal tone at 100 kHz and the bandwidth of interest is 10 MHz. The highest frequency of interest is typically set to the Nyquist frequency of the input, which is twice the bandwidth of the input, that is, 20 MHz. The lowest frequency of interest is the desired frequency resolution and is set to 5 kHz in this example. Note that if the lower frequency bound is overly low, the simulation time needed to generate the required time-domain response of the ADC will be exceedingly long as the step size is set in accordance with the highest frequency. On the other hand, if the lower frequency bound is too high, frequency resolution will be poor. In order to have a sufficient frequency resolution so as to capture the input tone and its harmonics while keeping the time-domain simulation time reasonably low, the bin size, that is, the frequency resolution, is set to f_{bin} = 5 kHz. The corresponding steady-state time-domain window where the time-domain response data are collected for FFT analysis is set to $T_w = 1/f_{bin}$ = 0.2 ms. Since most SPICE-based simulators use a variable step size to compute time-domain response in order to minimize simulation time while FFT requires that the time-domain response data be spaced with an equal distance, in order to force the simulator to calculate the time-domain response at equally-spaced time points, the strobe option available in most SPICE simulators to calculate the time-domain response of the ADCs at equally spaced time instants should be set with strobe period T_w/pts where *pts* is the number of data samples for FFT analysis. Often *pts* is set to the multiples of K = 1024, for example, 64K. Further, in order to minimize spectral leakage, the time-domain response of the start point and that of the ending point should be chosen in such a way that the selected block of time-domain response data for FFT analysis possesses periodicity.

It is well understood that the Fourier transform of a sinusoidal signal of frequency ω_{in} is a single-tone at ω_{in} (if we neglect the tone at the negative frequency), revealing that all the energy of the signal is concentrated at frequency ω_{in}. When an FFT analysis is performed on a nonperiodic signal, spectrum leakage results in the signal energy smearing out over a broad frequency range. As the data to be FFT-analyzed may contain unknown frequency components. They are normally not periodic in the predefined data block time period, a spectral leakage will occur. In order to minimize the spectral leakage, a window function that has a band-pass spectral characteristic is typically applied to the data samples prior to an FFT analysis. The window function takes a zero value at the start and end points of the window to ensure the periodicity of the window-weighted data. The choice of window functions depends upon a number of considerations such as the characteristics of the data (random or sinusoidal), bandwidth, the desired frequency resolution, and amplitude accuracy. Windows with a narrow passband and sharp stop band attenuation are preferred. Hanning window is the most widely used FFT window for ADCs [1].

The resolution of an ADC can be depicted using the DR of the ADC with both noise and nonlinearity considered. Alternatively, it can also be quantified using the ENOBs of the ADC. When the effect of distortion is not considered, the ENOBs of the ADC can be obtained from Equation 6.18 directly

$$\text{ENOB} = \frac{\text{SNR} - 1.76}{6.02}. \tag{6.18}$$

For oversampling ADCs, from Equation 6.16, we have

$$\text{ENOB} = \frac{\text{SNR} - 1.76 - 10 \log(\text{OSR})}{6.02}. \tag{6.19}$$

When the effect of the nonlinearity of the ADC is accounted for, Equation 6.18 becomes

$$\text{ENOB} = \frac{\text{SNDR} - 1.76}{6.02}. \tag{6.20}$$

Similarly, for oversampling ADCs, Equation 6.20 becomes

$$\text{ENOB} = \frac{\text{SNDR} - 1.76 - 10 \log(\text{OSR})}{6.02}. \tag{6.21}$$

6.1.5 DIFFERENTIAL NONLINEARITY

The differential nonlinearity (DNL) of an ADC is the normalized difference between the step width of the ADC and that of an ideal ADC. The nonuniform step size of the ADC is caused by the nonlinearity of the ADC. DNL is a measure of the amount of the deviation of the step size of the ADC from the step size of an ideal ADC. The DNL in the kth step, denoted by $\text{DNL}(k)$, is obtained from

$$\text{DNL}(k) = \frac{\Delta_k - \Delta}{\Delta}, \tag{6.22}$$

where Δ and Δ_k are the step size of an ideal ADC and that of the actual ADC in step k, respectively. The root-mean-square (rms) value of the DNL is obtained from

$$\text{DNL}_{rms} = \sqrt{\frac{1}{2^N - 2} \sum_{k=1}^{2^N - 2} \text{DNL}^2(k)}. \tag{6.23}$$

It is seen from Equation 6.22 that DNL provides an effective means to quantify the effect of the nonlinearity on the step width of ADCs. The larger the DNL, the larger is the quantization error. If DNL is less than 1 LSB, the ADC will have a monotonic transfer function with no missing code. If DNL exceeds 1 LSB, the converter will become nonmonotonic and a missing code will occur.

6.1.6 INTEGRAL NONLINEARITY

The integral nonlinearity (INL) of an ADC is the accumulation of the normalized deviation of the step size of the ADC from that of an ideal ADC over the input range of the ADC. Since both the static gain error and offset error of the ADC contribute to the accumulated step size deviation, INL is measured after the effect of the static gain error and offset error is removed. The INL in step k is obtained from

$$\text{INL}(k) = \sum_{i=1}^{k-1} \text{DNL}(i). \tag{6.24}$$

It can be shown that if $i = 1, 2, 3, ..., 2^N - 1$ where N is the number of the bits, then

$$\sum_{i=1}^{2^N-1} \mathrm{DNL}(i) = 0. \tag{6.25}$$

It should be noted that both DNL and INL are static design specifications of ADCs as they quantify the effect of the nonlinearity of ADCs in a DC steady state. SNR and SNDR, on the other hand, are dynamic design specifications as they depict the performance of ADCs in an AC steady state. For noise-shaping modulators, frequency-domain measures such as SNR and SNDR are widely favored over time-domain measures such as DNL and INL (Figure 6.5). This is because SNR and SNDR provide an explicit comparison of signal, quantization noise, and harmonics of ADCs. The spectrum plot of ADCs not only explicitly shows the location and strength of harmonics, it also quantifies the profile of the noise-shaping provided by the noise-shaping mechanisms of the ADCs.

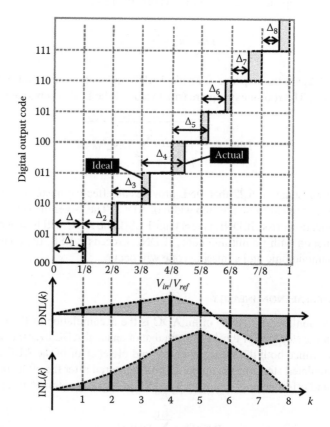

FIGURE 6.5 Differential nonlinearity and integral nonlinearity of analog-to-digital converters.

6.1.7 FIGURE-OF-MERITS

It is often desirable to compare the performance of ADCs of different architectures or design specifications. The following FOM quantifies the overall performance of an ADC by evaluating the amount of the power consumption of the ADC per conversion step

$$\text{FOM} = \frac{P}{2^{\text{ENOB}} f_s},\tag{6.26}$$

where
f_s is the sampling frequency
ENOB is the effective number of the bits of the ADC
P is the power consumption of the ADC

Clearly for ADCs with the same sampling frequency and power consumption, the higher the SNDR, the larger is the ENOB, and the better is the FOM.

6.2 VCO PHASE QUANTIZERS

6.2.1 PRINCIPLE OF VCO PHASE QUANTIZERS

An input voltage can be digitized by first converting it to a time variable using a voltage-to-time converter. The resultant time variable can then be quantized using a time-to-digital converter [2,3]. Although straightforward, the performance of these ADCs is largely dominated by the nonlinearity and speed of the voltage-to-time converter. Itawa et al. showed that an input voltage can also be quantized using a multistage voltage-controlled ring oscillator [4]. Specifically, the input voltage to be digitized is sampled and held by a clocked S/H block, a multistage voltage-controlled ring oscillator whose control voltage is the sampled-and-held value of the input voltage to be digitized, and a counter that records the number of the oscillation cycles of the oscillator per sampling period, as shown in Figure 6.6 [3,5–9]. Since the phase of ring oscillators is typically a monotonic function of the control voltage, a one-to-one mapping between the sampled-and-held control voltage and the phase of the oscillator exists. The number of the oscillation cycles of the oscillator per sampling period T_s thus yields the digital representation of the sampled-and-held input voltage.

The counter that records the number of the oscillation cycles of the oscillator per sampling period functions as a phase quantizer that quantifies the phase of the oscillator with a quantization error of 2π. One important characteristic of the phase of the VCO is that the phase of the VCO is continuous from one sampling period to the next, attributive to the continuity of the output voltage of the oscillator. As a result, the residual phase of the VCO in $(k-1)$th sampling period, denoted by $e_f(k-1)$, is carried over to the next sampling phase and becomes the initial phase of kth sampling period, denoted by $e_i(k)$, that is,

$$e_i(k) = e_f(k-1),\tag{6.27}$$

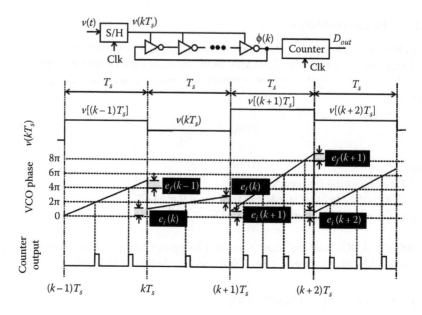

FIGURE 6.6 Voltage-controlled ring oscillator phase quantizers.

as shown graphically in Figure 6.6. The net phase accumulation in kth sampling period, denoted by $\phi(k)$, is obtained from

$$\phi(k) = K_{VCO}v_{in}(k) + \left[e_f(k) - e_i(k)\right], \tag{6.28}$$

where K_{VCO} is the voltage-to-phase gain of the VCO. If we define $e(k)$ as the residual phase, that is, phase quantization noise, in phase k, that is, $e(k) = e_f(k)$, then

$$e_i(k) = e_f(k-1) = e(k-1). \tag{6.29}$$

Equation 6.28 therefore becomes

$$\phi(k) = K_{VCO}v_{in}(k) + \left[e(k) - e(k-1)\right]. \tag{6.30}$$

Since the content of the counter increments every 2π phase accumulation, the output of the counter, denoted by $D_{out}(k)$, is obtained from

$$D_{out}(k) = \frac{\phi(k)}{2\pi}. \tag{6.31}$$

In z-domain, it becomes

$$D_{out}(z) = \frac{1}{2\pi}\left[K_{VCO}V_{in}(z) + \left(1 - z^{-1}\right)E(z)\right]. \tag{6.32}$$

Equation 6.32 reveals that the VCO quantizer performs first-order noise-shaping on the phase quantization noise. This unique and attractive characteristic of VCO phase quantizers does not exist in voltage-mode quantizers.

The quantization error of a counter-based phase quantizer is 2π as the content of the counter increments once the oscillator completes an oscillation cycle. The quantization error can be lowered from 2π to π/N where N is the number of the delay stages of the VCO if the output of each stage of the VCO is utilized [5,10]. It can be further decreased to $\pi/(MN)$ where M is the degree of interpolation between the neighboring delay stages of the VCO [11].

6.2.2 CHARACTERISTICS OF VCO PHASE QUANTIZERS

As compared with voltage-mode quantizers, VCO phase quantizers possess the following intrinsic characteristics, some are desirable while others are not. We examine them in detail.

6.2.2.1 First-Order Noise-Shaping

It was shown in the preceding section that VCO phase quantizers perform first-order noise-shaping on phase quantization noise. If a VCO phase quantizer is employed in an open-loop ADC, as shown in Figure 6.7, since the sampling frequency f_s is much higher than the signal frequency f, the counter that functions as a phase quantizer with its transfer function given by

$$H_c(z) = \frac{1 - z^{-1}}{2\pi} \tag{6.33}$$

can be approximated with utilization of $z^{-1} = e^{-sT_s} \approx 1 - sT_s$ where $T_s = 1/f_s$. We arrive at

$$H_c(s) \approx \frac{sT_s}{2\pi}, \tag{6.34}$$

as shown in Figure 6.7. It becomes evident that the signal experiences no loss while the quantization noise is first-order shaped. The first-order noise-shaping provided by the VCO phase quantizer improves the SNR of the ADC without employing an explicit noise-shaping block.

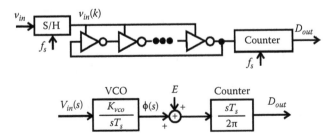

FIGURE 6.7 VCO phase quantizer with first-order noise-shaping.

If a VCO phase quantizer is employed in $\Delta\Sigma$ modulators, the first-order noise-shaping provided by the quantizer will allow the order of the loop filter to be lowered by 1. To demonstrate this, let us first consider the $\Delta\Sigma$ modulator with a first-order loop filter whose transfer function is given by

$$H_{LF}(s) = \frac{K_{LF}}{s} \tag{6.35}$$

and a voltage-mode quantizer shown in Figure 6.8a. It can be shown that

$$D_{out}(s) = \frac{K_{LF}}{K_{LF} + sT_s} V_{in}(s) + \frac{sT_s}{K_{LF} + sT_s} E(s). \tag{6.36}$$

Since for oversampling $sT_s \ll 1$ holds in the signal band, with a large in-band loop gain, we have

$$D_{out}(s) \approx V_{in}(s) + \frac{sT_s}{K_{LF}} E(s). \tag{6.37}$$

Equation 6.37 reveals that the modulator with a first-order loop filter and a voltage-mode quantizer provides first-order noise-shaping on the quantization noise.

Let us now consider Figure 6.8b and c where the order of the loop filter remains to be 1. Quantization is now performed by a VCO phase quantizer that consists of

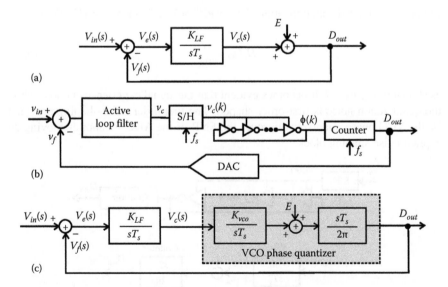

FIGURE 6.8 $\Delta\Sigma$ modulators with a first-order loop filter and (a) a voltage-mode quantizer and (b, c) a VCO phase quantizer.

a ring oscillator performing voltage-to-phase conversion and a counter performing phase quantization. It can be shown that

$$D_{out}(s) = \frac{K_{LF}K_{VCO}}{2\pi(sT_s) + K_{LF}K_{VCO}} V_{in}(s) + \frac{(sT_s)^2}{2\pi(sT_s) + K_{LF}K_{VCO}} E(s). \quad (6.38)$$

If the loop gain is sufficiently large, Equation 6.38 can be simplified to

$$D_{out}(s) \approx V_{in}(s) + \frac{(sT_s)^2}{K_{LF}K_{VCO}} E(s). \quad (6.39)$$

Equation 6.39 shows that the $\Delta\Sigma$ modulator provides second-order noise-shaping on the quantization noise even though the order of the loop filter is only 1.

The loop filter of $\Delta\Sigma$ modulators is often implemented using active components such as operational amplifiers in order to achieve a large in-band loop gain critical to achieving the adequate suppression of the nonlinearities in the forward path and quantization noise. These active components are the primary source of the power consumption of the modulators due to their large bias currents needed for achieving a large slew rate so as to satisfy the timing constraints. Reducing the order of the loop filter effectively lowers the overall power consumption of the modulators.

6.2.2.2 Large Oversampling Ratio

The OSR of ADCs with a VCO phase quantizer is defined as the ratio of the sampling frequency of the quantizer, that is, the frequency of the clock of the S/H block and counter to the Nyquist frequency of the input. Since the performance of ring oscillators, in particular, oscillation frequency, scales well with technology, ring oscillators implemented in the nanometer domain can oscillate at tens of GHz. As a result, the sampling frequency of VCO quantizers can be made sufficiently large while retaining the same number of the oscillation cycles of the oscillator per sampling period. A large OSR is therefore permitted.

6.2.2.3 Multibit Quantization

The relation between the number of the bits of a VCO phase quantizer, denoted by n, and the number of the delay stages of the oscillator of the quantizer, denoted by N, is given by: $N = 2^n - 1$. For example, to implement a 3-bit VCO phase quantizer, a ring oscillator with a total of seven stages is needed. Similarly, to implement a 5-bit VCO phase quantizer, a ring oscillator with a total of 31 stages is needed. In comparison, if a 5-bit voltage-mode quantizer is to be constructed, the number of the comparators required to implement the quantizer will be 31. As each voltage comparator consists of a large number of transistors, a reference voltage, and a mismatch compensation network, both the silicon and power cost of implementing these voltage-mode comparators will be prohibitively large. In addition to soaring power and silicon cost, mismatch between comparators also severely affects the performance of the quantizer.

6.2.2.4 Implicit Dynamic Element Matching

Although $\Delta\Sigma$ modulators are effective in suppressing nonidealities present in the forward path of the modulators, they are ineffective in suppressing the effect of any nonideality present in the feedback path. This drawback does not become an issue if the order of $\Delta\Sigma$ modulators is only 1, because only a single-bit DAC, which is intrinsically linear, is needed in these modulators. When a multibit quantizer is employed, a corresponding multibit DAC is needed in the feedback path to convert the output of the multibit quantizer back to an analog quantity, that is, the feedback voltage with which the input voltage compares. Almost exclusively current-mode DACs are used for this purpose. Figure 6.9 shows the architecture of a 3-bit current-steering DAC. To minimize mismatch, multiple unit current-steering blocks whose tail current is I are used for current-steering blocks with large tail currents. For example, the total tail current of the current-steering block for D_2 is $2I$. This current-steering block consists of two identical unit current-steering blocks whose tail current is I. Similarly, the current-steering block for D_3 is made of four identical unit current-steering blocks, each with a tail current I.

Process spread inevitably gives rise to the mismatch of the DAC that is typically modeled using a voltage source known as the mismatch noise source, as shown in Figure 6.10, where the block diagram of a first-order $\Delta\Sigma$ modulator is shown. This

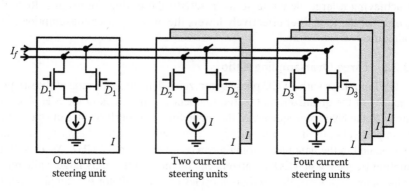

One current   Two current   Four current
steering unit   steering units   steering units

FIGURE 6.9   Three-bit current-steering digital-to-analog converter.

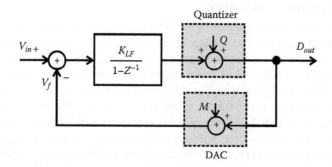

FIGURE 6.10   Mismatch of digital-to-analog converter is represented by the mismatch noise source residing in the feedback path.

mismatch noise source consists of the tones set by the pattern of the output code of the quantizer. It can be shown that the transfer function from the mismatch noise source to the output of the modulator is given by

$$\frac{D_{out}(z)}{M(z)} = \frac{K_{LF}}{K_{LF} + (1 - z^{-1})}.$$

(6.40)

If the in-band loop gain is sufficiently larger and noting that $1 - z^{-1} \approx 1 - (1 - sT_s) \approx 0$ because $f_s \gg f$, we will have $D_{out}(z) \approx M(z)$, that is, the mismatch noise will propagate to the output of the modulator without attenuation. As a result, mismatch noise will be sampled by the quantizer and will manifest themselves in the spectrum of the modulator as both in-band tones and a rising noise floor, deteriorating the performance of the modulator. For example, 1% component mismatches of the DAC will completely wipe out the noise-shaping of a third-order $\Delta\Sigma$ modulator [12].

Since the element mismatch of a DAC is deterministic, the degradation of SNDR caused by the mismatch noise is due to the concentration of the energy of the feedback signal at few frequencies. Clearly if the energy of the feedback signal can be spread over a broad frequency range, the effect of the mismatch noise in the signal band will be reduced. This can be achieved by permuting the unit elements of the DAC dynamically by the output of the quantizer. This is known as DEM. DEM is widely used and most effective in reducing the effect of the nonidealities of the DAC of $\Delta\Sigma$ modulators [13].

The randomness of the residue phase of the ring oscillator of VCO phase quantizers provides the inherent randomization of the mismatch noise. If, in addition to the output of the counter, the output of all the delay stages of the ring oscillator of the quantizer is used to generate the feedback signal, a built-in DEM that randomizes the mismatch noise of the DAC is also provided [10,14].

6.2.2.5 Low Power Consumption

As pointed out earlier, the absence of a large number of power-greedy voltage comparators, their mismatch compensation, and reference voltages in time-mode ADCs with VCO phase quantizers greatly lower the power consumption of these ADCs. The reduction of the order of the loop filter due to the extra order of noise-shaping provided by VCO phase quantizers also contributes to power reduction. Both make ADCs with VCO phase quantizers attractive candidates for low-power applications. For example, in the multirate 1-1 MASH $\Delta\Sigma$ modulator with the first $\Delta\Sigma$ modulator clocked at 100 MHz and implemented in a conventional first-order voltage-mode configuration and the second $\Delta\Sigma$ modulator clocked at 1.2 GHz and implemented using a VCO quantizer, the power consumption of the first modulator is 12.7 mW while that of the second modulator is only 1.1 mW despite its high sampling rate [15].

6.2.2.6 Technology Compatibility

The removal of voltage comparators in time-mode ADCs greatly improves the compatibility of these ADCs with technology scaling. Due to the nonavailability of time integrators with a large gain, most recently reported VCO quantizer ADCs such as those in [10,16] still rely on active loop filters implemented using operational amplifiers to obtain a large in-band loop gain needed to suppress the nonlinearities of VCO phase quantizers and provide the desired order of noise-shaping. The performance of these active loop filters scales poorly with technology. They also consume a significant amount of power.

6.2.2.7   Nonlinearity

Although there is a one-to-one mapping between the sampled-and-held control voltage of the oscillator and the phase of the oscillator, the relation between them is far from linear, especially when the variation of the control voltage is large. The nonlinear voltage-phase characteristics of voltage-controlled ring oscillators are rooted to the fact that the transistors of the voltage-controlled ring oscillators go through all modes of the operation, specifically, cut-off, subthreshold, triode, and saturation, each having its distinct characteristics. To illustrate this, let us examine the widely used current-starved voltage-controlled ring oscillator shown in Figure 6.11. When v_{in}^+ is at logic-1, C_1 is discharged via M5 initially in saturation when v_{DS5} is large, that is, the initial part of the discharging process, and then in triode when v_{DS5} is small, that is, the remaining part of the discharging process. Similarly, when v_{in}^+ is at logic-0, C_1 is charged via M1 and M2 with M1 is saturation when *on* and M2 in triode when v_{DS} is low, that is, initial charging process, and in saturation when v_{DS} is large, that is, the remaining part of the charging process.

The nonlinear voltage-phase characteristic of VCO phase quantizers manifests itself in the spectrum of modulators as both harmonic tones [10,17] and an arising noise floor [13,18]. The degree of the nonlinearity of VCO phase quantizers will improve when the variation of the input voltage is small and worsen when it is large. If the control voltage only toggles between two fixed voltages, the frequency of the oscillator will correspondingly vary between two fixed frequencies. In this case, the voltage-phase relation of the oscillator will become intrinsically linear. In fact, this characteristic has been widely used in design of time-mode ADCs with the input voltage to be digitized mapped to a PWM signal first. The pulse-width-modulated version of the input is then used as the control voltage of the oscillator of VCO phase quantizers [19–21].

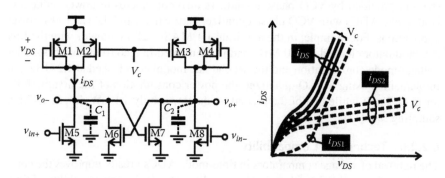

FIGURE 6.11   Delay stage of voltage-controlled ring oscillators.

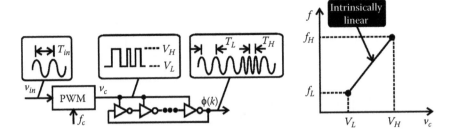

FIGURE 6.12 Input signal is pulse-width modulated before applying to voltage-controlled ring oscillator.

6.2.2.8 Latency

As mentioned earlier, the counter of a VCO phase quantizer functions as a phase quantizer. Although the performance of the voltage-controlled ring oscillator scales well with technology, the latency of the counter is typically much larger as compared with the propagation delay of the delay stage of the ring oscillator. As a result, the counter is the bottleneck of the VCO phase quantizer (Figure 6.12).

6.3 VCO FREQUENCY QUANTIZERS

It was shown in the previous section that VCO phase quantizers provide first-order noise-shaping on phase quantization noise. The first-order noise-shaping is stemmed from the continuity of the phase of the output of the ring oscillator of the quantizer. The latency of the counter that performs phase quantization greatly limits the sampling frequency of the quantizer. As improving the speed of counters is rather difficult, it is therefore highly desirable from the sampling frequency point of view to replace the counter with an alternative quantization mechanism. Since the output of ring oscillators is phase and the derivative of phase is frequency, that is, $\omega = d\phi/dt$ [22], if a digital differentiator is employed at the output of the oscillator, the output of the differentiator will yield the frequency of the oscillator. As the frequency of the oscillator is typically a monotonic function of its control voltage, the output of the differentiator thus yields the digital representation of the control voltage of the oscillator.

Digital differentiation can be conveniently realized using two D flip-flops and a XOR2 gate, as shown in Figure 6.13a. As pointed out earlier, the voltage-phase relation of a voltage-controlled oscillator (VCO) is given by

$$H_{VCO}(z) = \frac{K_{VCO}}{1-z^{-1}}. \tag{6.41}$$

In order to allow the input voltage, which is the control voltage of the oscillator, to be quantified without attenuation while still suppressing in-band quantization noise, a first-order digital differentiator with transfer function

$$H_d(z) = 1-z^{-1} \tag{6.42}$$

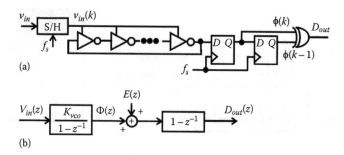

FIGURE 6.13   VCO frequency quantizer. (a) Simplified schematic and (b) transfer function block diagram.

is needed, as shown in Figure 6.13b [23,24]. From the block diagram of the quantizer, we have

$$D_{out}(z) = K_{VCO}V_{in}(z) + \left(1 - z^{-1}\right)E(z). \tag{6.43}$$

As compared with VCO phase quantizers investigated earlier, although both offer first-order noise-shaping, VCO frequency quantizers offer a speed advantage due to the absence of the counter. Similar to VCO phase quantizers, since no mechanism exists to suppress the effect of the nonlinear voltage-phase characteristics of the ring oscillator of the VCO phase quantizer, large harmonic tones exist in the spectrum of the output of the ADC, especially when the range of the variation of the input voltage is large.

6.4   OPEN-LOOP ANALOG-TO-DIGITAL CONVERTERS WITH VCO QUANTIZERS

As pointed out earlier, a VCO phase quantizer maps the control voltage of the VCO to a digital code. Thus, a VCO phase quantizer is an open-loop ADC with first-order noise-shaping. The open-loop architecture of these ADCs allows them to rapidly digitize an input voltage so as to cope with the speed constraint of many applications. The non-linearity of the voltage-phase relation of voltage-controlled ring oscillators, however, gives rise to both harmonic tones in the spectrum of ADCs and a rising noise floor. These harmonic tones degrade the SNDR and limit the DR of the ADCs. It becomes evident that the most important design issue encountered in design of open-loop ADCs with a VCO phase quantizer is how to minimize the effect of the nonlinearity of the VCO phase quantizer on the DR of ADCs. A number of novel design techniques emerged recently to achieve this. We examine them in detail in this section.

6.4.1   ADCs with Differential VCO Phase Quantizers

Fully differential configurations or operations are an effective means to eliminate the effect of the nonlinearity to the first order. Examples include differential pairs of the input stage of operational amplifiers to reject common-mode disturbances and double correlating sampling to eliminate the fixed-pattern noise of CMOS image sensors. To utilize a differential configuration in VCO phase quantizers to suppress the effect of the nonlinearity of these quantizers, two identical VCO phase quantizers, one driven by v_{in}^+ and the other by v_{in}^- where v_{in}^+ and v_{in}^- are the differential inputs to be quantized, are employed, as shown in Figure 6.14 [7,25–27]. To demonstrate this, we let

$$v_{in}^+ = v_{in,cm} + \frac{\Delta v}{2}, \tag{6.44}$$

and

$$v_{in}^- = v_{in,cm} - \frac{\Delta v}{2}, \tag{6.45}$$

where

$v_{in,cm}$ is the common-mode voltage of the input
Δv is the amplitude of the differential input voltage

The relation between the control voltage of the oscillator and the phase of the oscillator $\phi = f(v_{in})$ is approximated using its Taylor series expansion at the common-mode voltage of the input to the third order

$$\phi(v_{in,cm} + \Delta v) \approx \phi(v_{in,cm}) + a_1 \Delta v + a_1 \left(\Delta v\right)^2 + a_3 \left(\Delta v\right)^3, \tag{6.46}$$

where a_1, a_2, and a_3 are the first-, second-, and third-order Taylor series expansion coefficients of $\phi = f(v_{in})$, respectively. Evaluating Equation 6.46 at v_{in}^+ and v_{in}^- and noting that v_{in}^+ and v_{in}^- are applied to two different VCO phase quantizers separately, we arrive at

$$\phi(v_{in}^+) \approx \phi(v_{in,cm}) + a_1 \left(\frac{\Delta v}{2}\right) + a_2 \left(\frac{\Delta v}{2}\right)^2 + a_3 \left(\frac{\Delta v}{2}\right)^3, \tag{6.47}$$

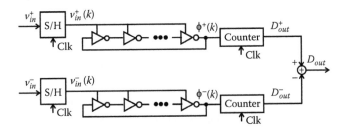

FIGURE 6.14 Differential VCO phase quantizer. (From Daniels, J. et al., A 0.02 mm² 65 nm CMOS 30 MHz BW all-digital differential VCO-based ADC with 64 dB SNDR, in *Symposium of VLSI Circuits Digest of Technical Papers*, 2010, pp. 155–156; Yoon, Y. et al., A linearization technique for voltage-controlled oscillator-based ADC, in *Proceedings of the IEEE International SoC Conference*, 2009, pp. 317–320; Daniels, J. et al., *IEEE Trans. Circuits Syst. I*, 57(9), 2404, 2010; Hamilton, J. et al., *IEEE Trans. Circuits Syst. II*, 57(11), 848, 2010.)

and

$$\phi(v_{in}^-) \approx \phi(v_{in,cm}) - \hat{a}_1\left(\frac{\Delta v}{2}\right) + \hat{a}_2\left(\frac{\Delta v}{2}\right)^2 - \hat{a}_3\left(\frac{\Delta v}{2}\right)^3, \tag{6.48}$$

where a_1 and $\hat{a}_1$, a_2 and $\hat{a}_2$, a_3 and $\hat{a}_3$ are the first-, second-, and third-order Taylor series expansion coefficients of $\phi = f(v_{in})$ of the two VCO phase quantizers, respectively. If $\Delta v = A\cos(\omega_{in}t)$, since $(\Delta v)^2$ only contains the second-order harmonic component of ω_{in} while $(\Delta v)^3$ contains both the fundamental and third-order harmonic components of ω_{in}, the differential output will not have the second-order harmonic, provided that $a_1 = \hat{a}_1$, $a_2 = \hat{a}_2$, and $a_3 = \hat{a}_3$, that is, the two VCO phase quantizers are perfectly matched. Differential VCO quantizers thus exhibit a better SNDR as compared with their single-ended counterparts due to the removal of the second harmonic. For example, in [25], the SNDR of the ADC with a differential VCO phase quantizer and the same ADC with a nondifferential VCO phase quantizer are 56.5 and 41.3 dB, respectively. The improvement obtained from using the differential VCO phase quantizer is approximately 15 dB.

The performance of ADCs with differential VCO phase quantizers will deteriorate if mismatches between the two VCO phase quantizers exist. To demonstrate this, let us assume that $\hat{a}_1 = a_1 + \Delta a_1$, $\hat{a}_2 = a_2 + \Delta a_2$, and $\hat{a}_3 = a_3 + \Delta a_3$ with $\Delta a_1 \ll a_1$, $\Delta a_2 \ll a_2$, and $\Delta a_3 \ll a_3$ quantifying the mismatches. In this case, we have

$$\phi(v_{in}^-) \approx \phi(v_{in,cm}) - (a_1 + \Delta a_1)\left(\frac{\Delta v}{2}\right) + (a_2 + \Delta a_2)\left(\frac{\Delta v}{2}\right)^2$$

$$-(a_3 + \Delta a_3)\left(\frac{\Delta v}{2}\right)^3, \tag{6.49}$$

while $\phi(v_{in}^+)$ is given by Equation 6.47. As a result,

$$\phi(v_{in}^+) - \phi(v_{in}^-) \approx (2a_1 + \Delta a_1)\left(\frac{\Delta v}{2}\right) - (\Delta a_2)\left(\frac{\Delta v}{2}\right)^2$$

$$+(2a_3 + \Delta a_3)\left(\frac{\Delta v}{2}\right)^3$$

$$\approx a_1(\Delta v) - (\Delta a_2)\left(\frac{\Delta v}{2}\right)^2 + (2a_3)\left(\frac{\Delta v}{2}\right)^3. \tag{6.50}$$

It is seen from Equation 6.50 that the severer the mismatches between the two VCO phase quantizers, the lesser the suppression of the second-order harmonic tone provided by the differential configuration.

6.4.2 ADCs with Digitally Calibrated VCO Phase Quantizers

The nonlinear characteristics of VCO phase quantizers are rooted to the nonlinear voltage-phase characteristics of voltage-controlled ring oscillators. Such a nonlinear characteristic will deteriorate if the variation of the control voltage of the oscillators is large. As the input voltage range is mainly dictated by applications, improving the linearity of the voltage-phase relation of ring oscillators over a large input voltage range is rather difficult. This is because most transistors in a ring oscillator go through all four regions of the operation of MOS transistors, namely, cutoff, subthreshold conduction, triode, and saturation, in every cycle of oscillation. One universal and effective way to linearize a nonlinear element is digital calibration. Specifically, the output of the nonlinear element is recorded in a calibration phase. The difference between the output of the nonlinear element and that of the desired linear element is obtained. This difference is the amount of the correction that needs to be imposed onto the response of the nonlinear element so as to make it linear.

To use digital calibration to remove the effect of the nonlinearity of a VCO phase quantizer, two identical VCO phase quantizers, one called calibration quantizer and the other called measurement quantizer, are used. In the calibration phase, a calibration signal often a ramping voltage that covers the entire range of the input voltage to be digitized is fed to the calibration quantizer. The calibration signal is digitized by the calibration quantizer and the results are recorded. The obtained results are compared with those of an ideal VCO phase quantizer, and their differences are computed and stored in a look-up table (LUT). In the measurement phase, the input to be digitized is routed to the measurement quantizer. The corrections obtained in the calibration phase are subtracted from the output of the measurement quantizer to remove the effect of the nonlinear voltage-phase characteristics of the measurement quantizer [6].

The effectiveness of digital calibration in compensation of the effect of the nonlinear voltage-phase relation of VCO phase quantizers was validated in a number of studies. For example, in [7], a 64 dB SNDR was obtained over 30 MHz bandwidth with 300 MHz sampling frequency. In [11], a 52.5 dB SNDR was obtained over 20 MHz bandwidth with 600 MHz sampling frequency. In [28], a peak 69 dB SNDR was obtained over 18 MHz bandwidth with sampling frequency 1.15 GHz.

Digital calibration is a universal and effective means to combat the effect of the nonlinear voltage-phase characteristics of VCO phase quantizers. It does not need the prior knowledge of the nonlinear characteristics of VCO phase quantizers. The need for complex logic to perform digital calibration inevitably results in high silicon and power consumption. Mismatches between the calibration quantizer and measurement quantizer also set the lower bound of the effectiveness of this approach.

6.4.3   ADCs with Pulse-Width-Modulation VCO Phase Quantizers

As pointed out earlier although the voltage-phase relation of ring oscillators is nonlinear, if the control voltage of a ring oscillator is only toggled between two different voltages V_L and V_H, the frequency of the oscillator will also vary between two different frequencies f_L and f_H corresponding to V_L and V_H, respectively. In this case, the voltage-phase relation of the ring oscillator becomes intrinsically linear [19–21]. To utilize this characteristic of VCO phase quantizers to minimize harmonic tones in the spectrum of ADCs with a VCO phase quantizer, the input voltage to be digitized is first converted to a pulse-width-modulated signal with pulse width directly proportional to the amplitude of the input. The modulated signal is then fed to the ring oscillator of the VCO phase quantizer as its control voltage, as shown in Figure 6.15.

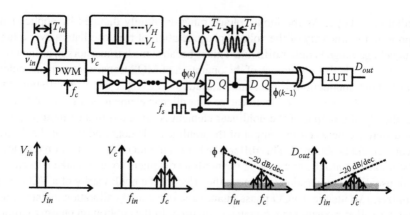

FIGURE 6.15 ADCs with pulse-width-modulation VCO phase quantizer. (From Rao, S. et al., A 71 dB SFDR open loop VCO-based ADC using 2-level PWM modulation, in *Symposium on VLSI Circuits Digest of Technical Papers*, 2011, pp. 270–271; Yoon, Y. et al., A time-based noise shaping analog-to-digital converter using a gated-ring oscillator, in *Proceedings of the IEEE International Microwave Workshop Intelligent Radio for Future Personal Terminals*, 2011, pp. 1–4; Gao, P. et al., Design of an intrinsically linear double-VCO-based ADC with 2nd-order noise shaping, in *Proceedings of the IEEE Design, Automation and Test in Europe Conference and Exhibition*, 2012, pp. 1215–1220.)

In order to maximize the DR of the ADC, V_H and V_L of the PWM-modulated signal should be set to the upper and lower bounds of the input voltage. Let the input to be digitized by the ADC be a sinusoid of frequency ω_{in}, that is, $v_{in} = V_m\sin(\omega_{in}t)$. v_{in} is PWM-modulated by a pulse width modulator. If the modulator is ideal, its spectrum will contain the input tone at ω_{in}, the tone of the modulating signal at ω_c, and their intermodulation tones at $\omega_{in} \pm n\omega_c$ where $n = 1, 2, 3, \ldots$. The frequency of the modulating signal should satisfy $\omega_c \gg \mathrm{BW}_{in}$ where BW_{in} is the bandwidth of the input such that all intermodulation tones of the input and modulating signals are located outside the Nyquist band of the input. In reality, the nonideality of the PWM modulator gives rise to unwanted tones in the Nyquist band of the input [19]. As a result, intermodulation frequency components of these unwanted tones and ω_c also exist in the spectrum of the output of the PWM modulator, as shown graphically in Figure 6.15. It is well understood that a single-tone signal at frequency ω_{in} present on the control voltage line of a VCO will generate tones at $\omega_o \pm \omega_{in}$ in the spectrum of the oscillator [22]. Both the unwanted tones and their intermodulation components with ω_c will generate tones at the output of the oscillator, corrupting its spectrum. It is therefore important to suppress these unwanted intermodulation tones existing in the spectrum of the oscillator. Since the voltage-phase relation of oscillators is given by [22]

$$\frac{\Phi(s)}{V_c(s)} = \frac{K_{VCO}}{s},\tag{6.51}$$

the oscillator functions as a low-pass filter that attenuates high-frequency components existing at the output of the oscillator. In order to adequately suppress the unwanted intermodulation frequency components of the PWM modulator, ω_c should be chosen in such a way that the intermodulation components of the PWM modulator are located outside the signal band such that they will be adequately attenuated by the low-pass characteristics of the oscillator [21].

In addition to the constraints imposed on the clock frequency of the PWM modulator, PWM-based ADCs with a VCO phase quantizer are also in need for a pulse width modulator where voltage comparators are typically needed. The inability to suppress unwanted tones in the signal band also limits SNDR.

Several designs utilizing this approach were reported recently. For example, in [19], a 59 dB SNDR over 8 MHz bandwidth with 640 MHz sampling frequency was achieved. A similar approach was proposed by Yoon et al. where the width of the pulse-width-modulated signal is digitized using a gated ring ADC [20]. Since the gated ring ADC provides first-order noise-shaping [29], the need for an explicit differentiator and a LUT-based phase-to-binary conversion is removed. This approach achieved 56 dB SNDR over 500 kHz bandwidth with 10 MHz sampling frequency.

6.5 FUNDAMENTAL OF ΔΣ MODULATORS

It was shown earlier that the open-loop configurations of ADCs with a VCO phase quantizer provide first-order noise-shaping on quantization noise, allowing the digitization of an input with a large bandwidth while achieving a good SNDR when techniques such as differential configurations, digital calibration, and PWM are utilized. To further improve the resolution of ADCs, not only the order of noise-shaping on quantization noise needs to be increased, the effect of the nonlinear voltage-phase characteristics of the voltage-controlled ring oscillator of the quantizers also needs to be adequately suppressed. Clearly this cannot be accomplished using an open-loop architecture. The closed-loop architectures of ADCs, specifically ΔΣ modulators, are known for their ability to provide high-order noise-shaping on quantization noise and the suppression of nonidealities present in the forward path of the modulators. This section reviews the fundamentals of ΔΣ modulators.

6.5.1 ΔΣ OPERATION

A voltage-mode ΔΣ modulator consists of an integrator also known as loop filter that accumulates the difference between the input voltage v_{in} of the modulator and the feedback voltage v_f, a voltage-mode quantizer typically realized using a voltage comparator for single-bit quantization and a total of $2^N - 1$ voltage comparators for N-bit quantization, and a DAC that converts the digital output of the modulator back to an analog voltage, specifically the feedback voltage v_f with which the input voltage compares. The name ΔΣ stems from the sequence of the Δ-operation and Σ-operation carried out by the modulator. The modulator first performs the subtraction of the feedback signal v_f from the input v_{in}, that is, $v_e = v_{in} - v_f$ (Δ-operation). It is then followed by an accumulation operation (Σ-operation) on v_e by the integrator. Both the integrator and the DAC are typically implemented using switched-capacitor networks to take the advantages of the superior performance of switched-capacitor networks, such as the high accuracy of the ratio of capacitances that a standard CMOS technology provides (0.1% typically), the fast settling time of switched-capacitor networks that permits a high sampling frequency and subsequently a large OSR, and the programmability of switched-capacitor networks. In order to ensure that the input signal to be digitized is unchanged during each sampling phase, the input signal is first sampled and held by an S/H block clocked by a sampling clock of frequency f_s. To eliminate the aliasing of unwanted signals into the desired signal band, an antialiasing filter (AAF) preceding the S/H block is required.

We use the switched-capacitor $\Delta\Sigma$ modulator shown in Figure 6.16a as an example to demonstrate the operation of switched-capacitor $\Delta\Sigma$ modulators. The integrator accumulates the difference between the input voltage v_{in} and the feedback voltage v_f. The accumulated difference between v_{in} and v_f is compared with a zero voltage reference in each phase. If

$$\sum_{k=1}^{K}\left[v_{in}(k)-v_f(k)\right]>0, \tag{6.52}$$

the quantizer outputs a logic-1, otherwise, it conveys a logic-0. If

$$\sum_{k=1}^{K}\left[v_{in}(k)-v_f(k)\right]=0, \tag{6.53}$$

the output of the quantizer provides a faithful digital representation of the input. Note that $k=1$ represents the start of the modulator, while $k=K$ specifies the time at which comparison takes place. The clocked operation of the switched-capacitor modulator and the fact that the equivalence of the digital output of the quantizer and the sampled input is built upon the condition of a zero accumulated difference between v_{in} and v_f reveal that $\Delta\Sigma$ modulators do not provide a one-to-one mapping between the input and its digital representation, that is, $v_{in}(k) - v_f(k) = 0$; rather, it establishes the equivalence of v_{in} and v_f in a statistic sense, that is, $\sum_{k=1}^{K}\left[v_{in}(k)-v_f(k)\right]=0$. This differs fundamentally from other ADCs such as successive approximation ADCs

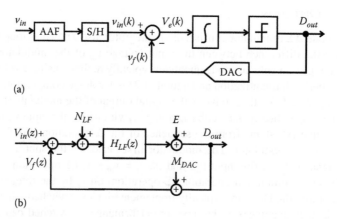

(a)

(b)

FIGURE 6.16 Basic configuration of singe-bit switched-capacitor $\Delta\Sigma$ modulator. (a) Configuration and (b) transfer function block diagram.

where a one-to-one mapping between an analog input and corresponding digital output exists. For $\Delta\Sigma$ modulators, as long as there is an accumulated difference between v_{in} and v_f at the end of each clock period, the output of the integrator will be nonzero, and the output of the quantizer will be adjusted accordingly with the objective to nullify the accumulated error between v_{in} and v_f.

6.5.2 CHARACTERISTICS OF $\Delta\Sigma$ MODULATORS

The block diagram of a first-order $\Delta\Sigma$ modulator is shown in Figure 6.16b. We assume that the loop filter is an ideal integrator of order 1 with its transfer function given by

$$H_{LF}(z) = \frac{K_{LF}}{1 - z^{-1}}. \tag{6.54}$$

Let the effect of the nonidealities of the integrator be represented by an input-referred voltage source N_{LF}, and the effect of the mismatch of the DAC be represented by an input-referred mismatch noise source M_{DAC}. It can be shown that

$$D(z) = \frac{K_{LF}}{K_{LF} + (1 - z^{-1})}\left[V_{in}(z) + N_{LF}(z) + M_{DAC}(z)\right]$$

$$+ \frac{(1 - z^{-1})}{K_{LF} + (1 - z^{-1})}E(z). \tag{6.55}$$

If the OSR of the modulator is sufficiently large, $z^{-1} = e^{-sT_s} \approx 1 - sT_s$ will hold in the signal band. As a result, Equation 6.55 can be simplified to

$$D(s) \approx V_{in}(s) + N_{LF}(s) + M_{DAC}(s) + \frac{sT_s}{K_{LF}}E(s). \tag{6.56}$$

It is seen from Equation 6.56 that the input passes through the modulator without attenuation, the effect of the mismatch of the DAC and that of the nonidealities of the loop filter also make their way to the output of the modulator without attenuation. The quantization noise, on the other hand, is first-order shaped and also suppressed by the in-band gain of the integrator.

To quantify the SNR of the $\Delta\Sigma$ modulator, we notice from Equation 6.55 that the noise transfer function in the signal band is given by

$$\text{NTF}(z) = \frac{D(z)}{E(z)} \approx \frac{1 - z^{-1}}{K_{LF}}. \tag{6.57}$$

Further, we recall that the PSD of oversampling ADCs without noise-shaping was derived in Equation 6.13 and is repeated here for convenience

$$S_{e,OSR} = \frac{\Delta^2}{12}\frac{1}{2f_B}\frac{1}{\text{OSR}}. \tag{6.58}$$

The power of the total in-band quantization noise of the $\Delta\Sigma$ modulator is therefore obtained from

$$P_{e,in\text{-}band} = \int_{-f_N/2}^{f_N/2} |\text{NTF}(jf)|^2 S_{e,OSR}(f)df.$$

(6.59)

Since

$$1 - z^{-1} = 1 - e^{-j\omega T_s} = 2je^{-j\frac{\omega T_s}{2}} \sin\left(\frac{\omega T_s}{2}\right)$$

(6.60)

and in the signal band, $\omega T_s \ll 1$, we have

$$\sin\left(\frac{\omega T_s}{2}\right) \approx \frac{\omega T_s}{2}.$$

(6.61)

As a result, Equation 6.59 is simplified to

$$P_{e,in\text{-}band} \approx \int_{-f_N/2}^{f_N/2} \frac{4}{K_{LF}^2} \left(\frac{\omega T_s}{2}\right)^2 \frac{\Delta^2}{12} \frac{1}{2f_B} \frac{1}{OSR} df$$

$$= \frac{\pi^3}{3} \frac{1}{OSR^3} \frac{\Delta^2}{12} \frac{1}{K_{LF}^2}.$$

(6.62)

Recall that the total in-band noise power of oversampling ADCs without noise-shaping was derived early in Equation 6.14. We rewrite it here for convenience

$$P_{e,OSR,in\text{-}band} = \int_{-f_B}^{f_B} S_{e,OSR}(f)df$$

$$= \frac{\Delta^2}{12} \frac{1}{OSR}.$$

(6.63)

A comparison of Equations 6.62 and 6.63 reveals that

$$P_{e,in\text{-}band} = \frac{\pi^3}{3} \frac{1}{OSR^2} \frac{1}{K_{LF}^2} P_{e,OSR,in\text{-}band}.$$

(6.64)

It is evident from Equation 6.64 that the power of the in-band noise of oversampling $\Delta\Sigma$ modulators is lower as compared with that of oversampling modulators without noise-shaping.

6.5.3 DECIMATION

The reduced quantization noise of $\Delta\Sigma$ modulators at low frequencies is at the expense of excessive quantization noise at high frequencies, as illustrated graphically in Figure 6.17. To extract the wanted signal from the output of the $\Delta\Sigma$ modulators, which is a bit stream made of 1s and 0s from the quantizer, since the wanted signal is not only at a low frequency but also band-limited, it can be fully recovered using a low-pass filter with its bandwidth set to the Nyquist frequency of the input, as shown graphically in Figure 6.18. This low-pass filter is typically realized using a digital filter so that its characteristics such as bandwidth can be made fully programmable. Since the input is buried in the quantization noise that is spread over the frequency domain $[-f_s, f_s]$ with most of the energy of the quantization noise concentrated at high frequencies, the low-pass operation provided by the low-pass digital filter removes the excessive quantization noise at high frequencies. As a result, the output of the digital low-pass filter will only contain the wanted signal and low quantization noise.

In order to extract the input from the quantization noise that is spread over $[-f_s, f_s]$, downsampling or decimation with sampling frequency set to the Nyquist frequency of the input, that is, f_N, is needed. For example, if the OSR of a $\Delta\Sigma$ modulator is 64, the sampling frequency of the modulator will be $f_{s,OSR} = 64f_N$. The output of the low-pass filter needs to be sampled at $f_{s,d} = f_N = f_{s,OSR}/64$, that is, decimated by 64.

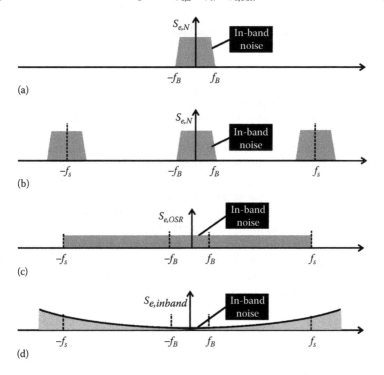

FIGURE 6.17 Noise shaping in oversampling $\Delta\Sigma$ modulators. (a) In-band quantization noise. (b) Oversampling replicates in-band quantization noise at multiples of sampling frequency. (c) Spread of quantization noise over $[-f_s, f_s]$ due to oversampling. (d) Shaped quantization noise.

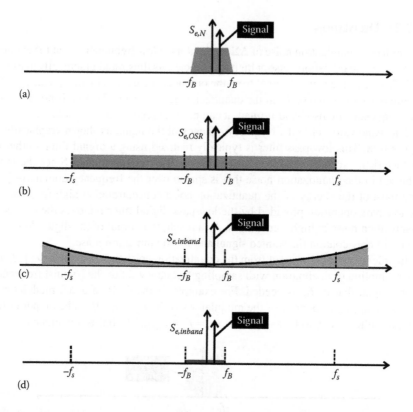

FIGURE 6.18 Noise shaping and signal extraction in oversampling ΔΣ modulators. (a) Signal with in-band quantization noise. (b) Signal with spread quantization noise over [−f_s,f_s]. (c) Signal with shaped quantization noise. (d) Output after decimation filter.

The subscript d identifies the downsampling frequency of the decimation. Because the spectrum to be decimated contains both the wanted input located in the signal band [−$f_N/2$, $f_N/2$] and quantization noise spread over [−f_s, f_s], the input is sampled at its Nyquist frequency while the quantization noise is undersampled with an undersampling ratio of 64. As a result, quantization noise will be aliased back to the signal band, deteriorating SNR, as illustrated graphically in Figure 6.19.

As mentioned earlier, the low-pass filter for removing the excessive quantization noise at high frequencies is typically implemented in digital domain to take advantage of the full programmability of digital filters. The simplest low-pass digital filters are comb filters. The transfer function of a comb filter of order N is given by

$$H_{comb}(z) = \frac{Y(z)}{X(z)} = \sum_{n=0}^{N-1} z^{-n}.$$ (6.65)

Writing Equation 6.65 in the time domain

$$y(n) = x(n) + x(n-1) + x(n-2) + + x(N-1)$$ (6.66)

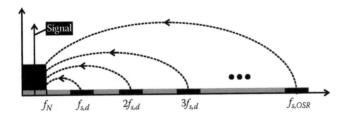

FIGURE 6.19 Fold-over of quantization noise to signal band due to undersampling of quantization noise in decimation.

Equation 6.66 shows that the comb filter of order N performs a moving average operation on a total of N consecutive samples. Since only the addition operation and delay are needed and no multiplication is required, comb filters are most economical in terms of silicon and power consumption. From Equation 6.65, we obtain the closed form of the transfer function of the comb filter of order N

$$Y(z) = \frac{1-z^{-N}}{1-z^{-1}} X(z)$$

$$= \left(\frac{1}{1-z^{-1}}\right)\left(1-z^{-N}\right)X(z). \tag{6.67}$$

Since the time-domain equivalence of $(1 - z^{-N})X(z)$ is $x(n) - x(n - N)$, that is, the difference between two samples that are N steps apart, $(1 - z^{-N})X(z)$, is the same as $x_d(n) - x_d(n - 1)$ where $x_d(n)$ is the output of the decimation block with decimation ratio N. In other words, the Nth-order differentiation $x(n) - x(n - N)$ at sampling rate $f_{s,OSR}$ is the same as first-order differentiation $x_d(n) - x_d(n - 1)$ at sampling rate $f_{s,d} = f_{s,OSR}/N$. The operation of Equation 6.67 can therefore be divided into three steps: (1) The first-order integration of the output of the $\Delta\Sigma$ modulators at $f_{s,OSR}$, (2) the decimation of the output of the integration by the OSR, and (3) the first-order differentiation on the output of the decimation operation at $f_{s,d}$, as shown in Figure 6.20. Performing integration first is important as the low-pass characteristics of the integrator remove the excessive quantization noise present at high frequencies before any aliasing. Note that since $f_{s,OSR} = \text{OSR} \times f_{s,d}$, the integration at $f_{s,OSR}$ consumes most of the power. At the output of the integrator, the noise floor will be low due to

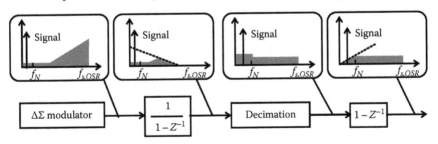

FIGURE 6.20 First-order comb filter and decimation.

the removal of the excessive quantization noise by the integrator. Since decimation samples the output of the integrator that contains both the wanted signal in the signal band $[-f_B, f_B]$ and the quantization noise of bandwidth $[-f_N \times \mathrm{OSR}, f_N \times \mathrm{OSR}]$, due to the undersampling of the quantization noise, the fold-over of the quantization noise to the signal band will take place. As a result, the in-band noise at the output of the decimation block will be worse as compared with that at the input of the decimation block, as shown in Figure 6.20. Fortunately, the differentiator following the decimation block reduces in-band quantization noise caused by the decimation.

To improve performance, the order to the comb filter can be increased. Clearly a high-order comb filter will remove more excessive quantization noise at high frequencies in the integration step and more decimation-induced noise in the differentiation step.

6.5.4 Nonidealities of ΔΣ Modulators

The performance of switched-capacitor ΔΣ modulators is affected by the nonidealities of the modulators. These nonidealities include the static error and dynamic errors of the loop filters; the former is caused by the parasitic capacitances at the input and output of the operational amplifiers of the loop filter and latter is rooted to the finite gain, finite bandwidth, finite slew rate, and gain nonlinearity of the operational amplifiers of the loop filter, the offset of the quantizer, clock jitter, and noise. There exists a long list of exhaustive study of the effect of the nonidealities of switched-capacitor ΔΣ modulators on their performance. A detailed examination of the effect of the nonidealities of ΔΣ modulators is clearly well beyond the scope of this book. This short section is by no means an attempt to provide an in-depth study on this topic; rather, our sole purpose of having this short section here is simply to bring this important issue to the attention of readers.

Although the performance of an ideal switched-capacitor integrator is determined by the ratio of capacitance of the sampling capacitor to that of the integration capacitor, the nonzero input and output parasitic capacitances of the operational amplifier of the switched-capacitor integrator give rise to a static error.

An ideal integrator has an infinite DC gain. The DC gain of practical operational amplifier-based integrator, however, is rather finite. The effect of the finite DC gain of the loop filter is twofold: It results in a signal loss and the rise of the noise floor. The larger the DC gain of the loop filter, the smaller the signal loss and the lesser the increase of the quantization noise.

If the gain of the operational amplifiers of the loop filter varies with the input voltage nonlinearly, harmonic distortion will appear in the spectrum of the modulator. The gain nonlinearity of the operational amplifier in the first-stage of a high-order loop filter has the greatest effect as the effect of the gain nonlinearity of later stages is reduced by the gain of the preceding stages.

For oversampling ΔΣ modulators, the high sampling rate imposes a stringent constraint on the slew rate of operational amplifiers of the loop filter. The slewing process of operational amplifiers is a nonlinear settling process that gives rise to input signal harmonics, which degrades the performance of the modulator in the form of both increasing quantization noise and harmonic distortion.

Voltage-mode quantizers are typically implemented using voltage comparators that are subject to the effect of device mismatches, among them, threshold mismatch, load mismatch, and dimension mismatch the most prominent. The effect of the mismatches of voltage comparators is typically represented by an input-referred voltage called input offset voltage. Since the location at which the input offset voltage is injected to the modulator is the same as the quantization noise, its effect will be attenuated in the same way as that of the quantization noise. As a result, its effect is not of a critical concern.

6.6 TIME-MODE ΔΣ MODULATORS

To improve resolution and stability, multibit quantization is highly desirable. As mentioned previously, multibit voltage-mode quantizers are typically realized using a total of $2^N - 1$ voltage comparators where N is the number of the bits of the quantization. When N is large, not only the silicon and power consumption of the voltage-mode quantizers become prohibitively large, the performance of these quantizers also deteriorates due to mismatches between the comparators. To eliminate the drawbacks of voltage-mode quantizers, VCO phase quantizers or VCO frequency quantizers studied in early chapters can be used. As the control voltage of the voltage-controlled ring oscillator of these quantizers must be kept unchanged during each sampling phase, an S/H block that samples and holds the control voltage of the voltage-controlled ring oscillator is required. In this case, the loop filter can be implemented using a continuous-time (CT) filter, such as a RC or g_m-C filter, rather than a switched-capacitor loop filter. ΔΣ modulators of this nature are known as continuous-time ΔΣ modulators. We use the ΔΣ modulator with a VCO phase quantizer shown in Figure 6.21a to demonstrate the operation of ΔΣ modulators with a VCO phase quantizer.

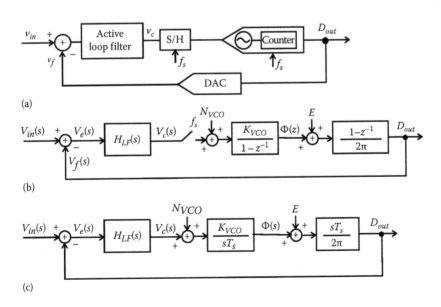

FIGURE 6.21 ΔΣ modulator with a VCO phase quantizer.

To derive the transfer function of the modulator, we note that the voltage-phase relation of the ring oscillator is given by

$$H_{VCO}(s) = \frac{K_{VCO}}{1 - z^{-1}} \qquad (6.68)$$

and the transfer function of the counter for phase-to-digital conversion is given by

$$H_c(z) = \frac{1 - z^{-1}}{2\pi}. \qquad (6.69)$$

We further assume that the DAC in the feedback path is ideal. Let the effect of the nonlinear voltage-phase characteristics of the voltage-controlled ring oscillator be represented by an input-referred voltage source N_{VCO}, as shown in Figure 6.21b. Since $f_s \gg f$ due to oversampling, we have $z^{-1} = e^{-sT_s} \approx 1 - sT_s$ in the signal band. Figure 6.21b can then be simplified to Figure 6.21c where all variables are in s-domain. If the loop gain in the signal band is sufficiently large, one can show that

$$D(s) \approx V_{in}(s) + \frac{1}{H_{LF}(s)} N_{VCO}(s) + \frac{sT_s}{H_{LF}(s)K_{VCO}} E(s). \qquad (6.70)$$

It is seen from Equation 6.70 that the input of the modulator passes through the modulator without attenuation. If the loop filter is an ideal integrator, that is, $H_{LF}(s) = K_{LF}/s$, the effect of the voltage-phase nonlinear characteristics of the voltage-controlled ring oscillator is suppressed by the in-band gain of the loop filter, and the in-band quantization noise is both second-order noise-shaped and suppressed by the in-band loop gain.

6.7   CHARACTERISTICS OF TIME-MODE ΔΣ MODULATORS

6.7.1   SUPPRESSION OF NONLINEARITY

It was shown in the previous section that the effect of the nonlinear voltage-phase characteristics of a voltage-controlled ring oscillator of the VCO phase quantizer is suppressed by the loop gain $H_{LF}(s)$. Consider the following three cases: (1) the loop filter is an ideal integrator, (2) the loop filter is a single-pole system, and (3) the loop filter is a double-pole system. The corresponding distortion transfer functions (DTFs) defined as the ratio of the output of the modulator to the distortion source of the ring VCO of the quantizer are tabulated in Table 6.1.

It is seen from Table 6.1 that the nonlinearity of the VCO phase quantizer is suppressed by the in-band gain of the loop filter. For example, if the loop filter is modeled as a single-pole system with its −3 dB frequency ω_b, harmonic tones that fall into $[0, \omega_b]$ will be attenuated by the DC gain K_{LF} of the loop filter. Harmonic tones at frequencies higher than ω_b, however, will be increased, as illustrated graphically in Figure 6.22. Clearly, loop filters with a large in-band loop gain are critical. To achieve this, the loop filter needs to be implemented using active configurations, typically using operational amplifiers [10,14].

The reduction of the effect of the nonlinear voltage-phase characteristics of the ring VCO provided by the modulator can also be viewed from a different perspective that will provide more insight of the operation of ΔΣ modulators. If we assume that

TABLE 6.1

Distortion Transfer Function of ΔΣ Modulators with a VCO Quantizer

Loop Filter Type	Loop Filter Transfer Function	Distortion Transfer Function
Ideal integrator	$H_{LF} = \dfrac{K_{LF}}{s}$	$DTF = \dfrac{1}{K_{LF}}$
First-order	$H_{LF} = \dfrac{K_{LF}}{1 + \dfrac{s}{\omega_b}}$	$DTF = \dfrac{1 + \dfrac{s}{\omega_b}}{K_{LF}}$
Second-order	$H_{LF} = \dfrac{K_{LF}}{\left(1 + \dfrac{s}{\omega_b}\right)^2}$	$DTF = \dfrac{\left(1 + \dfrac{s}{\omega_b}\right)^2}{K_{LF}}$

a large change in v_{in} takes place, a large change in the control voltage v_c of the ring VCO of the quantizer will occur. This will in turn deteriorate the nonlinear voltage-phase behavior of the oscillator. If the feedback v_f is generated sufficiently fast, it will be subtracted from v_{in} in time. As a result, the net variation of the control voltage of the ring VCO of the quantizer will be small. This will in turn result in an improved voltage-phase linearity of the ring VCO.

A downside of deploying an active integrator is its finite bandwidth and slew rate. The finite bandwidth sets the upper bound of the loop dynamics of the modulator and subsequently the maximum sampling frequency of the modulator. The slew rate, on the other hand, sets the maximum rate of change that the output of the active integrator can provide. To increase the slew rate, a large DC bias current is needed. This is because for the differentially configured input stage of an operational amplifier, the maximum slew rate is obtained when the total tail current of the differential pair is steered to one of the two arms of the differential pair and is given by [30]

$$\text{Slew rate} = \frac{I_{ss}}{C_L}, \tag{6.71}$$

where I_{ss} and C_L are the tail current and load capacitor of the differential input stage of the operational amplifier, respectively. A large tail current will inevitably lead to a high level of power consumption.

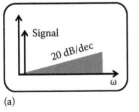
(a)

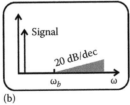
(b)

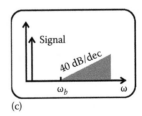
(c)

FIGURE 6.22 The effect of loop filter of ΔΣ modulators in suppression of nonlinear characteristics of VCO phase quantizers. (a) Ideal loop filter, (b) first-order loop filter, and (c) second-order loop filter.

If the required slew rate exceeds the maximum slew rate of the operational amplifier of the integrator, the output of the operational amplifier will not follow the rate of the change of the input. Slew-induced distortion will occur.

6.7.2 High-Order Noise-Shaping

It was shown in Equation 6.70 that the noise transfer function is given by

$$\text{NTF}(s) = \frac{D(s)}{E(s)} \approx \frac{sT_s}{H_{LF}(s)K_{VCO}}. \tag{6.72}$$

If the loop filter is an ideal first-order integrator, that is, $H_{LF}(s) = K_{LF}/s$, then

$$\text{NTF}(s) \approx \frac{s^2 T_s}{K_{LF}K_{VCO}}. \tag{6.73}$$

Similarly, if the loop filter is an ideal second-order integrator, that is, $H_{LF}(s) = K_{LF}/s^2$, then the noise transfer function is given by

$$\text{NTF}(s) \approx \frac{s^3 T_s}{K_{LF}K_{VCO}}. \tag{6.74}$$

The aforementioned results show that the use of the VCO phase quantizer effectively increases the order of noise-shaping provided by the modulator by one. To realize second-order noise-shaping, only a first-order active loop filter is needed. Lowering the order of the loop filter will greatly reduce its power consumption and subsequently the overall power consumption of the modulator as most of the power of $\Delta\Sigma$ modulators is consumed by the loop filter.

6.7.3 Continuous-Time $\Delta\Sigma$ Modulation

It was shown in Figure 6.21 that the sampling of the control voltage of the ring VCO of the quantizer takes place at the S/H block, which is located in the forward path of the modulator. The input to the $\Delta\Sigma$ modulator is continuous rather than sampled-and-held. This differs fundamentally from conventional switched-capacitor $\Delta\Sigma$ modulators where the input signal must first be sampled and held by an S/H block preceding the $\Delta\Sigma$ modulator. The S/H operation of the S/H block of SC $\Delta\Sigma$ modulators requires the deployment of an AAF in front of the S/H block in order to eliminate the aliasing of unwanted out-of-band signals into the signal band.

In the time-mode $\Delta\Sigma$ modulator of Figure 6.21, sampling takes place inside the loop of the modulator, specifically after the loop filter. Since the input signal must first pass through the loop filter before reaching the S/H block where sampling takes place, the loop filter can be implemented using a continuous-time filter rather than a switched-capacitor filter. In addition, the loop filter limits the bandwidth of the signal to be sampled by the S/H block and therefore in essence functions as an AAF. As a result, the need for an explicit AAF could be removed [31,32]. Further, because sampling frequency is much higher than the loop bandwidth of the modulator due to oversampling, the fold-over effect associated with oversampling is also largely eliminated by the loop dynamics of the modulator.

Also observed from Figure 6.21 is the fact that sampling and quantization take place at the same location. The former gives rise to aliasing, while the latter is the root to quantization noise. Aliasing is therefore attenuated in a similar way as quantization noise. This is another unique and desirable characteristic of continuous-time $\Delta\Sigma$ modulators that switched-capacitor $\Delta\Sigma$ modulators do not possess [33,34].

Since the loop filter is continuous-time, the DAC in the feedback path of time-mode $\Delta\Sigma$ modulators can now be implemented using a current-steering DAC. Although current-steering DACs suffer from mismatch, the output of these DACs has no abrupt but rather incremental changes. Since the output of the DAC is fed to the loop filter directly, the bandwidth and slew rate constraints imposed on the operational amplifiers of the loop filter are therefore greatly relaxed [34]. If a switched-capacitor DAC was used, the large change of the output of the DAC would impose a stringent constraint on the bandwidth and slew rate of the operational amplifiers of the loop filter. This inevitably leads to poor performance and high power consumption.

6.7.4 INHERENT DYNAMIC ELEMENT MATCHING

As pointed out earlier that $\Delta\Sigma$ modulators are effective in suppressing nonidealities residing in the forward path of the modulators, they are, however, ineffective in suppressing the effect of nonidealities, specifically, the mismatches of the DAC present in the feedback path. This is not a concern if the order of $\Delta\Sigma$ modulators is only one, because only a single-bit DAC is needed in this case and single-bit DACs are intrinsically linear. When a multibit quantizer is employed, a corresponding multibit DAC is needed in the feedback path to convert the output of the multibit quantizer back to an analog quantity, specifically the feedback voltage with which the input voltage compares. Process spread gives rise to the mismatch of the DAC that is typically modeled using a voltage source known as the mismatch noise source. Mismatch noise consists of the tones set by the pattern of the output code of the quantizer. These tones are sampled by the quantizer and manifest themselves in the spectrum of the modulator as both in-band tones and a rising noise floor, deteriorating the performance of the modulator.

Since the mismatch of DACs is deterministic, the degradation of SNDR caused by the mismatch noise of the DACs is due to the concentration of the energy of the feedback signal at few frequencies. Clearly if the energy of the feedback signal can be spread over a broad frequency range, the effect of the mismatch noise in the signal band will be reduced. This can be achieved by permuting the unit elements of the DACs dynamically by the output of the quantizer. This technique is known as DEM. DEM is widely used to reduce the effect of the mismatch of the DAC of $\Delta\Sigma$ modulators [13], and will be studied in Section 6.13. Unlike voltage-mode $\Delta\Sigma$ modulators, when a VCO phase quantizer is used, the random nature of the residue phase of the ring oscillator of the quantizer in each sampling period provides the built-in randomization of the mismatch noise. An inherent DEM therefore exists in $\Delta\Sigma$ modulators with a VCO phase quantizer [10,14].

6.8 $\Delta\Sigma$ MODULATORS WITH VCO PHASE QUANTIZERS

This section investigates $\Delta\Sigma$ modulators with VCO phase quantizers. A time-mode $\Delta\Sigma$ modulator with a VCO phase quantizer is shown in Figure 6.23a where M_{DAC} represents the mismatch of the DAC while N_{VCO} and E have their usual meanings. Following the same arguments as those used in derivation of Figures 6.21c and 6.23a can be converted to Figure 6.23b where all variables are in s-domain. If the in-band loop gain of the modulator $H_{LF}(s)K_{VCO}$ is sufficiently large, one can show that in the signal band

$$D(s) \approx V_{in}(s) + M_{DAC}(s) + \frac{1}{H_{LF}(s)}N_{VCO}(s) + \frac{sT_s}{H_{LF}(s)K_{VCO}}E(s). \qquad (6.75)$$

It is seen from Equation 6.75 that the input passes through the modulator without attenuation, the effect of the voltage-phase nonlinear characteristics of the voltage-controlled ring oscillator of the quantizer is suppressed by the gain of the loop filter, the quantization noise is both noise-shaped and suppressed by the loop gain, and no attenuation of the effect of the mismatch of the DAC exists. The order of noise-shaping depends upon the order of the loop filter.

A drawback of $\Delta\Sigma$ modulators with a VCO phase quantizer is the speed limitation of the counter. Although the frequency of the voltage-controlled ring oscillator scales well with technology, the latency of the counter is much larger as compared with the propagation delay of the delay stage of the ring oscillator. As a result, the speed of the counter is the bottleneck of the modulator that limits the sampling

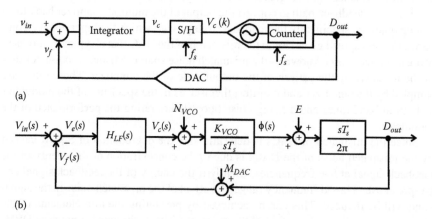

(a)

(b)

FIGURE 6.23 $\Delta\Sigma$ modulator with a VCO phase quantizer. (a) Configuration and (b) transfer function block diagram.

frequency of the modulator. To exemplify this, let us assume that the bandwidth of the input to be digitized is 20 MHz. The Nyquist frequency is therefore 40 MHz. If the OSR is set to 8, the sampling frequency is given by: $f_s = 8 \times 40$ MHz $= 320$ MHz. Assume that in order to ensure an adequate raw resolution, in each sampling period, the ring oscillator of the VCO quantizer needs to complete 10 cycles. The oscillation frequency of the oscillator will therefore be 3.2 GHz. The counter must therefore operate at 3.2 GHz in order to capture the phase of the oscillator.

6.9 ΔΣ MODULATORS WITH VCO FREQUENCY QUANTIZERS

It was shown in the last section that the latency of the counter that performs phase quantization limits the sampling frequency of ΔΣ modulators with VCO phase quantizers. As improving the speed of counters is rather difficult, an approach to avoid this difficulty so as to increase the sampling frequency of ΔΣ modulators is needed. We notice that the output of the voltage-controlled ring oscillator of the quantizer is phase and the derivative of phase is frequency [22]. If a digital differentiator is employed at the output of the ring oscillator, it will differentiate the phase of the oscillator with respect to time. The output of the differentiator will be the frequency of the oscillator. As the frequency of ring oscillators is typically a monotonic function of their control voltage, the output of the differentiator will yield the digital representation of the control voltage of the oscillator. This quantizer is clearly different from the VCO phase quantizer studied earlier. We term it VCO frequency quantizer.

A first-order digital differentiator can be realized by utilizing the characteristics of XOR2 operation, as shown in Figure 6.24. It is seen from the truth table given in the figure that when $D(k)$ and $D(k-1)$ differs, the differentiator yields a nonzero output, that is, a difference between $D(k)$ and $D(k-1)$ exists. Otherwise, it outputs a zero, that is, no difference between $D(k)$ and $D(k-1)$ exists.

The ΔΣ modulator proposed in [10,14] and shown in Figure 6.25 employs a VCO frequency quantizer that consists of a voltage-controlled ring oscillator and a set of first-order digital differentiators at the output of each stage of the oscillator. The first-order digital differentiators perform phase-to-frequency conversion. The feedback signal is obtained via a current-mode DAC to take the advantage of its high-speed operation and low switching noise. Buffers are typically inserted between the output of the delay stages of the ring oscillator and the differentiators to sharpen the transition edges of the output voltages of the oscillator.

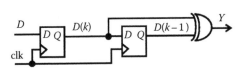

$D(k-1)$	$D(k)$	Y
0	0	0
0	1	1
1	0	1
1	1	0

FIGURE 6.24 First-order digital differentiator.

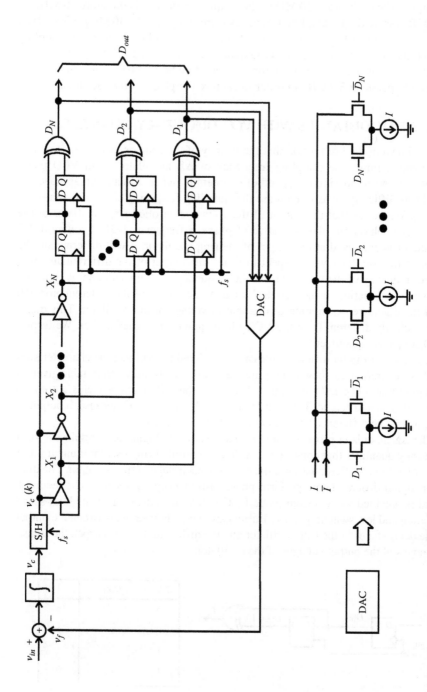

FIGURE 6.25   ΔΣ modulator with a VCO frequency quantizer. (From Straayer, M. and Perrott, M., *IEEE J. Solid-State Circuits*, 43(4), 805, 2008.)

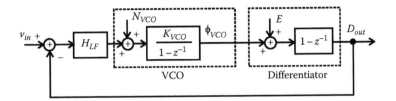

FIGURE 6.26 Block diagram of $\Delta\Sigma$ modulator with a VCO frequency quantizer.

The system diagram of the modulator is shown in Figure 6.26 with the mismatch of the DAC neglected. It can be shown that

$$D_{out}(z) = \frac{1-z^{-1}}{D(z)} E(z) + \frac{K_{VCO}}{D(z)} N_{VCO}(z) + \frac{K_{VCO}H_{LF}(z)}{D(z)} V_{in}(z), \qquad (6.76)$$

where

$$D(z) = 1 + K_{VCO}H_{LF}(z). \qquad (6.77)$$

Consider the case where the loop filter is an ideal integrator, that is,

$$H_{LF}(z) = \frac{K_{LF}}{1-z^{-1}}. \qquad (6.78)$$

Assume that a large in-band loop gain exists, that is, $K_{VCO}K_{LF} \gg 1$, Equation 6.76 can be simplified to

$$D_{out}(z) \approx V_{in}(z) + \frac{(1-z^{-1})^2}{K_{VCO}K_{LF}} E(z) + \frac{N_{VCO}(z)}{K_{LF}}. \qquad (6.79)$$

It is evident from Equation 6.79 that the effect of the nonlinear voltage-phase characteristics of the ring VCO is suppressed by the in-band gain of the loop filter, the quantization noise is both second-order noise-shaped and suppressed by the in-band loop gain, and the input signal passes through the modulator with no loss.

Because K_{VCO} is typically small, the loop filter must provide most of the required loop gain. To achieve this, active loop filters realized using operational amplifiers are typically used at the cost of high power consumption. The loop bandwidth and slew rate of the operational amplifiers must also be large enough to cope with high sampling rates. This imposes a stringent constraint on the bandwidth and slew rate of the operational amplifiers. As an example, the loop filter in the time-mode $\Delta\Sigma$ modulators reported in [10,14] has a DC gain over 50 dB and 2–3 GHz unity gain frequency.

We model the loop filter using a single-pole model with a DC gain 50 dB and unity gain frequency 3 GHz, that is,

$$H_{LF}(s) = \frac{A_o}{1 + \dfrac{s}{\omega_b}}, \tag{6.80}$$

with $\omega_T = A_o\omega_b$ where ω_b and ω_T are the −3 dB frequency and unity gain frequency, respectively, and A_o is the DC gain. Clearly $A_o = 50$ dB and $f_T = 3 \times 10^9$. Since the gain of the loop filter rolls off at the rate of −20 dB/dec, the −3 dB frequency of the loop filter is $f_b = f_T/A_o = 94.8$ MHz.

Another important factor that limits the suppression of harmonic tones in the spectrum of the modulator is the fact that the feedback signal of $\Delta\Sigma$ modulators with a VCO frequency quantizer is the frequency rather than the phase of the ring oscillator. Since a voltage-controlled ring oscillator is a voltage-to-phase converter rather than a voltage-to-frequency converter, in order to confine the control voltage of the oscillator to a small range so as to minimize the effect of the nonlinear voltage-phase characteristics of the ring oscillator on the DR of the modulator, phase feedback should be used. Although one can replace the differentiator-based frequency quantizer with a counter-based phase quantizer without sacrificing first-order noise-shaping on quantization noise, the latency of the counter will set the upper bound of the sampling frequency of the modulator, making the digitization of inputs with a large bandwidth rather difficult.

6.10 $\Delta\Sigma$ MODULATORS WITH PHASE FEEDBACK

As pointed out in the last section, phase feedback is preferred over frequency feedback when VCO-based multibit quantizers are used in $\Delta\Sigma$ modulators. A number of $\Delta\Sigma$ modulators with phase feedback emerged recently. We examine these modulators in this section.

6.10.1   TAILLEFER–ROBERTS $\Delta\Sigma$ MODULATOR

The block diagram of the $\Delta\Sigma$ modulator proposed by Taillefer and Roberts is shown in Figure 6.27a [35]. The modulator employs a voltage-to-time integrator realized using two voltage-controlled delay units (VCDUs) and a static inverter. If we assume that the VCDUs are linear, then the delay of the VCDU controlled by the input v_{in} and that controlled by the feedback v_f are given by $\tau_{in} = K_{in}v_{in}$ and $\tau_f = K_f v_f$, respectively, where K_{in} and K_f are the gain of the VCDU controlled by the input and that controlled by the feedback, respectively. The total propagation delay of the loop, denoted by τ_{VCO}, is obtained from

$$\tau_{VCO} = K_{in}v_{in} + K_f v_f + \tau_{inv}, \tag{6.81}$$

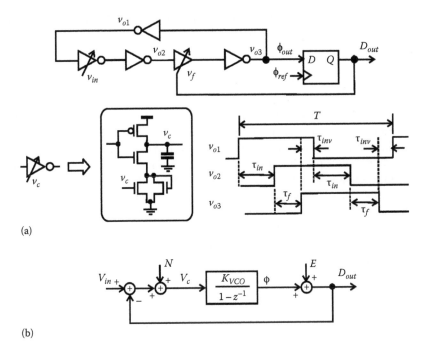

(a)

(b)

FIGURE 6.27 ΔΣ modulator with phase feedback. (a) Configuration and (b) transfer function block diagram. (From Taillefer, C. and Roberts, G., *IEEE Trans. Circuits Syst. I*, 56(9), 1908, 2009.)

where τ_{inv} is the delay of the static inverter. The two VCDUs and the static inverter form a ring oscillator with oscillation frequency that is given by

$$T = 2\left(K_{in}v_{in} + K_f v_f + \tau_{inv}\right). \tag{6.82}$$

It is evident from Equation 6.82 that the oscillation period of the ring oscillator is controlled by both v_{in} and v_f. For a given input voltage v_{in}, a corresponding feedback voltage v_f can be generated to counteract v_{in} such that T remains unchanged. To generate the required feedback signal, a single-bit phase quantizer implemented using a D flip-flop is employed to perform bang-bang phase detection between ϕ_{out} of the sensing oscillator and ϕ_{ref} of a reference oscillator of a constant oscillation frequency, and adjust v_f in accordance with

$$v_f = \text{sign}[D_{out}], \tag{6.83}$$

where sign[x] is defined as

$$\text{sign}[x] = \begin{cases} 1, & \text{if } x > 0 \\ 0, & \text{if } x < 0. \end{cases} \tag{6.84}$$

The block diagram of the modulator is shown in Figure 6.27b. In the signal band, $1 - z^{-1} \approx sT_s \ll 1$. If $K_{VCO} \gg 1$, we will have

$$D_{out} \approx V_{in}(z) + N_{VCO}(z) + \frac{1 - z^{-1}}{K_{VCO}} E(z). \qquad (6.85)$$

Equation 6.85 shows that the modulator provides first-order noise-shaping on the phase quantization noise and no attenuation on both the nonlinearity of the VCDUs and input. As pointed out in early chapters on voltage-to-time converters, the VCDU used in this design suffers from a poor linearity. As a result, harmonic tones will exist in the spectrum of the oscillator. This will also raise the noise floor. Although the absence of a loop filter in the modulator lowers the power consumption of the modulator, it also eliminates the ability of the modulator to suppress the harmonic tones caused by the nonlinear characteristics of the VCDUs. As a result, although the modulator can achieve 48.9 SNR in digitizing an input of bandwidth 400 kHz with a 140 MHz sampling clock, the SNDR is only 38.6 dB due to the existence of a large second-order harmonic tone.

6.10.2   LIN–ISMAIL ΔΣ MODULATOR

The ΔΣ modulator proposed by Lin and Ismail and shown in Figure 6.28 employs a voltage-controlled delay line-based architecture [36–38]. A reference clock ϕ_{ref} is fed to both the voltage-controlled delay line and the phase detector to perform voltage-to-time conversion. The resultant phase difference $\Delta\phi$ is sampled by a D flip-flop that functions as a single-bit phase quantizer. The quantizer is controlled by a quantization

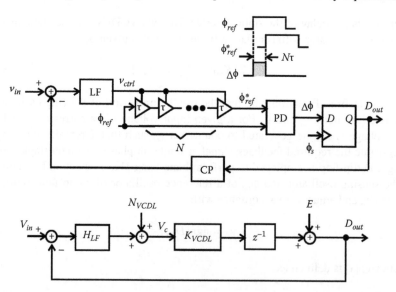

FIGURE 6.28 ΔΣ modulator with phase feedback. (From Lin, Y. et al., A phase-based singe-bit delta-sigma ADC architecture, in *Proceedings of the IEEE New Circuits and Systems*, 2011, pp. 406–409; Lin, Y. and Ismail, M., *Analog Integr. Circuits Signal Process.*, 73, 801, 2012.)

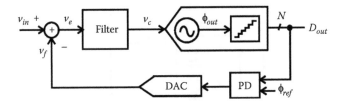

FIGURE 6.29 ΔΣ modulator with phase feedback. (From Park, M. and Perrott, M., A single-slope 80 Ms/s ADC using two-step time-to-digital conversion, in *IEEE International Symposium on Circuits and Systems*, 2009, pp. 1125–1128.)

clock ϕ_s. The charge pump in the feedback path acts as a 1-bit DAC that maps the digital output of the quantizer to an analog voltage with which the input voltage compares.

Lin–Ismail ΔΣ modulator offers a number of unique characteristics. Let us examine them in detail. First, a loop filter can be inserted in the forward path. This enables the realization of high-order ΔΣ modulators with a large loop gain, both are critical to achieve high-order noise-shaping and the suppression of the harmonic tones caused by the nonlinearity of the voltage-controlled delay line. Second, voltage-to-time conversion is performed using a voltage-controlled delay line rather than a VCO. As a result, the drawback of jitter accumulation inherent to oscillators is avoided. This, of course, is at the cost of losing the first-order noise-shaping provided by counter-based VCO phase quantizers. Third, similar to Taillefer–Roberts ΔΣ modulator, no VCO phase quantizers are used. As a result, the difficulty associated with the latency of counters is avoided. Finally, since the power consumption of voltage-controlled delay lines is significantly lower as compared with that of ring oscillators, Lin–Ismail ΔΣ modulator is attractive for low-power applications.

6.10.3 PARK–PERROTT ΔΣ MODULATOR

The ΔΣ modulator proposed in [16] and shown in Figure 6.29 employs a multibit phase detector to detect phase lead/lag of the VCO quantizer with respect to a reference phase Φ_{ref}. Since the output of the VCO is phase rather than frequency, the use of phase as the feedback signal allows the modulator to track the input signal more closely. As a result, v_e is confined to a very small region. This in turn reduces the range of v_c and subsequently the effect of the nonlinear voltage-phase characteristics of the VCO of the quantizer. A better suppression of the nonlinearity of voltage-phase characteristics of the VCO can therefore be achieved. For example, the third-order time-mode ΔΣ modulator in [14] utilized a frequency feedback and achieved SNDR of 55 dB over 20 MHz bandwidth with 950 MHz sampling frequency. The SNDR of the fourth-order time-mode ΔΣ modulator in [16] where a phase feedback was used is 78 dB over 20 MHz bandwidth with 900 MHz sampling frequency.

6.11 ΔΣ MODULATORS WITH PULSE-WIDTH-MODULATION

An effective means to overcome the effect of the nonlinear voltage-phase characteristics of VCO quantizers is to use a pulse width modulator to convert the input signal to a PWM signal with pulse width proportional to the amplitude of the input.

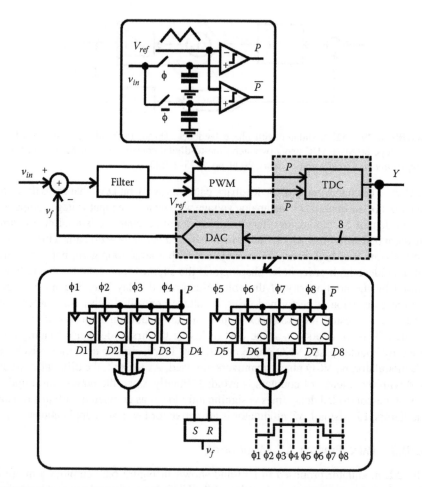

FIGURE 6.30 $\Delta\Sigma$ modulator with pulse with modulation. (From Dhanasekaran, V. et al., A 20 MHz BW 68 dB DR CT $\Delta\Sigma$ ADC based on a multi-bit time-domain quantizer and feedback element, in *IEEE International Solid-State Circuits Conference Digest of Technical Papers*, 2009, pp. 174–175; Dhanasekaran, V. et al., *IEEE J. Solid-State Circuits*, 46(3), 639, 2011.)

The PWM signal toggles the frequency of the VCO between two fixed frequencies f_H and f_L, as shown in Figure 6.30 [19,39–42]. The PWM-signal is then digitized using the TDC that samples the pulse-width-modulated signal eight times per sampling clock period. The multibit digital output of the TDC is converted to a single-bit signal for feedback. It was shown earlier that the spectrum of PWM contains the input tone at f_{in}, the modulating signal tones centered at f_c, and intermodulation tones at $f_{in} + nf_c$. The nonidealities of the PWM also give rise to tones other than $nf_c + f_{in}$. The frequency of the modulating signal f_c should satisfy $f_c \gg \mathrm{BW}_{in}$ such that intermodulation tones and modulating signals are outside the signal band. It is also critical to choose the sampling frequency f_s to be far away from the sideband tones of the output of the pulse width modulator. This will ensure that the intermodulation tones of f_s and f_c are outside the signal band.

As compared with open-loop VCO quantizers with PWM, the closed-loop configuration with PWM allows in-band noise arising from both quantization and aliasing effect of TDC sampling to be suppressed by the loop gain and noise-shaped. For example, the open-loop PWM ADC in [19] achieved a 59 dB SNDR over 8 MHz bandwidth with a 640 MHz sampling rate. The closed-loop ADC with PWM in [39] achieved a 60 dB SNDR over 20 MHz bandwidth with 950 MHz sampling frequency.

6.12 MASH TIME-MODE ΔΣ MODULATORS

Another effective way to minimize the effect of the nonlinear voltage-phase characteristics of VCO quantizers while achieving a large SNDR is to use a multistage configuration with the first-stage implemented using a voltage-mode modulator called signal modulator to perform the coarse quantization of the input signal and the second-stage implemented using a time-mode ΔΣ modulator with a VCO quantizer called residue modulator to perform the fine quantization of the residue error of the signal modulator, as shown in Figure 6.31 [43–46]. Since voltage-mode modulators can easily handle a large input especially when the constraints on quantization noise are relaxed while ΔΣ modulators with a VCO phase quantizer exhibit superior performance if the amplitude of their input is small, the cascade of a voltage-mode modulator and a time-mode modulator elegantly minimizes the effect of the nonlinear voltage-phase characteristics of VCO quantizers encountered in time-mode ΔΣ modulators.

The coarse modulator is often implemented using voltage-mode flash ADCs to take the advantages of their high speed [44,45]. The reduced quantization error constraint is another justification of doing so. Voltage-mode ΔΣ modulators can also be used for the signal modulator. Also, since the input of the residue modulator is the quantization noise of the signal modulator, which is much smaller as compared with the input, the linearity of the residue modulator is greatly improved. The random nature of the quantization error of the signal modulator also effectively prevents the nonlinearity of the VCO of the residue modulator from producing harmonic tones [15].

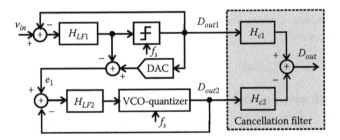

FIGURE 6.31 Single-rate MASH ΔΣ modulators. (From Gupta, A. et al., *IEEE Trans. Circuits Syst. II*, 58(11), 734, 2011; Reddy, K. et al., A 16 mW 78 dB SNDR 10 MHz-BW CT-ΔΣ ADC using residue-canceling VCO-based quantizer, in *IEEE International Solid-State Circuits Conference Digest of Technical Papers*, 2012, pp. 152–154; Reddy, K. et al., *IEEE J. Solid-State Circuits*, 47(12), 1–12, 2012; Bos, L. et al., *IEEE J. Solid-State Circuits*, 45(6), 1198, 2010.)

6.12.1   Single-Rate Time-Mode MASH $\Delta\Sigma$ Modulators

In a single-rate MASH $\Delta\Sigma$ modulators, both the signal modulator and residue modulator are clocked by the same clock, as shown in Figure 6.31. As a result, the quantization error of the signal modulator can be directly fed to the residue modulator. Let us examine how this single-rate $\Delta\Sigma$ operates. Because

$$D_{out1}(z) = H_1(z)V_{in}(z) + N_1(z)E_1(z) \tag{6.86}$$

and

$$D_{out2}(z) = H_2(z)E_1(z) + N_2(z)E_2(z), \tag{6.87}$$

where $H(z)$ and $N(z)$ are signal and noise transfer functions, respectively, $E(z)$ is the quantization noise, and subscripts 1 and 2 identify the signal modulator and residue modulator, respectively, and we have

$$\begin{aligned} D_{out} = H_{c1}(z)H_1(z)V_{in1}(z) \\ + \left[H_{c1}(z)N_1(z) - H_{c2}(z)H_2(z) \right] E_1(z) \\ - H_{c2}(z)N_2(z)E_2(z), \end{aligned} \tag{6.88}$$

where $H_{c1}(z)$ and $H_{c2}(z)$ are the transfer functions of the noise cancellation filter, which is implemented in the digital domain, as shown in Figure 6.31. To remove $E_1(z)$ from D_{out}, we impose the cancellation condition [13]

$$H_{c1}(z)N_1(z) - H_{c2}(z)H_2(z) = 0. \tag{6.89}$$

A natural choice to satisfy the cancellation condition is

$$H_{c1}(z) = kH_2(z) \quad \text{and} \quad H_{c2}(z) = kN_1(z), \tag{6.90}$$

where k is a constant whose value is chosen to achieve a unity signal gain. Equation 6.88 thus becomes

$$D_{out} = kH_1(z)H_2(z)V_{in}(z) - kN_1(z)N_2(z)E_2(z). \tag{6.91}$$

If the noise-shaping provided by the signal and residue modulators is first-order, that is, $N_1(z) = N_2(z) = 1 - z^{-1}$ while no attenuation of the signal exists, that is, $H_1(z) = H_2(z) = z^{-1}$, the modulator will provide the second-order noise-shaping on the quantization noise of the residue modulator and no attenuation on the input. Since the quantization noise of the residue modulator can be made small by increasing OSR and multibit quantization, the quantization noise at the output of the MASH modulator can be made quite low.

It is also evident from Equation 6.88 that if the cancellation condition is not met due to errors in estimating H_2 and N_1, $E_1(z)$ will leak to the output, jeopardizing the performance of the modulator as the signal modulator is only a coarse modulator with a high level of quantization noise. If we assume that

$$H_{c1} = k(H_2 + \Delta H), \tag{6.92}$$

and

$$H_{c2} = k(N_1 + \Delta N), \tag{6.93}$$

where ΔH and ΔN denote the error from the desired value of H_2 and N_1, respectively, it can be shown that the amount of the leakage of $E_1(z)$ is quantified by $(N_1 \Delta H - H_2 \Delta N)E_1(z)$. To minimize the effect of the leakage of $E_1(z)$, $E_1(z)$ itself should be minimized. This can be achieved by using a second-order voltage-mode $\Delta\Sigma$ modulator without concerning stability [47]. Since

$$H_{c1}(z) = kH_2 = kz^{-1}, \tag{6.94}$$

and

$$H_{c2}(z) = kN_1 = k(1 - z^{-1}), \tag{6.95}$$

$H_{c1}(z)$ is an all-pass while $H_{c2}(z)$ provides noise-shaping. As a result, error in estimating $N_1(z)$ dominates that in estimating $H_2(z)$ as the former will pass through $H_{c1}(z)$ without attenuation while the latter will be suppressed by $H_{c2}(z)$ [13].

6.12.2 MULTIRATE TIME-MODE MASH $\Delta\Sigma$ MODULATORS

The power consumption of a MASH ADC is largely dominated by that of the signal modulator [46,48]. This is because the voltage swing of the input of the signal modulator is much larger as compared with that of the residue modulator, resulting in a large amount of charge flowing through the integrator of the signal modulator in each clock cycle. As a result, a large slew rate and subsequently a high level of the power consumption of the integrator of the signal modulator is needed. To improve the performance of MASH ADCs without sacrificing power consumption, the sampling frequency of the signal modulator can be intentionally reduced to relax timing constraints, while the performance loss is compensated by improving the performance of the residue modulator [49,50]. Fortunately, the high sampling frequency of time-mode $\Delta\Sigma$ modulators and their low power consumption make them an ideal candidate for the residue modulator [15,18,51].

The single-loop multirate $\Delta\Sigma$ modulator proposed by Colodro and Torralba requires a decimator in the feedback path. Any nonideality of the decimator will not be suppressed by the loop dynamics and will therefore have a detrimental effect on the overall performance of the modulator [50]. Bos et al. showed that this difficulty can be removed by cascading two $\Delta\Sigma$ modulators of different sampling rates with a upsampler bridging the signal modulator that has a lower sampling rate and the residue modulator that has a higher sampling rate, that is, multirate MASH $\Delta\Sigma$ modulators, as shown in Figure 6.32 [46]. Zaliasl et al. further showed that the upsampler

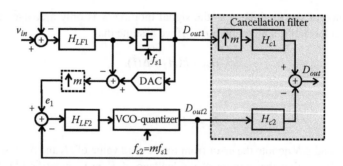

FIGURE 6.32   Multirate MASH $\Delta\Sigma$ modulators. The sampling frequency of residue modulator is m times that of signal modulator. (From Zaliasl, S. et al., *IEEE Trans. Circuits Syst. I*, 59(8), 1604, 2012; Bos, L. et al., *IEEE J. Solid-State Circuits*, 45(6), 1198, 2010.)

bridging the signal and residue modulators can be removed as the residue modulator operates at a higher sampling rate [15]. In the digital cancellation filter, D_{out1} needs to be upsampled by m first and then scaled by H_{c1}, while D_{out2} needs to be scaled by H_{c2} only. The principle of the digital cancellation filter is the same as that of the single-rate MASH $\Delta\Sigma$ modulator detailed earlier.

6.13 DYNAMIC ELEMENT MATCHING

VCO quantizers are multibit quantizers. A multibit DAC is required to generate the feedback voltage with which the input of the modulator compares. DACs are typically implemented using an array of unit elements such as unit capacitors or unit current sources to improve component matching accuracy so as to minimize mismatch. Typical CMOS technologies provide 0.1% for capacitor ratio and 1% for MOS transistor dimension [52]. The effect of the mismatches can be represented by an additive mismatch noise source M_{DAC}, as shown in Figure 6.33.

Since the DAC is located in the feedback path, the mismatch noise will not be attenuated by the loop dynamics. M_{DAC} consisting of tones set by the pattern of

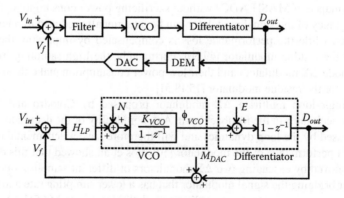

FIGURE 6.33 $\Delta\Sigma$ ADC with VCO quantizer and DEM.

the modulator output codes will be sampled by the quantizer and manifest itself in the frequency domain as both in-band tones and an increasing noise floor, both deteriorating the performance of the modulator. For example, 1% component mismatches of the DAC will completely wipe out the noise-shaping of a third-order $\Delta\Sigma$ modulator [12].

Since the element mismatch of a DAC is deterministic, the degradation of SNDR caused by the mismatch noise is due to the concentration of the energy of the feedback signal at few frequencies. Clearly, if the energy of the feedback signal can be spread over a broad frequency range evenly, the effect of the mismatch noise in the signal band will be minimized. This can be achieved by permuting the unit elements of the DAC dynamically in accordance with the output of the quantizer. This technique is known as DEM. DEM is one of the most studied topics in mixed-mode circuits. Many effective DEM techniques emerged; each has its pros and cons. In the following, we briefly examine these DEM techniques.

Conventional clocked averaging (CLA) rotates the elements of the DAC periodically. It suffers from the correlation between averaging action inherent in $\Delta\Sigma$ modulators and clocked averaging action [53]. Random averaging (RDA) permutes the elements of the DAC randomly, thereby whitening the mismatch noise at the cost of the increased baseband noise floor [54]. Individual level averaging (ILA) removes the drawback of CLA by decorrelating the averaging actions [55,56].

Data-weighted averaging (DWA) selects the element of the DAC sequentially with the selection process controlled by a pointer. The selection process ensures that each element is selected by the same number of times over a long period of time so that statistically, the mismatch noise is whitened. A distinct characteristic of DWA is its first-order noise-shaping on mismatch noise [12,48,57]. Tonal behavior where large tones exist in the baseband was observed in modulators with DWA if the input signal is slowly time-varying [12] or if the amplitude of the input is relatively small [58,59]. Both are due to the repetition of the output codes of the modulator. The former can be minimized by dithering (dithering-DWA) prior to the output of the modulator to randomize the output code of the modulator, however, at the cost of degrading SNR [12], while the latter can be reduced by adding a small offset to the input (offset-DWA) so that the output code of the modulator will not be confined to a small portion of the bits but rather cycling through most if not all of the bits [60]. Rotated data-weighted averaging (RDWA) minimizes the signal-dependent in-band tones of DWA by switching randomly between different patterns for element selection with a deteriorating SNR [52]. Similarly, randomized data-weighted averaging (RnDWA) combats the tonal behavior of generic DWA by randomizing the selection of the element of DAC [60]. Bidirectional data-weighted averaging (Bi-DWA) combats the tonal issue of DWA by alternately changing the direction of the cyclic selection of the element of the DAC at the price of a rising noise floor [61]. Partial data-weighted averaging (Partial-DWA) utilizes the fact that unwanted in-band tones are mostly generated when the input of the modulator is small (excluding dc inputs) and permuting the upper digits of the DAC input that correspond to a large input of the modulator contributes little to removing the unwanted tone. Fujimori et al. showed that a partial permutation of the DAC input, more specifically, the lower digits of the DAC input, will effectively remove the unwanted in-band tones at a reduced cost and the increased frequency of the use of each element subsequently better whitening of mismatch noise [62]. A similar approach was proposed in [63]. Incremental data-weighted averaging (Incremental DWA) removes the baseband tones of DWA by adding more unit elements in DACs that effectively removes the repetitive codes of the modulator [59,64]. The relocated tones at high frequencies can then be removed in a postprocessing step. A similar approach was proposed in [65].

GLA, dithering-DWA, rotated-DWA, Bi-DWA, RnDWA, P-DWA, and pseudo-DWA are tone-suppressing DWA, while offset-DWA, partial-DWA, and incremental DWA are tone-transferring DWA [66]. The former suppress unwanted in-band tones by whitening the input of the DAC at the cost of the increased noise floor, while the latter move mismatch-induced tones outside the signal band so that they can be removed later. They have a little impact on the noise floor. The joint effect of the first-order noise-shaping of DWA and the reduction of in-band tones provided by various improved versions of DWA provides a large SNDR adequate for most applications. The text by Geerts et al. [67] provides a detailed explanation of the principle and an in-depth comparison of the performance of the preceding DEM. Readers are referred to this reference for further information on DEM.

Digital correction is another effective means to eliminate the effect of the nonlinearities of the feedback DAC [68–70]. As demonstrated in [68], digital correlation not only removes harmonic tones caused by the nonlinearity of the DAC, it also lowers the noise floor. The improved performance of digital circuits from technology scaling makes digital correction very attractive.

6.14   PERFORMANCE COMPARISON

Table 6.2 compares the performance of some recently reported closed-loop time-mode modulators. It is seen that closed-loop configurations yield a good SNDR without the need for a large OSR or digital calibration. $\Delta\Sigma$ modulators with voltage-to-time integration offer the advantage of low power consumption but suffer from a poor SNDR due to the absence of a loop filter with a large in-band gain.

TABLE 6.2

Performance Comparison of Closed-Loop Time-Mode $\Delta\Sigma$ Modulators

Reference	Tech. (nm)	f_s (MHz)	BW (MHz)	SNDR (dB)	Power (mW)	FOM (fJ/step)
Taillefer-Roberts [35]	180	140	0.4	38.6	0.8	14381.9
Straayer-Perrott [14]	130	950	23.75	55	38.4	2089
Straayer-Perrott [10]	130	950	10	72	40	614.7
Park-Perrott [16]	130	900	20	78.1	87	331.2
Asl et al. [51]	130	1200	4	77	13.8	298.1
Reddy et al. [45]	90	600	10	78.3	16	119.0
Dhanasekaran et al. [39]	65	250	20	60	10.5	321.2

TABLE 6.3

Performance Comparison of Time-Mode Modulators with Large Signal Bandwidth

Reference	f_s (MHz)	BW (MHz)	OSR	SNDR (dB)	Power (mW)	FOM (fJ/step)
Gao et al. [21]	560	20	14	67.3	3.1	40.6
Jang et al. [11]	600	20	15	52.5	14.3	1043.8
Daniels et al. [26]	300	30	5	64	11.4	150.7
Straayer-Perrott [14]	950	20	23.75	55	38.4	2089
Park-Perrott [16]	900	20	22.5	78.1	87	331.2
Dhanasekaran et al. [39]	250	20	6.25	60	10.5	321.2

$\Delta\Sigma$ modulators with frequency feedback provide an additional order of noise-shaping, allowing the use of a low-order loop filter to achieve the same order of noise-shaping. $\Delta\Sigma$ modulators with phase feedback provide a better SNDR as compared with those with frequency feedback due to the reduced range of the variation of the control voltage of the VCO and subsequently improved voltage-phase linearity. $\Delta\Sigma$ modulators with PWM must carefully address issues such as modulating frequency and sampling frequency so as to minimize the aliasing of the tones from the pulse width modulator back to the signal band of the modulator. MASH $\Delta\Sigma$ modulators are effective in minimizing the effect of the nonlinear voltage-phase characteristics of VCO quantizers at the cost of multiple modulators and the need for a precision digital cancellation filter.

Table 6.3 compares the performance of some recently reported open-loop and closed-loop time-mode modulators for digitizing inputs of a large bandwidth ($\geq$20 MHz). It is seen that the open-loop 1-1 MASH modulator with PWM in [21] offers the lowest FOM due to its low power consumption and large SNDR. Open-loop time-mode modulators perform well if digital calibration is employed to eliminate the effect of the nonlinear voltage-phase characteristics of VCO quantizers. The $\Delta\Sigma$ modulator with phase feedback, multibit quantization, DEM, and a fourth-order active loop filter provides the highest SNDR and a moderate FOM [16]. Digital calibration is the most robust and effective means in combating the effect of nonlinear voltage-phase characteristics of VCO quantizers. The improved performance of digital circuits from technology scaling makes digital correction particularly attractive.

6.15 SUMMARY

This chapter started with a close examination of the fundamentals of ADCs with a special attention to the design specifications of ADCs. An in-depth examination of the principles, advantages, and challenges of VCO phase quantizers was provided. We showed that VCO phase quantizers possess the intrinsic advantages of first-order noise-shaping, a large OSR, multibit quantization, implicit DEM, low power consumption, and full compatibility with technology scaling but suffer from the effect of the nonnegligible voltage-phase nonlinearity of ring oscillators and a large latency due to the use of counter-based phase quantizers. Open-loop techniques that combat the nonlinear voltage-to-phase characteristics of VCO quantizers such as differential VCO phase quantizers, digital calibration, and PWM were investigated and their pros and cons were studied. Due to the existence of only first-order noise-shaping, the SNDR of open-loop time-mode ADCs is generally low unless a large OSR or digital calibration is used at the cost of more power consumption. ADCs with digital calibration consume more power but yield an excellent SNDR. PWM-based ADCs also yield an excellent SNDR and consume a moderate amount of power. The open-loop configuration of these ADCs allows the digitization of inputs with a large bandwidth. PWM and digital calibration are the most effective techniques to reduce the effect of nonlinear voltage-phase characteristics of VCO quantizers to achieve a large SNDR over a large signal bandwidth. The need for a pulse width modulator makes it less attractive with technology scaling.

The fundamentals of $\Delta\Sigma$ modulators including $\Delta\Sigma$ operation, the characteristics of $\Delta\Sigma$ modulators, decimation, and the nonidealities of $\Delta\Sigma$ modulators were examined in detail. Time-mode $\Delta\Sigma$ modulators, specifically, $\Delta\Sigma$ modulators with a VCO quantizer, were introduced. The characteristics of time-mode $\Delta\Sigma$ modulators such as the suppression of nonlinear voltage-phase characteristics of VCO quantizers, the extra order of noise-shaping, continuous-time characteristics, and inherent DEM were studied. $\Delta\Sigma$ modulators with a VCO phase or a VCO frequency quantizer were investigated in detail. We showed that $\Delta\Sigma$ modulators with a VCO phase quantizer outperform those with a VCO frequency quantizer in achieving a better SNDR at the price of a low sampling frequency. $\Delta\Sigma$ modulators with phase feedback were also explored. We showed that an external reference phase from a timing oscillator is needed in these modulators and the choice of the frequency of the reference oscillator must be made in consideration of the frequency of the oscillator of the quantizer. $\Delta\Sigma$ modulators with PWM utilize the intrinsic linear voltage-phase characteristics of VCOs under the condition that the control voltage is only toggled between two fixed values, preferably the lowest and highest values of the control voltage of the oscillator. An attention must be given to the tones generated in the pulse width modulator so as to avoid the aliasing of these tones back to the signal band of the modulator. Another effective means to achieve a better SNDR using time-mode modulators is to use MASH $\Delta\Sigma$ modulators with the signal modulator realized using a conventional voltage-mode modulator and the residue modulator implemented using a time-mode modulator with a VCO quantizer. The former permits a large input, while the latter yields superior performance. Single-rate MASH $\Delta\Sigma$ modulators feature simple configurations and operation, while multirate MASH $\Delta\Sigma$ modulators consume less power by deploying a frequency translation block to coordinate the slow operation of the signal modulator and the fast sampling clock of the residue modulator. Finally, a close look at various DEM techniques for minimizing tones generated by the mismatch of the DAC was given. The chapter ended with a comparison of the performance of some recently reported time-mode $\Delta\Sigma$ modulators.

REFERENCES

1. P. Wickramrachi. Effects of windowing on the spectral content of a signal. *Sound and Vibration*, 10–11, January 2003.
2. G. Li, Y. Tousi, A. Hassibi, and E. Afshari. Delay-line-based analog-to-digital converters. *IEEE Transactions on Circuits and Systems II*, 56(6):464–468, June 2009.
3. Y. Tousi and E. Afshari. A miniature 2 mW 4 bit 1.2 GS/s delay-line-based ADC in 65 nm CMOS. *IEEE Journal of Solid-State Circuits*, 46(10):2312–2325, October 2011.
4. A. Iwata, N. Sakimura, M. Nagata, and T. Morie. The architecture of delta sigma analog-to-digital converters using a voltage-controlled oscillator as a multi-bit quantizer. *IEEE Transactions on Circuits and Systems II*, 46(7):941–945, July 1999.
5. T. Watanabe, T. Mizuno, and Y. Makino. An all-digital analog-to-digital converter with 12-μV/LSB using moving-average filtering. *IEEE Journal of Solid-State Circuits*, 38(1):120–125, January 2003.
6. J. Kim and S. Cho. A time-based analog-to-digital converter using a multi-phase voltage controlled oscillator. In *Proceedings of the IEEE International Symposium on Circuits and Systems*, Island of Kos, Greece, 2006, pp. 3934–3937.
7. J. Daniels, W. Dehaene, M. Steyaert, and A. Wiesbauer. A 0.02 mm^2 65 nm CMOS 30 MHz BW all-digital differential VCO-based ADC with 64 dB SNDR. In *Symposium of VLSI Circuits Digest of Technical Papers*, 2010, pp. 155–156.
8. J. Kim, T. Jang, Y. Yoon, and S. Cho. Analysis and design of voltage-controlled oscillator based analog-to-digital converter. *IEEE Transactions on Circuits and Systems I*, 57(1):18–30, January 2010.
9. R. Si, F. Li, and C. Zhang. A 100 MH/s, 7 bit VCO-based ADC which is used in time interleaved ADC architecture. In *Proceedings of the International Consumer Electronics, Communications and Networks Conferences*, Yichang, 2012, pp. 4–7.
10. M. Straayer and M. Perrott. A 12-bit, 10-MHz bandwidth, continuous time ΛΣ ADC with a 5-bit, 950-MS/s VCO-based quantizer. *IEEE Journal of Solid-State Circuits*, 43(4):805–814, April 2008.
11. T. Jang, J. Kim, Y. Yoon, and S. Cho. A highly-digital VCO-based analog-to-digital converter using phase interpolator and digital calibration. *IEEE Transactions on VLSI Systems*, 20(8):1368–1372, August 2012.
12. R. Baird and T. Fiez. Linearity enhancement of multibit ΔΣ A/D and D/A converters using data weighted averaging. *IEEE Transactions on Circuits and Systems II*, 42(12):753–762, December 1995.
13. R. Schreier and G. Temes. *Understanding Delta-Sigma Data Converters*. John Wiley & Sons, Hoboken, NJ, 2005.
14. M. Straayer and M. Perrott. A 10-bit 20 MHz 38 mW 950 MHz CT ΣΔ ADC with a 5-bit noise-shaping VCO-based quantizer and DEM circuit in 0.13 μm CMOS. In *Symposium VLSI Circuits Digest of Technical Papers*, Honolulu, HI, 2007, pp. 246–247.
15. S. Zaliasl, S. Saxena, P. Hanumolu, K. Mayaram, and T. Fiez. A 12.5-bit 4 MHz 13.8 mW MASH ΔΣ modulator with multirate VCO-based ADC. *IEEE Transactions on Circuits and Systems I*, 59(8):1604–1613, August 2012.
16. M. Park and M. Perrott. A single-slope 80 Ms/s ADC using two-step time-to-digital conversion. In *IEEE International Symposium on Circuits and Systems*, Taipei, Taiwan, 2009, pp. 1125–1128.
17. M. Allam, H. Aboushady, and M. Louerat. Continuous-time ΣΔ modulators with VCO-based voltage-to-phase and voltage-to-frequency quantizers. In *Proceedings of the IEEE Mid-West Symposium on Circuits and Systems*, Seattle, WA, 2010, pp. 676–679.
18. T. He, Y. Du, Y. Jiang, S. Sin, U. Seng-Pan, and R. Martin. A DT 0-2 MASH ΣΔ modulator with VCO-based quantizer for enhanced linearity. In *Proceedings of the IEEE Asia-Pacific Conference on Circuits and Systems*, Kaohsiung, Taiwan, 2012, pp. 33–36.

19. S. Rao, B. Young, A. Elshazly, W. Yin, N. Sasidhar, and P. Hanumolu. A 71 dB SFDR open loop VCO-based ADC using 2-level PWM modulation. In *Symposium on VLSI Circuits Digest of Technical Papers*, Kyoto, 2011, pp. 270–271.
20. Y. Yoon, S. Park, and S. Cho. A time-based noise shaping analog-to-digital converter using a gated-ring oscillator. In *Proceedings of the IEEE International Microwave Workshop Intelligent Radio for Future Personal Terminals*, Daejeou, Korea, 2011, pp. 1–4.
21. P. Gao, X. Xing, J. Cranincks, and G. Gielen. Design of an intrinsically linear double-VCO-based ADC with 2nd-order noise shaping. In *Proceedings of the IEEE Design, Automation and Test in Europe Conference and Exhibition*, Dresden, Germany, 2012, pp. 1215–1220.
22. B. Razavi. *RF Microelectronics*, 2nd edn. Prentice Hall, Upper Saddle River, NJ, 2011.
23. M. Hovin, A. Olsen, T. Lande, and C. Toumazou. Delta-sigma modulators using frequency-modulated intermediate values. *IEEE Journal of Solid-State Circuits*, 32(1):13–22, January 1997.
24. U. Wismar, D. Wisland, and P. Andreani. A 0.2 V 0.44 µW 20 kHz analog to digital ΔΣ modulator with 57 fJ/conversion FoM. In *Proceedings of the IEEE European Solid-State Circuits Conference*, Montreux, Switzerland, 2006, pp. 187–190.
25. Y. Yoon, M. Cho, and S. Cho. A linearization technique for voltage-controlled oscillator-based ADC. In *Proceedings of the IEEE International SoC Conference*, Belfast, 2009, pp. 317–320.
26. J. Daniels, W. Dehaene, M. Steyaert, and A. Wiesbauer. A/D conversion using asynchronous ΔΣ modulation and time-to-digital conversion. *IEEE Transactions on Circuits and Systems I*, 57(9):2404–2412, September 2010.
27. J. Hamilton, S. Yan, and T. Viswanathan. A discrete-time input ΔΣ ADC architecture using a dual-VCO-based integrator. *IEEE Transactions on Circuits and Systems II*, 57(11):848–852, November 2010.
28. G. Taylor and I. Galton. A mostly-digital variable-rate continuous-time delta-sigma modulator ADC. *IEEE Journal of Solid-State Circuits*, 45(12):2634–2646, December 2010.
29. M. Straayer and M. Perrott. A multi-path gated ring oscillator TDC with first-order noise shaping. *IEEE Journal of Solid-State Circuits*, 44(4):1089–1098, April 2009.
30. P. Gray, P. Hust, S. Lewis, and R. Meyer. *Analysis and Design of Analog Integrated Circuits*, 4th edn. John Wiley & Sons, New York, 2001.
31. D. Tuite. Continuous-time delta-sigma ADCs. *Electronic Design*, October 2005.
32. P. Whytock. Get to know continuous-time ADCs. *Electronic Design*, June 2009.
33. R. Schreier. Understanding continuous-time, discrete-time sigma-delta ADCs and Nyquist ADCs. *Electronic Design*, February 2009.
34. A. Gupta. What is the difference between continuous-time and discrete-time delta-sigma ADCs? *Electronic Design*, May 2014.
35. C. Taillefer and G. Roberts. Delta-sigma A/D converter via time-mode signal processing. *IEEE Transactions on Circuits and Systems I*, 56(9):1908–1920, Sept. 2009.
36. Y. Lin, D. Liao, C. Hung, and M. Ismail. A phase-based singe-bit delta-sigma ADC architecture. In *Proceedings of the IEEE New Circuits and Systems*, Bordeaux, France, 2011, pp. 406–409.
37. Y. Lin and M. Ismail. A delta-sigma modulator with a phase-to-digital and digital-to-phase quantizer. In *Proceedings of the IEEE New Circuits and Systems*, Bordeaux, France, 2011, pp. 209–212.
38. Y. Lin and M. Ismail. Time-based all-digital sigma-delta modulators for nanometer low voltage CMOS data converters. *Analog Integrated Circuits and Signal Processing*, 73:801–808, 2012.

39. V. Dhanasekaran, M. Gambhir, M. Elsayed, E. Sanche-Sinencio, J. Silva-Martinez, C. Mishra, L. Chen, and E. Pankratz. A 20 MHz BW 68 dB DR CT ΔΣ ADC based on a multi-bit time-domain quantizer and feedback element. In *IEEE International Solid-State Circuits Conference Digest of Technical Papers*, 2009, pp. 174–175.
40. V. Dhanasekaran, M. Gambhir, M. Elsayed, E. Sanche-Sinencio, J. Silva-Martinez, C. Mishra, L. Chen, and E. Pankratz. A continuous time multibit ΣΔ ADC using time domain quantizer and feedback element. *IEEE Journal of Solid-State Circuits*, 46(3):639–650, March 2011.
41. M. Elsayed, V. Dhanasekaran, M. Gambhir, J. Silva-Martinez, and E. Pankratz. A 0.8 ps DNL time-to-digital converter with 250 MHz event rate in 65 nm CMOS for time-mode-based ΣΔ modulator. *IEEE Journal of Solid-State Circuits*, 46(9):2084–2098, September 2011.
42. L. Hernandez, S. Paton, and E. Prefasi. VCO-based sigma delta modulator with PWM pre-coding. *IET Electronics Letters*, 47(10):588–589, May 2011.
43. A. Gupta, K. Nagaraj, and T. Viswanathan. A two-stage ADC architecture with VCO-based second stage. *IEEE Transactions on Circuits and Systems II*, 58(11):734–738, November 2011.
44. K. Reddy, R. Sachin, R. Inti, B. Young, A. Elshazly, M. Talegaonkar, and P. Hanumolu. A 16 mW 78 dB SNDR 10 MHz-BW CT-ΔΣ ADC using residue-canceling VCO-based quantizer. In *IEEE International Solid-State Circuits Conference Digest of Technical Papers*, 2012, pp. 152–154.
45. K. Reddy, S. Rao, R. Inti, B. Young, A. Elshazly, M. Talegaonkar, and P. Hanumolu. A 16-mW 78-dB SNDR 10-MHz BW CT ΔΣ ADC using residue-canceling VCO-based quantizer. *IEEE Journal of Solid-State Circuits*, 47(12):1–12, December 2012.
46. L. Bos, G. Vandersteen, P. Rombouts, A. Geis, A. Morgado, G. Van der Plas, and J. Ryckert. Multirate cascaded discrete-time low-pass ΔΣ modulator for GSM/Bluetooth/UMTS. *IEEE Journal of Solid-State Circuits*, 45(6):1198–1208, June 2010.
47. J. de la Rosa and R. del Rio. *CMOS Sigma-Delta Converters: Practical Design Guide*. John Wiley & Sons, New York, 2013.
48. O. Nys and R. Henderson. A 19-bit low-power multibit sigma-delta ADC based on data weighted averaging. *IEEE Journal of Solid-State Circuits*, 32(7):933–942, July 1997.
49. F. Colodro, A. Torralba F. Munoz, and L. Franquelo. New class of multibit sigma-delta modulators using multirate architecture. *IEE Electronics Letters*, 36(9):783–785, April 2000.
50. F. Colodro and A. Torralba. Multirate single-bit ΣΔ modulators. *IEEE Transactions on Circuits and Systems II*, 49(9):629–634, September 2002.
51. S. Asl, S. Saxena, P. Hanumolu, K. Mayaram, and T. Fiez. A 77 dB SNDR, 4 MHz MASH ΔΣ modulator with a second-stage multi-rate VCO-based quantizer. In *Proceedings of the IEEE Custom Integrated Circuits Conference*, San Jose, CA, 2011, pp. 1–4.
52. R. Radke, A. Eshraghi, and T. Fiez. A 14-bit current-mode ΣΔ DAC based upon rotated data weighted averaging. *IEEE Journal of Solid-State Circuits*, 35(8):1074–1084, August 2000.
53. R. van der Plassche and D. Goedhart. A monolithic 14-bit D/A converter. *IEEE Journal of Solid-State Circuits*, SC-14(3):552–556, June 1979.
54. L. Carley. A noise-shaping coder topology for 15+ bit converters. *IEEE Journal of Solid-State Circuits*, 24(2):267–273, April 1989.
55. B. Leung and S. Sutarja. Multibit Σ—*Delta* A/D converter incorporating a novel class of dynamic element matching techniques. *IEEE Transactions on Circuits and Systems II*, 39(1):35–51, January 1992.
56. F. Chen and B. Leung. A high resolution multibit sigma-delta modulator with individual level averaging. *IEEE Journal of Solid-State Circuits*, 30(4):453–460, April 1995.

57. O. Nys and R. Henderson. An analysis of dynamic element matching techniques in sigma-delta modulation. In *IEEE International Symposium on Circuits and Systems*, Atlanta, 1996, pp. 213–234.
58. F. Chen and B. Leung. Some observations on tone behavior in data weighted averaging. In *Proceedings of the IEEE International Symposium on Circuits and Systems*, Monterey, CA, 1998, Vol. 1, pp. 500–503.
59. K. Chen and T. Kuo. An improved technique for reducing baseband tones in sigma-deltas modulators employing data weighted averaging algorithm without adding dither. *IEEE Transactions on Circuits and Systems II*, 46(1):63–68, January 1999.
60. M. Vadipour. Techniques for preventing tonal behavior of data weighted averaging algorithm in $\Sigma\Delta$ modulators. *IEEE Transactions on Circuits and Systems II*, 47(11):1137–1144, November 2000.
61. I. Fujimori, L. Longo, A. Hairapetian, K. Seiyama, S. Kasic, J. Cao, and S. Chan. A 90-dB SNR 2.5 MHz output-rate ADC using cascaded multibit delta-sigma modulation at 8× oversampling ratio. *IEEE Journal of Solid-State Circuits*, 35(12):1820–1828, December 2000.
62. I. Fujimori, A. Nogi, and T. Sugimoto. A multi-bit delta-sigma audio DAC with 120-dB dynamic range. *IEEE Journal of Solid-State Circuits*, 35(8):1066–1073, August 2000.
63. K. Vleugels, S. Rabii, and B. Wooly. A 2.5 V sigma-delta modulator for broadband communications applications. *IEEE Journal of Solid-State Circuits*, 36(12):1887–1899, December 2001.
64. T. Kuo, K. Chen, and H. Yeng. A wideband CMOS sigma-delta modulator with incremental data weighted averaging. *IEEE Journal of Solid-State Circuits*, 37(1):11–17, January 2002.
65. A. Hamoui and K. Martin. High-order multibit modulators and pseudo data-weighted-averaging in low-oversampling $\Delta\Sigma$ ADCs for broad-band applications. *IEEE Transactions on Circuits and Systems I*, 51(1):72–85, January 2004.
66. D. Lee and T. Kuo. Advancing data weighted averaging technique for multibit sigma-delta modulators. *IEEE Transactions on Circuits and Systems II*, 54(4):838–842, October 2007.
67. Y. Geerts, M. Seyaert, and W. Sansen. *Design of Multi-Bit Delta-Sigma A/D Converters*. Kluwer Academic Press, Secaucus, NJ, 2002.
68. M. Sarhang-Nejad and G. Temes. A high-resolution multibit $\Sigma\Delta$ ADC with digital correction and relaxed amplifier requirements. *IEEE Journal of Solid-State Circuits*, 28(6):648–660, June 1993.
69. S. Ali, S. Tanner, and P. Farine. A background calibration method for DAC mismatch correction in multibit sigma-delta modulators. In *Proceedings of IEEE International SoC Design Conference*, Jeju, Korea, 2012, pp. 427–430.
70. S. Ali, S. Tanner, and P. Farine. A DAC mismatch calibration technique for multibit $\Sigma\Delta$ modulators. In *Proceedings of the IEEE New Circuits and Systems*, Paris, 2013, pp. 1–4.

7 Time-Mode Delta-Sigma Converters

*Soheyl Ziabakhsh, Ghyslain Gagnon,
and Gordon W. Roberts*

CONTENTS

The aim of this chapter is to *describe* delta-sigma (ΔΣ) converters that adopt time-mode signal processing (TMSP) techniques. A key advantage of time-mode ΔΣ converters is that they are realized using digital circuits and process information in the form of time-difference intervals. As a consequence of using digital circuits, this technique benefits from low voltage operation without concern for reduced signal swings, sensitivities to thermal noise effects, or switching noise sensitivity. Recently, several studies on time-mode ΔΣ converters have been conducted showing that such methodology has high potential in low-voltage design. The noise-shaping behavior demonstrated by this technique can be implemented and extended in various ways, including voltage-controlled delay unit (VCDU) or gated-ring oscillator (GRO)-based implementations of TMΔΣ converters. In this chapter, after a brief review of ΔΣ ADC specifications, we will discuss different architectures of TMΔΣ converters that have been recently proposed.

7.1   INTRODUCTION

A large number of research works over the last decade has shown that a decrease in the voltage supply will seriously impact the operation of analog circuits implemented in advancing CMOS technologies [1,2]. Some of these limitations include: increased leakage current through the gate of individual transistors, reduction in voltage node swings, increases in switching noise, and consumption of more power. Another constraint in designing high-precision analog circuits in mixed-signal fully integrated systems is the reduced intrinsic gain of MOS transistors [3]. Intrinsic gains of less than 10 V/V are not uncommon with very fine-line MOS transistors. In contrast, digital circuits have proven to be quite amenable in advanced CMOS technologies working with transistors with such low gains. Billions of transistors are now being integrated on a single chip in the form of digital logic where they are used to realize reliable computing and signal-processing algorithms. It is therefore the goal of this work to develop analog signal-processing techniques that use digital logic gates as their basic building blocks. The emphasis of this chapter will be on the intuitive understanding of the various architectural approaches rather than their empirical performance, as it is expected that their performance will advance significantly in the next few years.

Recently, several studies have explored the realization of wideband, high-resolution analog-to-digital converters (ADCs) under low supply voltage conditions [4]. As a result of these studies, delta-sigma modulators ($\Delta\Sigma$Ms) used in the realization of high-resolution ADCs were deemed better suited to low-voltage operation than other types of ADC architectures on account of their insensitivity to individual component values. Within this class of $\Delta\Sigma$-type ADC, three techniques emerged that improved their conversion accuracy: direct feed-forward compensation [5], multibit quantization [3], and increasing the order of the $\Delta\Sigma$ modulator [2]. Although the direct feed-forward compensation technique improves the linearity of $\Delta\Sigma$ modulators [6], it suffers from a timing problem between the feed-forward path and feedback path that increases the design complexity. In the case of a multibit quantization implementation approach, large power consumption and silicon area is required to realize a multibit digital-to-analog converter (DAC) in the feedback path with a linearity that matches that of the overall ADC [4]. While increasing the order of a $\Delta\Sigma$ modulator will improve the overall performance of an ADC, it does so at the expense of an exponential increase in chip area.

One promising technique to overcome the aforementioned challenges is to process signal information in the time-domain where sampled data is represented by discrete time-differences. A major advantage of this method is that the performance of the $\Delta\Sigma$Ms is not directly related to the supply voltage. Shifting the signal processing from the voltage or current domains to the time domain enables the use of digital circuits instead of analog elements (capacitors, op-amps, etc.). By doing so, simple digital circuits can be used to replace otherwise complicated analog circuit elements. As an example, a high-speed, high-precision analog quantizer can be replaced by a 1-bit digital comparator (D-type flip-flop) as reported in [7]. Due to this attractive property, several efforts on time-domain $\Delta\Sigma$Ms have been recently reported. Such efforts are exemplified in a VCDU-based $\Delta\Sigma$M [8], a modulator realized using a voltage-controlled gated-ring oscillator (VCGRO) [9], a voltage-controlled oscillator (VCO)-based multibit quantizer used in a $\Delta\Sigma$M [10–12], a delay-locked-loop (DLL)-based and voltage-controlled delay line (VCDL) technique as a $\Delta\Sigma$M quantizer [13], and a time-to-digital converter (TDC)-based $\Delta\Sigma$M [7,14]. One of the significant trends in these research works is to uncover the means in which to realize analog signal-processing techniques with digital circuits.

Given the aforementioned motivation, this chapter will continue in Section 7.2 by providing background on $\Delta\Sigma$ modulators and their performance parameters. The time-mode first-order single-bit $\Delta\Sigma$ modulator is then presented in Section 7.3. Also provided in this section is a description of several research works that appeared in the literature over the last few years. Following this in Section 7.4, high-order TM$\Delta\Sigma$ modulators will be described and in Section 7.5, the realization of several multibit TM$\Delta\Sigma$ modulators will be outlined. A performance comparison of the published realizations of various TM$\Delta\Sigma$ modulators versus their voltage-mode counterparts will be provided in Section 7.6. Some of the design limitations of TM$\Delta\Sigma$ modulators will be outlined in Section 7.7. Finally, in Section 7.8, conclusions are drawn.

7.2 THEORY OF ΔΣ MODULATORS

The basic architecture of a conventional ΔΣ modulator is shown in Figure 7.1. It consists of a difference amplifier, loop filter and a one-bit quantizer in the feed-forward path, and a one-bit DAC in the feedback signal path. The front-end difference amplifier and loop filter are often realized using a high-performance switched-capacitor circuit, and the quantizer is often realized using a high-speed latched comparator circuit. As the DAC provides only one-bit conversion, a simple set of analog switches is used to realize this element. Through the negative-feedback action provided by the feedback path, the effects of nonlinearities in the feed-forward signal path are reduced by the loop gain. This action has come to be known in the ΔΣ modulator literature as noise-shaping [15–18], as the quantization noise introduced by the quantizer is pushed or *shaped* away in frequency from the signal band.

In multibit ΔΣ modulators, the quantizer and the DAC operate on more than two levels and, correspondingly, require a more complicated circuit realization. An important implementation issue with the multibit DAC is the requirement for high linearity. To achieve this result over multiple manufacturing runs, very good element matching is required. However, as the dimensions of CMOS transistors scale downward, matching becomes more difficult to achieve [12].

In the following, the basic principles of a ΔΣ modulator will be described along with its various performance metrics.

7.2.1 Basic Principles of a ΔΣ Modulator

The process of converting an analog continuous-time signal $x(t)$ into a sequence of digital numbers $y[n]$ requires a front-end antialiasing filter circuit, a sample-and-hold (S/H) circuit, and the corresponding multilevel quantizer or ADC circuit as illustrated in Figure 7.2. The front-end low-pass filter is used to minimize the potential threat of unwanted high-frequency signals from aliasing into the base-band frequency region that the desired signal occupies. The antialiasing filter is designed to have a bandwidth equal to the incoming desired signal and a stop-band region very near to one-half the sampling frequency, f_s—also referred to the Nyquist frequency. In contrast, an ADC constructed using a ΔΣ modulator takes on a slightly different realization. While the front-end requires an antialiasing filter and an S/H circuit, the quantizer is realized using a ΔΣ modulator, followed by a low-pass brick-wall digital filter used to remove the noise-shaped quantization noise.

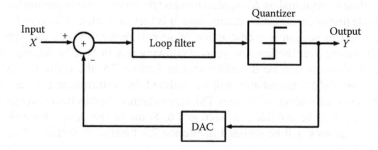

FIGURE 7.1   Basic architecture of a conventional ΔΣ modulator.

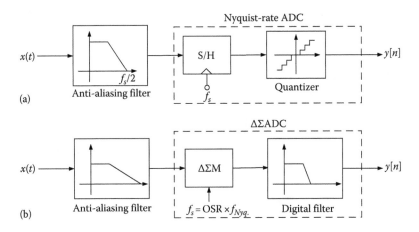

FIGURE 7.2 Two types of analog-to-digital conversion schemes: (a) Nyquist-rate implementation, and (b) oversampling or $\Delta\Sigma$-based implementation.

Unlike a Nyquist-rate ADC that samples the input signal at twice its signal bandwidth, that is, $f_s = 2 \times f_{BW}$, a $\Delta\Sigma$-based ADC oversamples the incoming signal at a rate much greater than twice the signal bandwidth. The ratio of one-half of the sampling rate to the signal bandwidth f_{BW} is defined as the oversampling ratio (OSR), that is,

$$\text{OSR} = \frac{f_s/2}{f_{BW}} \tag{7.1}$$

An important advantage of oversampling the incoming signal is that it relaxes the requirements on the antialiasing filter circuit. In fact, the antialiasing filter is typically implemented with a simple low-order filter that requires little power [16]. As a means to compare the filter requirements for the two types of analog-to-digital conversion schemes, an illustration of the antialiasing filter magnitude response is provided in Figure 7.3. In Figure 7.3a, the magnitude response for the Nyquist-rate ADC filter is shown, and Figure 7.3b, the corresponding response for the $\Delta\Sigma$-based ADC implementation is shown. While each filter has the same analog bandwidth, as it is assumed that each ADC will see the same incoming signal, the transition region of the filter is quite different. The stop-band region for the $\Delta\Sigma$-based ADC would be much higher than that required for the Nyquist-rate ADC. This greatly reduces the complexity of the antialiasing filter as mentioned earlier.

The error between the information carried by the input analog signal $x(t)$ and the information carried by an ideal quantized output digital signal $y[n]$ is defined as the quantization error. Such a situation is depicted in Figure 7.4 where the transfer characteristic of an ideal quantizer is shown in Figure 7.4a. Here we see the quantizer has a staircase-like shape with the width of each staircase equal to Δ, also referred to as the least-significant bit (LSB) of the analog-to-digital conversion process. When subtracted from a perfect conversion process (one without error, as depicted by the dashed line in Figure 7.4a), the quantization error curve results as

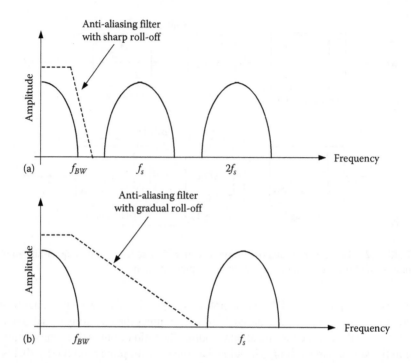

FIGURE 7.3 Filter requirement for (a) Nyquist rate $(f_s > 2f_{BW})$ and (b) $\Delta\Sigma$ modulator $(f_s \gg 2f_{BW})$.

shown in Figure 7.4b. The quantization error is bounded between $-\Delta/2$ and $+\Delta/2$ for any level of analog input. Also we see that the average error is zero and it has an RMS error defined by

$$e_{rms}^2 = \int_0^1 \left[e(v_{in}) \right]^2 dv_{in} = \frac{\Delta^2}{12} \tag{7.2}$$

Through careful construction, one can show that under certain conditions, this quantization noise power is uniformly distributed in the frequency range $(0, f_s/2)$. Based on this criterion, the single-sided quantization noise power spectral density expressed in terms V²/Hz can be defined simply as

$$N_q = \frac{\Delta^2}{6f_s} \tag{7.3}$$

Figure 7.5a illustrates the uniform power-spectral density (PSD) of the ideal quantizer with $\Delta = 1$. Here the total noise power is equal to 1/12 V², which is equivalent to an RMS noise voltage of $1/\sqrt{12}$ V.

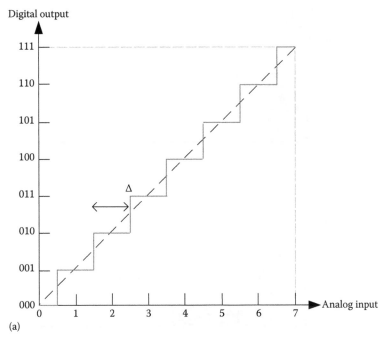

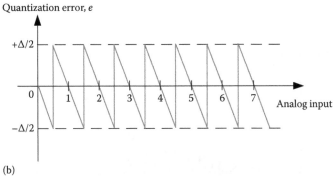

FIGURE 7.4 Illustrating the quantization process: (a) transfer characteristic of input and output signal and (b) quantization error.

From Equation 7.3, the magnitude of the RMS quantization error can be reduced by decreasing the step size of the quantizer. Another approach would be to increase the sampling rate of the quantizer, say by a factor of K, and pass the quantizer output through a low-pass filter with a bandwidth equal to the signal bandwidth f_{BW}. By doing so, according to Equation 7.3, the PSD of the quantization noise will be spread over a larger frequency range $(0, Kf_s/2)$ with magnitude

$$N'_q = \frac{\Delta^2}{6} \frac{1}{Kf_s} \tag{7.4}$$

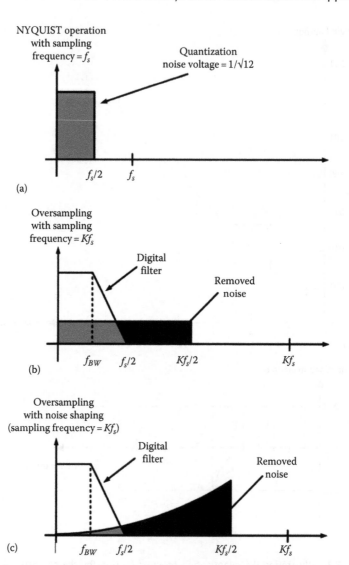

FIGURE 7.5 Illustrating the effect of oversampling and noise-shaping on the PSD of an ideal quantizer: (a) PSD reference quantizer with $\Delta = 1$, (b) PSD of an oversampled quantizer by factor K with digital filter response overlaid, and (c) noise-shaped PSD of quantizer with digital filter response overlaid.

Since the noise power is uniform over the frequency range $(0, f_{BW})$, and since the factor K is equivalent to the OSR parameter introduced earlier, then the total output quantization noise power reduces to

$$e'^2_{rms} = N'_q \cdot \frac{f_{BW}}{K \cdot f_s/2} = N'_q \cdot \frac{f_{BW}}{\text{OSR} \cdot f_s/2} \tag{7.5}$$

For a digital filter with transfer function $H_d(z)$, the in-band noise power that passes through the filter would be more correctly represented by the integral equation,

$$n_0^2 = \int_{f_1}^{f_2} |H_d(f)|^2 N_q' df \tag{7.6}$$

This situation is depicted in Figure 7.5b where the quantization noise PSD is spread over a bandwidth of Kf_s, but only a portion of the noise passes through the digital filter.

Equation 7.5 reveals two important facts related to oversampling, followed by low-pass filtering: (1) The higher the oversampling factor OSR, the smaller is the output RMS error, and (2) the smaller the signal bandwidth, the smaller is the RMS noise error. To gain, say, a 10 dB improvement in noise reduction would require a 10-fold increase in the sampling rate. Any further noise power reduction would come at the expense of impractical increases in the sampling rate of the quantizer.

$\Delta\Sigma$-based analog-to-digital conversion provides an alternative means in which to reduce the magnitude of the quantization noise at the ADC output. Through the application of noise-shaping, the feedback loop established around the quantizer reduces the amount of quantization that makes its way to the output based on the amount of gain in the feedback loop. Mathematically, this effect can be quantified by writing an expression for the output signal $Y(z)$ in the z-domain in terms of the input signal $X(z)$ and the quantization error signal $E(z)$, according to

$$Y(z) = \text{STF}(z)X(z) + \text{NTF}(z)E(z), \tag{7.7}$$

where $\text{STF}(z)$ and $\text{NTF}(z)$ are denoted as the signal and noise transfer functions, respectively. For a first-order low-pass $\Delta\Sigma$ modulator, the output signal in the z-domain is expressed as

$$Y_{LP}(z) = z^{-1}X(z) + (1 - Z^{-1})E(z) \tag{7.8}$$

leading to STF $= z^{-1}$ and NTF $= 1 - z^{-1}$. The signal transfer function (STF) simply signifies that the output will contain a one-clock period delay of the input signal, essentially with the input information unchanged. In the case of the noise transfer function (NTF), a transmission zero appears at DC, corresponding to a gain of 0. Likewise, at the Nyquist frequency, that is, $f = f_s/2$, the NTF has a gain of 2. For frequencies between DC and Nyquist, the NTF will have a high-pass behavior. Consequently, the quantization noise injected by the quantizer will have little effect on the incoming signal at low frequencies but doubles up for frequencies close to the Nyquist frequency. This situation is illustrated in Figure 7.5c whereby noise shaping combined with digital filtering greatly reduces the level of output quantization noise.

For higher-order modulators, whereby the NTF is of order greater than 1, even less quantization noise power will appear at the output. This is illustrated in Figure 7.6 where the quantization noise PSD appears at the output of the $\Delta\Sigma$ modulator for orders of 1, 2, and 3. As is evident from this plot, the higher the modulator order, the lower is the quantization noise PSD at frequencies close to DC.

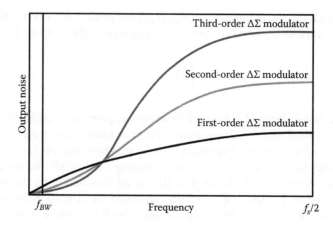

FIGURE 7.6 High-order noise shaping and its effect on the output quantization noise PSD.

7.2.2 SOME ADC PERFORMANCE METRICS

The most commonly quoted ADC performance parameters are sampling frequency, dynamic range (DR), resolution, distortion, power dissipation, chip area, and stability. These metrics will be defined and discussed in the following.

7.2.2.1 Sampling Frequency

Generally speaking, the sampling frequency is defined as the rate at which the sampling process takes place. This parameter in Nyquist-rate ADCs is set to twice the signal bandwidth and in $\Delta\Sigma$ADCs is set to higher values to significantly decrease the in-band quantization noise. On the other hand, higher sampling frequencies lead to higher power consumption and increased sensitivities to clock jitter issues. High-order and multibit $\Delta\Sigma$ modulators are two techniques that can be used to compensate for these effects and lead to higher performance ADCs.

7.2.2.2 Dynamic Range

DR is defined as the ratio between the maximum applied sinusoidal input signal to the smallest that is discernible at the output of the ADC from any other unwanted or undesirable signal created by the ADC, for example, quantization and thermal noise, and distortion. The simplest manner in which to extract the DR metric is to measure the signal-to-noise-plus-distortion ratio (SNDR) as a function of the input signal level, such as that shown in Figure 7.7. The DR metric expressed in dB would then be defined as the difference in the maximum and minimum input levels in dB for which the SNDR = 0 dB, that is,

$$DR\big|_{dB} = V_{in,\max,dB} - V_{in,\min,dB} \tag{7.9}$$

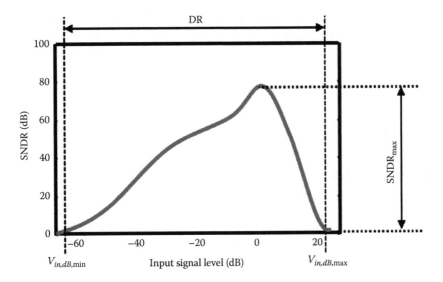

FIGURE 7.7 An example SNDR plot as a function of input level and several important performance metrics.

Often, in practice, the maximum input level will be limited by the power supply or some other upper limit instead of the SNDR value falling back to the 0 dB level. Nonetheless, Figure 7.7 conveys the general idea behind the DR metric.

As CMOS technology advances to smaller dimensions, one observes that the DR of ADCs tends to decrease, a result attributed to smaller transistor gate oxide, lower supply voltages, and greater transistor thermal noise levels. Novel signal-processing methods like TMSP may help to resolve the DR issue in advanced CMOS processes [8].

7.2.2.3 Resolution

The resolution of an N-bit ADC is defined as the smallest change in the analog input voltage that leads to a corresponding consistent change in the digital code output. In terms of our previous discussion related to Figure 7.4, the resolution of an ideal quantizer would simply be equal to the width of the staircase specified by Δ. In the ideal case, Δ would be defined in terms of the full-scale output range (FSR) and the total number of bits specified by the architecture, N, given by the following expression

$$\Delta = \frac{\text{FSR}}{2^N - 1} \tag{7.10}$$

A related metric, but one that is easier to extract in practice, is called the effective resolution of the ADC, Δ_{eff}, and is defined in terms of the effective number of bits (ENOBs) of the ADC in the following way:

$$\Delta_{eff} = \frac{\text{FSR}}{2^{ENOB} - 1} \tag{7.11}$$

Here ENOB represents the maximum value of the SNDR plot (see Figure 7.7) converted from dBs to equivalent bits [19] using the following formula

$$\text{ENOB} = \frac{\text{SNDR}\big|_{max} - 1.76 \text{ dB}}{6.02} \tag{7.12}$$

The effective ADC resolution is always less than the ideal ADC resolution, that is, $\Delta_{eff} < \Delta$.

7.2.2.4   Distortion

Signal distortion occurs with increasing input signal level on account of the general nonlinear nature of semiconductor devices, in addition to mismatches between ADC elements. One attractive technique to compensate for component mismatch effects is to include circuits that average out any component mismatch. Some of these approaches are: offset cancellation [20], delayed-lock loop (DLL) [21], dynamic element matching (DEM) [22], and laser trimming during postpackaging manufacturing.

7.2.2.5   Power Dissipation and Chip Area

In the design of ADC circuits, there are two important requirements that must be taken into account. This includes both the ability to achieve good performance and low power operation, as well as a small silicon footprint. With the ongoing scaling-down of CMOS technology, power consumption is expected to decrease as the supply voltage must be reduced. On the other hand, reducing the supply voltage causes several drawbacks in the corresponding analog circuits (such as reducing DR, increasing the switching noise and gate leakage, etc.) especially when the performance requirements are to be maintained, for example, build a 12-bit ADC. One way to overcome these limitations in low-voltage design is to shift the analog design to the digital domain in an attempt to be more compatible with modern CMOS technologies. To this end, in this chapter, we offer several novel techniques that perform analog-to-digital conversion in the time-domain without consuming large amounts of power or die area—hopefully, fulfilling the goal of achieving high-resolution analog circuits in nanometer CMOS technologies.

7.2.2.6   Stability Considerations

It is well known by ADC designers that high-order $\Delta\Sigma$ modulators can become unstable. In [2], the stability of a sixth-order modulator is evaluated. They show that changing the gain of quantizer can change the location of closed-loop poles and cause instability. The main reason is that for some unique values of quantizer gain, the poles move outside of the unit circuit in the z-domain and cause the modulator to go unstable. Several works have followed this line of reasoning and have attempted to model this effect in $\Delta\Sigma$ modulators. The basic model is to replace the quantizer with a two-input adder, having one input from the quantizer and the other connected to a source of additive noise modeling the quantization process. While this model defines the noise-shaping process described earlier in the usual way, it places emphasis on the lack of knowledge related to the noise properties of the quantizer and the fact that the stability of the modulator cannot be determined without this knowledge. Research is presently ongoing and further details can be found in [23–25].

7.2.3 ΔΣ MODULATOR ARCHITECTURAL CHOICES

There are two very different ways in which to construct a ΔΣ modulator for analog-to-digital conversion: one based on a continuous-time (CT) implementation and the other based on a discrete-time (DT) implementation.

7.2.3.1 Discrete-Time versus Continuous-Time ΔΣ Modulators

Discrete-time ΔΣADCs are constructed with an S/H stage at the front-end that converts the continuous-time input signal into a sampled-data signal. This signal is then processed by a discrete-time ΔΣ modulator as shown in Figure 7.8a. The S/H is an important bottleneck in the design. As the S/H sits outside the feedback loop of the ΔΣ modulator as depicted by Figure 7.8a, its performance must be equal or better than that required by the ΔΣADC. Also, as explained earlier, as the S/H implements the sampling process, an antialiasing filter is required to precede the S/H circuit. This is necessary to eliminate potential unwanted signal appearing at the input from being aliased into the signal band. In contrast, the sampling process in a CTΔΣADC occurs within the feedback loop of the ΔΣ modulator as shown in Figure 7.8b. By the low-pass nature of the loop at this injection point, the antialiasing filter requirements can be greatly relaxed, thus saving on power and silicon area. Moreover, the performance of the S/H stage can also be greatly relaxed, as its nonlinearities will be suppressed by the loop gain of the feedback path.

A DTΔΣADC is generally constructed with switched-capacitor (SC) type circuits. SC circuits enable the realization of an integrated circuit with precise pole-zero

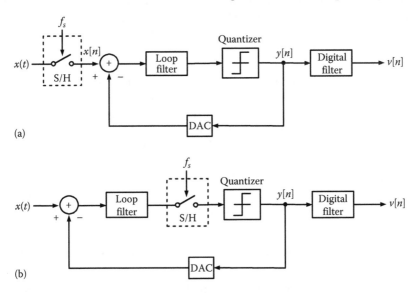

FIGURE 7.8 Block diagram of a ΔΣ modulator: (a) discrete-time and (b) continuous-time.

placement, even under larger manufacturing variations. This is made possible by realizing the poles and zeros as the ratio of two well-matched capacitors. In the case of the CT$\Delta\Sigma$ADC, the poles and zeros are realized from the product of R and C elements. As these two elements are constructed in very different ways, their product is quite sensitive to manufacturing variations. As a result, to tune the poles and zeros to their desired locations, a tuning circuit is required [26]. This significantly complicates the design of a CT$\Delta\Sigma$ADC.

An additional advantage of a CT$\Delta\Sigma$ADC is that they can operate at much higher sampling rates than a DT$\Delta\Sigma$ADC. This is due to the limited speed of operation of SC circuits. An important high-speed design issue that one must be aware when implementing a CT$\Delta\Sigma$ADC is its susceptibility to sampling clock jitter effects. Special circuit types are necessary to mitigate these effects [27,28].

In addition to the architectural issue of a discrete time versus continuous-time realization, a designer is faced with several additional architectural choices. These are: (1) selection of the order of the modulator, (2) number of levels used by the quantizer, and (3) $\Delta\Sigma$ modulator frequency response behavior. Each of these issues will be discussed in the following.

7.2.3.2 High-Order $\Delta\Sigma$ Modulators

The order of the NTF has a significant impact on the quantization noise that appears at the output of the ADC. The higher the order, the lower is the RMS output noise level as depicted in Figure 7.6. In much the same way, for a fixed modulator order, increasing the OSR can reduce the output RMS noise level. Thus, the overall $\Delta\Sigma$ modulator performance can be traded off with respect to modulator order and/or OSR.

An attractive feature of high-order $\Delta\Sigma$ modulators is that they can achieve high SNR for modest OSRs, unlike that required for low-order $\Delta\Sigma$ modulators that would require high OSRs. An additional benefit of a high-order implementation is that it is less prone to idle tones behavior, that is, unwanted oscillations. However, this comes at a price of not knowing for certain whether the modulator is stable for all possible inputs.

This uncertainty prevents its wide spread use, although different system level strategies are often used to maintain stability: for example, if an internal node voltage exceeds a pre-set limit, the $\Delta\Sigma$ modulator is reset to a known good state. This will clearly interrupt the normal operation and cause a noticeable change at the output.

Another drawback to the stability uncertainty is that it prevents the optimization of the design by overly restricting the input range to the $\Delta\Sigma$ modulator.

A method that realizes high-order NTFs and one that is inherently stable is the multistage noise-shaping (MASH) approach. This approach involves the cascade of multiple low-order, stable, $\Delta\Sigma$ modulators, combined to realize a high-order NTF as shown in Figure 7.9 for a third-order case. Three individual first-order $\Delta\Sigma$ modulators are placed in cascade and their corresponding outputs combined with a complex digital filter network, implementing the following equation

$$Y(z) = H_1(z)V_1(z) + H_2(z)V_2(z) + H_3(z)V_3(z) \qquad (7.13)$$

MASH $\Delta\Sigma$ converters work by canceling the quantization error of each preceding stage. The remaining signal after signal processing is the quantization error of

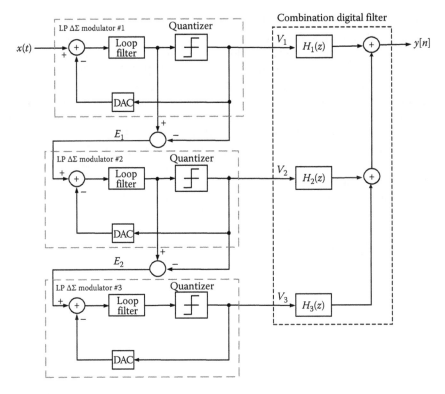

FIGURE 7.9   Block diagram of a third-order MASH $\Delta\Sigma$ modulator.

the final delta-sigma stage passed through a discrete differentiator chain—hence, providing the high-pass noise shaping effect. Note that matching between stages is critical for the successful implementation of the MASH architecture as it depends on cancellation processes.

7.2.3.3   Multibit $\Delta\Sigma$ Modulators

Replacing the single-bit quantizer by a multibit quantizer in any $\Delta\Sigma$ modulator will reduce the quantization noise level added in series with the feed-forward signal. This has beneficial consequences as it requires a lower OSR to achieve the same noise performance, or it provides higher performance for the same OSR, and it eliminates idle tones [2]. However, mismatch-induced nonlinearity and nonlinearity of the DAC in the feedback path will become the performance limitation of the design [29]. DEM or autocalibration is typically inserted in the digital feedback path between the multibit quantizer and the DAC to compensate for the mismatching issue. However, this approach can complicate the circuit design as well as add more delay in the feedback loop, limiting its high-frequency operation [26]. There are various approaches for improving DAC nonlinearity such as data-weighted averaging, mismatch-shaping, and digital correction schemes. These are discussed in great detail in [15].

7.2.3.4   Band-Pass ΔΣ Modulators

When a signal is composed entirely of AC signals whose frequencies are centered at some frequency, say f_o with bandwidth B, then a more efficient analog-to-digital conversion method is one involving a band-pass ΔΣ (BPΔΣ) modulator. Such a signal arises with RF transmitting methods that make use of the two-step heterodyning frequency translation principle known as superheterodyning. Here a baseband or low-pass signal is shifted up to an intermediate-frequency (IF) and then shifted again to the RF band. When either the IF or RF signal is to be recovered, say by a receiver, then the band-pass nature of the IF or RF signal is best converted into digital form using a BPΔΣ ADC as opposed to the low-pass ΔΣ ADC seen previously.

The critical difference between a low-pass and band-pass ΔΣ modulator lies with the frequency-selectivity of the loop filters. A BPΔΣ modulator uses a band-pass loop filter as opposed to a low-pass filter, as depicted in Figure 7.10. The rest of the elements that make up the modulator are the same as those described earlier. It is interesting to note that any Mth-order low-pass ΔΣ modulator can be transformed to a $2M$th-order BPΔΣ modulator by performing the following frequency transformation

$$z \rightarrow -z^2$$

Consequently, this transformation causes the zeros and poles to shift from DC to $f_s/4$ [16,30]. Figure 7.11 compares the PSD associated with a fourth-order BPΔΣ modulator to the PSD which was derived from a second-order low-pass ΔΣ modulator. As can be seen from this figure, the in-band noise for the order BPΔΣ modulator is nearly zero around $f_s/4$ and increases rapidly outside this region, whereas for the low-pass ΔΣ modulator, its PSD is nearly zero at DC and rises rapidly beyond this.

The primary difference between the realization of a CT and DT band-pass ΔΣADCs is in the structure of the resonators. Different topologies are presented for DT band-pass ΔΣ modulators such as forward-Euler (FE), lossless discrete integrator (LDI), and double delay (DD) resonators. In contrast, Gm-C resonator, active-RC resonator, and LC tank resonators are used as CT resonators [15].

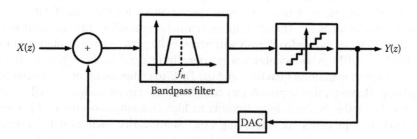

FIGURE 7.10   Block diagram of a BPΔΣ modulator.

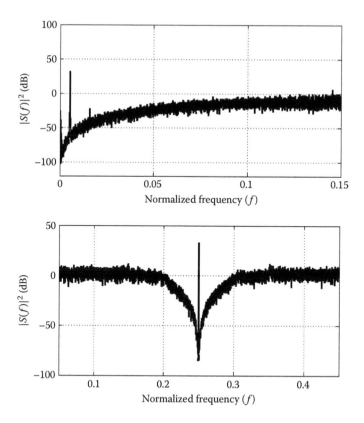

FIGURE 7.11 Comparing the PSD associated with a second-order low-pass and a fourth-order band-pass $\Delta\Sigma$ modulator.

7.3   FIRST-ORDER SINGLE-BIT TIME-MODE $\Delta\Sigma$ MODULATORS

It is the intent of this section to learn about various $\Delta\Sigma$ modulators implemented using TMSP techniques that have been published in recent years. The various architectures have been classified according to either their key building block or some distinguishing architectural feature. This will include a discussion on a VCDU-based $\Delta\Sigma$ modulator, an open-loop $\Delta\Sigma$ modulator design approach, a DLL-based $\Delta\Sigma$ modulator, and a TDC-based $\Delta\Sigma$ modulator.

7.3.1   VCDU-BASED TM$\Delta\Sigma$ MODULATORS

Let us begin by considering a first-order $\Delta\Sigma$ modulator with sampled-data input signal, $v_{in}[n]$, whereby the loop filter is implemented with a first-order discrete-time integrator, an analog summer, a quantizer with a 1-linear bit ADC with output $D_{out}[n]$, and a 1-bit DAC in the feedback path as shown in Figure 7.12a. Assuming a linear model for the quantizer with error $e[n]$, a first-order $\Delta\Sigma$ modulator can be described with the block diagram shown in Figure 7.12b. Writing the output

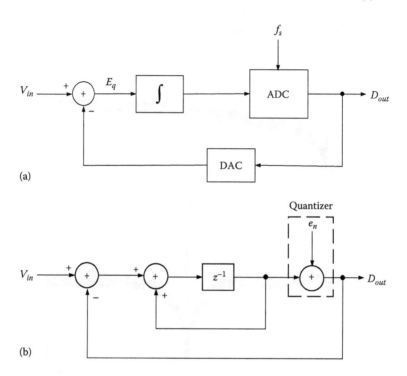

FIGURE 7.12   (a) Block diagram of a first-order $\Delta\Sigma$ modulator, (b) Linear model of operation.

signal $D_{out}[n]$ in terms of the two inputs $v_{in}[n]$ and $e[n]$, one obtains the following time-difference equation

$$D_{out}\left[n\right] = v_{in}\left[n-1\right] + \left(e\left[n\right] - e\left[n-1\right]\right) \tag{7.14}$$

The aforementioned equation reveals the error feedback nature of the $\Delta\Sigma$ modulator whereby the digital output is a sum of the input and the first-order time difference of the quantizer error. If the quantizer errors $e[n]$ and $e[n-1]$ are similar, that is, through oversampling, their contribution will be small and the output D_{out} will be a very good approximation to the input v_{in}.

An approach introduced in [8] that implements a similar first-order difference equation is shown in Figure 7.13. Here a noise-shaped error behavior is realized by two voltage-controlled ring oscillators that perform phase integration followed by a D-type flip-flop and some digital inverters. The two ring oscillators are constructed using two sets of VCDUs whose output is fed back to its input via a single inverter circuit.

The D-type flip-flop is employed as a one-bit quantizer.

In this design, the period of oscillation of the bottom-most ring oscillator is governed by the reference voltage, V_{ref}. By keeping this quantity fixed to a constant value, the period or frequency of this oscillator is held constant. The period of the top-most integrator is controlled by the input voltage $v_{in}[n]$ and the digital output

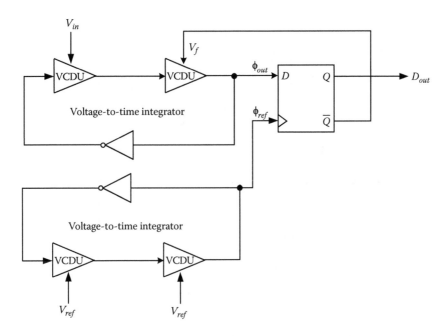

FIGURE 7.13   Time-mode single-ended $\Delta\Sigma$ modulator with VCDU. (From Taillefer, C.S. and Roberts, G.W., *IEEE Trans. Circuits Systems I: Regular Papers*, 56(9), 1908, 2009.)

of the modulator through the two input control terminals via the VCDUs. The difference in phase of the leading edge of the two ring oscillators is compared by the D-type flip-flop. If the phase of the reference oscillator output denoted by ϕ_{ref} lags the phase of the output of the input-controlled oscillator ϕ_{out}, the flip-flop will output logic 1, otherwise it will output logic 0. As the output of the flip-flop is also fed back to control the oscillation frequency of the top oscillator, a noise-shaping action occurs. Following the mathematical development given in [8], the difference equation between the input and output of the $\Delta\Sigma$ modulator is found to be

$$D_{out}\left[n\right] = v_{in}\left[n-1\right] + \frac{T_e\left[n\right] - T_e\left[n-1\right]}{G_\varphi} \tag{7.15}$$

where

G_φ is the voltage-to-time conversion factor of the VCDUs
$T_e[n]$ is the error sequence made by D-type flip-flop

When compared to the difference equation derived for the first-order modulator provided in Equation 7.14, one finds that they have similar form.

The single-ended first-order TM$\Delta\Sigma$ modulator of Figure 7.13 can easily be extended to a fully differential implementation [8] as shown in Figure 7.14. Here the input signal is fully differential where the positive half is applied to the top ring oscillator, and the negative half is applied to the other. In addition, a differential signal is fed back from the Q and Q-bar output of D-type flip-flop. A fully differential implementation

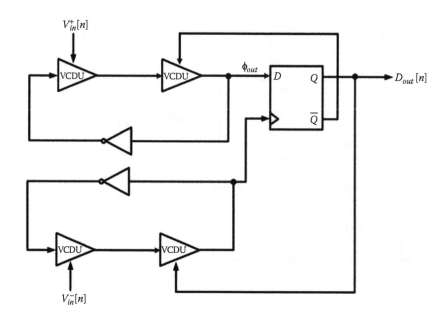

FIGURE 7.14 Differential-input first-order time-mode $\Delta\Sigma$ modulator. (From Taillefer, C.S. and Roberts, G.W., *IEEE Trans. Circuits Syst. I: Reg. Papers*, 57(9), 2404, 2010.)

offers some advantages over the single-ended approach, namely: reduced sensitivity to common-mode noise, SNR improvement by doubling of the signal amplitude, improved linearity, and elimination of the reference voltage [8]. However, a disadvantage of the fully differential implementation is an increase in the amount of nonuniform sampling on account of the elimination of the reference voltage [8].

7.3.2   OPEN-LOOP TM$\Delta\Sigma$ MODULATORS

Another approach to perform analog-to-digital conversion is based on the application of a VCO or a gated-ring oscillator (GRO) running in an open-loop manner followed by a digital counter or filter circuit [9,10,32–34]. These time-based architectures have some interesting features. For one, they all employ noise shaping but in a very simple and direct manner. Because of their simplicity, they require little power and are easily adapted to advanced CMOS processes. In this section, four types of open-loop approaches for TM $\Delta\Sigma$ modulation will be described: (1) VCO-based Open-Loop TM$\Delta\Sigma$ modulator, (2) GRO-based Open-Loop TM$\Delta\Sigma$ modulator, (3) vernier GRO-Based $\Delta\Sigma$ modulator, and (4) switched-ring oscillator (SRO)-based $\Delta\Sigma$ modulator.

7.3.2.1   VCO-Based Open-Loop TM$\Delta\Sigma$ Modulator

A VCO-based open-loop $\Delta\Sigma$ADC is shown in Figure 7.15, which consists of a front-end S/H stage, followed by a ring-VCO and a counter, register, and some logic [12].

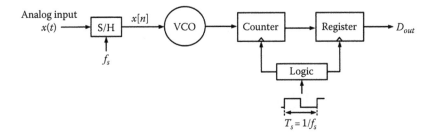

FIGURE 7.15 VCO-based $\Delta\Sigma$ADC. (From Yuan, F., *Analog Integr. Circuits Signal Process.*, 79(2), 191, 2014; Straayer, M.Z. and Perrott, M.H., *IEEE J. Solid State Circuits*, 43(4), 805, 2008.)

Here the output digital count $D_{out}[n]$ represents discrete samples of the input signal of $x[n]$ in some general form such as

$$D_{out}\left[n\right] = a \cdot x\left[n\right] + b \tag{7.16}$$

where a and b are arbitrary coefficients.

The basic principle of this system is that the oscillation frequency of the VCO is set at each sampling instant based on the sampled input value $x[n]$. The counter then counts the number of rising edge transitions associated with the VCO output in the sampling period, T_s. At the end of each sampling phase, the total transition count is latched into the output register and presented as the digital output $D_{out}[n]$. The counter is then reset at the start of the next sampling phase to begin the count all over again.

A timing diagram illustrating the internal action of the VCO-based open-loop $\Delta\Sigma$ADC is shown in Figure 7.16, where the input signal and the VCO input and output voltage signals are seen as on an oscilloscope. Also in the plot are the count pulses that correspond to the VCO output crossing the phase threshold of 2π, whereby the total number of count pulses minus one indicates the number of cycles the VCO output completes during the sampling period; we assign this value as the output digital value $D_{out}[n]$.

Also shown in this diagram is the instantaneous frequency and phase of the VCO output as a function of time. During any sampling phase, with the input to the VCO held constant by the input sample $x[n]$, its output will be a clock signal with a specific, but constant, frequency value given by

$$f_{VCO}\left[n\right] = K_{VCO} \cdot x\left[n\right] + f_{FR} \tag{7.17}$$

where
   K_{VCO} represents the voltage-to-frequency gain coefficient expressed in units of Hz/V
   f_{FR} represents the free-running oscillation frequency of the VCO expressed in Hz

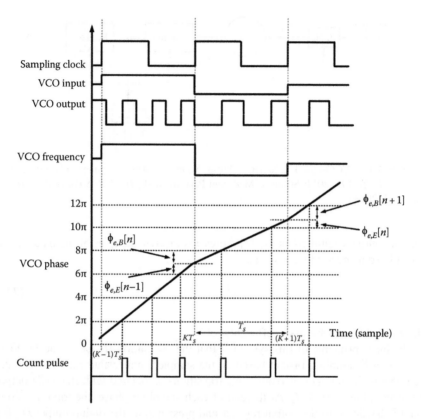

FIGURE 7.16 Timing diagram for the VCO-based $\Delta\Sigma$ADC.

The instantaneous VCO frequency $f_{VCO}(t)$ and phase $\phi_{VCO}(t)$ are related through the derivative operation:

$$f_{VCO}(t) = \frac{1}{2\pi} \frac{d\phi_{VCO}(t)}{dt} \tag{7.18}$$

Therefore, the change in the instantaneous phase over the sampling period T_s at any sampling instant can be approximated as

$$\Delta\phi_{VCO}[n] = 2\pi f_{VCO}[n]T_s \tag{7.19}$$

Substituting Equation 7.17 into the aforementioned equation allows the discrete-time phase change to be written as

$$\Delta\phi_{VCO}[n] = 2\pi K_{VCO}T_s \cdot x[n] + 2\pi f_{FR}T_s \tag{7.20}$$

Referring back to the VCO output phase behavior shown in Figure 7.16, one can also write the same phase change $\Delta\phi_{VCO}[n]$ as a number of full cycles 2π phases changes and a small phase error $\phi_e[n]$ as follows

$$\Delta\phi_{VCO}\left[n\right] = 2\pi \cdot D_{out}\left[n\right] + \phi_e\left[n\right] \qquad (7.21)$$

Equating Equations 7.20 and 7.21 leads to

$$D_{out}\left[n\right] = K_{VCO} \cdot T_s \cdot x\left[n\right] + f_{FR} \cdot T_s - \frac{1}{2\pi}\phi_e\left[n\right] \qquad (7.22)$$

The phase error $\phi_e[n]$ at any sampling instant consists of two components: a start and stop phase error, or what we shall refer to in this chapter as the begin and end phase error. To distinguish each component from one another, we shall denote the start or begin phase error with an additional subscript B appended to the phase error term and write it as $\phi_{e,B}[n]$. Correspondingly, the stop or end phase error will be described with an additional subscript E and write $\phi_{e,E}[n]$. These phase error terms can be seen in Figure 7.16, allowing one to write during any sampling instant

$$\phi_e\left[n\right] = \phi_{e,B}\left[n\right] + \phi_{e,E}\left[n\right] \qquad (7.23)$$

In addition, we also observe from the phase plot that the sum of the stop/end phase error during the sampling instant $[n-1]$ and the start/begin phase error at sampling instant $[n]$ must equal 2π. Hence, we can write

$$\phi_{e,E}\left[n-1\right] + \phi_{e,B}\left[n\right] = 2\pi \qquad (7.24)$$

Substituting this back into the total phase error expression of Equation 7.23 leads to

$$\phi_e\left[n\right] = 2\pi + \phi_{e,E}\left[n\right] - \phi_{e,E}\left[n-1\right] \qquad (7.25)$$

Here we see how the total phase error $\phi_e[n]$ depends on the difference in the stop or end phase errors at adjacent sampling instants. This should remind the reader of the difference equations related to $\Delta\Sigma$ modulation and the corresponding noise-shaping effect.

Armed with earlier result, the output digital code $D_{out}[n]$ can be rewritten as

$$D_{out}\left[n\right] = K_{VCO} \cdot T_s \cdot x\left[n\right] + f_{FR} \cdot T_s - 1 - \frac{1}{2\pi}\left(\phi_{e,E}\left[n\right] - \phi_{e,E}\left[n-1\right]\right) \qquad (7.26)$$

If we define the reference count $D_{out,Ref}$ at $x[n] = 0$, then

$$D_{out,Ref} = f_{FR} \cdot T_s - 1 \qquad (7.27)$$

and the output count relative to the reference can be written as

$$\Delta D_{out}\left[n\right] \equiv D_{out}\left[n\right] - D_{out,Ref} = K_{VCO} \cdot T_s \cdot x\left[n\right] - \frac{1}{2\pi}\left(\phi_{e,E}\left[n\right] - \phi_{e,E}\left[n-1\right]\right) \quad (7.28)$$

The corresponding z-transform of the calibrated output count becomes

$$\Delta D_{out}\left(z\right) = K_{VCO} \cdot T_s \cdot X\left(z\right) - \frac{1}{2\pi}\left(1 - z^{-1}\right)\Phi_{e,E}\left(z\right) \quad (7.29)$$

where $\Phi_{e,B}(z)$ is the z-transform of the forward phase error sequence, $\phi_{e,E}[n]$. The aforementioned equation reveals the first-order noise shaping, as the phase error term is weighted by the frequency dependent term, $(1 - z^{-1})$.

As the step size of the counter/quantizer is 2π, the PSD of the forward phase error sequence in rad²/Hz can be estimated as

$$N_{\Phi_{e,F}} = \frac{2\pi^2}{3f_s} \quad (7.30)$$

The phase error spread is inversely proportional to the sampling frequency f_s. Therefore, to maximize the noise-shaping benefit, the bandwidth of the incoming signal $x(t)$ should be small in comparison to f_s.

7.3.2.1.1 Improving the Nonlinear Behavior

A major issue associated with the VCO-based $\Delta\Sigma$ADC design is its nonlinear operation. While the VCO was assumed to be linear with respect to its voltage input, a more accurate representation is to assume the VCO has the following transfer characteristic

$$f_{VCO}\left[n\right] = f_{FR} + K_{VCO,1} \cdot x\left[n\right] + K_{VCO,2} \cdot x^2\left[n\right] + K_{VCO,3} \cdot x^3\left[n\right] + \cdots \quad (7.31)$$

where the coefficients, $K_{VCO,1}$ ... $K_{VCO,3}$, ..., represent the terms of the power series expansion around the operating point of the input–output behavior of the VCO. Substituting the aforementioned equation into Equation 7.20, one can write the output phase-difference $\Delta\phi_{VCO}[n]$ as

$$\Delta\phi_{VCO}\left[n\right] = 2\pi f_{FR} T_s + 2\pi T_s K_{VCO,1} \cdot x\left[n\right]$$
$$+ 2\pi T_s K_{VCO,2} \cdot x^2\left[n\right] + 2\pi T_s K_{VCO,3} \cdot x^3\left[n\right] + \cdots \quad (7.32)$$

By transforming the input $x[n]$ into a positive and negative version and applying each one to a separate VCO leads to the following two output phase difference terms, denoted as $\Delta\phi^+_{vco}[n]$ and $\Delta\phi^-_{vco}[n]$, one for the positive input as

$$\Delta\phi^+_{VCO}\left[n\right]=2\pi f_{FR}T_s+2\pi T_s K_{VCO,1}\cdot x\left[n\right]+2\pi T_s K_{VCO,2}\cdot x^2\left[n\right]$$
$$+\;2\pi T_s K_{VCO,3}\cdot x^3\left[n\right]+\cdots \tag{7.33}$$

and the other for the negative input as

$$\Delta\phi^-_{VCO}\left[n\right]=2\pi f_{FR}T_s+2\pi T_s K_{VCO,1}\cdot\left(-x\left[n\right]\right)+2\pi T_s K_{VCO,2}\cdot\left(-x\left[n\right]\right)^2$$
$$+\;2\pi T_s K_{VCO,3}\cdot\left(-x\left[n\right]\right)^3+\cdots \tag{7.34}$$

Adding a stage that takes the difference in these two phases, that is,

$$\Delta\phi_{VCO}\left[n\right]=\Delta\phi^+_{VCO}\left[n\right]-\Delta\phi^-_{VCO}\left[n\right] \tag{7.35}$$

the corresponding output phase difference depends only on the odd-order terms, thereby reducing the overall distortion level, that is,

$$\Delta\phi_{VCO}\left[n\right]=0+2\cdot2\pi T_s K_{VCO,1}\cdot x\left[n\right]+0+2\cdot2\pi T_s K_{VCO,3}\cdot x^3\left[n\right]+\cdots \tag{7.36}$$

Counting the corresponding 2π phase changes, one obtains the system shown in Figure 7.17 where the digital count value $D_{out}[n]$ becomes

$$D_{out}\left[n\right]=2\cdot K_{VCO,1}\cdot T_s\cdot x\left[n\right]+2\cdot K_{VCO,3}\cdot T_s\cdot x^3\left[n\right]$$
$$+\;f_{FR}\cdot T_s-1-\frac{1}{2\pi}\left(\phi_{e,E}\left[n\right]-\phi_{e,E}\left[n-1\right]\right) \tag{7.37}$$

Further balancing in the signal path can be performed with digital calibration techniques [41] implemented in the logic block of Figure 7.17. Figure 7.18 displays the output PSD for a VCO $\Delta\Sigma$ modulator implemented with both single-ended and differential configurations, with and without digital calibration. As is evident from Figure 7.18, the differential implementation reduced the even-order distortion terms below the noise level of the $\Delta\Sigma$ modulator, while the digital calibration further reduces the third-order distortion component.

A main drawback of this approach is of course related to the increase in hardware and power, as two parallel VCOs and a phase differencing circuit are required.

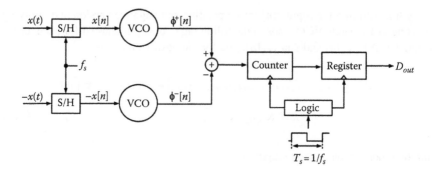

FIGURE 7.17 Differential configuration of VCOΔΣADC. (From Yuan, F., *Analog Integr. Circuits Signal Process.*, 79(2), 191, 2014; Daniels, J. et al., All-digital differential VCO-based A/D conversion, *Proceedings of 2010 IEEE International Symposium on Circuits and Systems (ISCAS)*, France, 2010.)

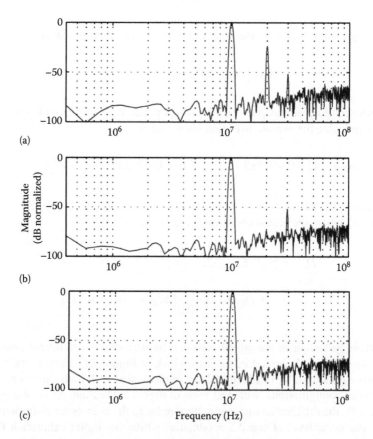

FIGURE 7.18 Measurement results of a VCOΔΣADC: (a) single-ended, (b) differential, and (c) differential with digital calibration. (From Daniels, J. et al., All-digital differential VCO-based A/D conversion, *Proceedings of 2010 IEEE International Symposium on Circuits and Systems (ISCAS)*, France, 2010.)

7.3.2.1.2 Increasing the Maximum Sampling Rate, f_s

The basic principle behind a VCO-based $\Delta\Sigma$ modulator is that an input signal $x[n]$ is encoded into the frequency of the VCO output, $f_{VCO}[n]$. By measuring the number of cycles that the VCO output completes during the sampling period T_s allows one to estimate the frequency of the VCO and hence to recover the input samples. As the output of the VCO completes $2\pi D_{out}[n]$ radians of phase change during the sampling period, T_s, the discrete-time VCO frequency is estimated from

$$f_{VCO}[n] = \frac{1}{2\pi}\frac{\Delta\phi_{VCO}[n]}{T_s} = \frac{D_{out}[n]}{T_s} \tag{7.38}$$

The ultimate speed of this operation depends on the speed at which the counter can count. In practice, this limits the sampling rate of the VCO-based $\Delta\Sigma$ modulator to relative low-frequency operation. Instead, one can use a phase discriminator and estimate a change in the VCO output phase in a shorter time, thereby increasing the maximum sampling rate of the VCO-based $\Delta\Sigma$ modulator. Figure 7.19a illustrates the block diagram of this arrangement. It is essentially the same as the implementation in Figure 7.15, except that a phase discriminator circuit shown in Figure 7.19b replaces the counter. The phase discriminator generates a pulse with width T_d that, when normalized by the sampling period T_s, is proportional to the VCO output frequency. The phase discriminator is made from two D-type flip-flops and an XOR gate. As the propagation delay of this logic gate combination is extremely short in comparison to that of an N-bit counter, this circuit can operate at much higher sampling rates than a counter-phase discriminator circuit.

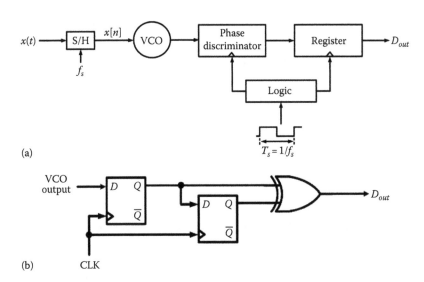

(a)

(b)   CLK

FIGURE 7.19 Using a phase discriminator instead of a counter to extract the VCO phase changes (a) block diagram, (b) phase discriminator circuit. (From Yuan, F., *Analog Integr. Circuits Signal Process.*, 79(2), 191, 2014.)

7.3.2.2   GRO-Based Open-Loop TMΔΣ Modulators

A GRO-based TMΔΣ modulator quantizes a time-difference interval, for example, time between a start and stop edge transition, rather than some input voltage. Before a discussion about GRO-based TMΔΣ modulator, consider the ring oscillator–based (RO-based) TDC [44,45] shown in Figure 7.20a. The RO-based TDC approach consists of four blocks: a ring oscillator, counter, register, and some logic gates. The operation of this design is to count the number of clock cycles of the ring oscillator between the time interval defined by the start and stop signals. Once again, the output count will be denoted as $D_{out}[n]$ and the input time-difference interval as $t_m[n]$. This design is essentially identical to that described for VCO-based open-loop ΔΣADC shown in Figure 7.15 with input signal $x[n] = 0$ as depicted in Figure 7.20b; albeit, there is a time-lag between the instant a count value is ready and the next input sample can be ready for conversion.

Following a similar mathematical development as for the VCO-based open-loop ΔΣADC, consider that the ring oscillator oscillates at some free-running frequency f_{FR}. As the instantaneous frequency of the ring oscillator output is equal to the derivative of the phase, the change in phase output over the duration of the start-stop interval $t_m[n]$ can be described as

$$\Delta\phi_{VCO}\left[n\right] = 2\pi f_{FR} \cdot t_m\left[n\right] \tag{7.39}$$

This phase change corresponds to the 2π multiples of the count $D_{out}[n]$, with some quantization error $\phi_e[n]$, which is equivalent to the sum of a start/begin and stop/end phase error component as shown in Figure 7.21, that is, $\phi_e[n] = \phi_{e,B}[n] + \phi_{e,E}[n]$. However, as derived earlier, $\phi_{e,B}[n] + \phi_{e,E}[n-1] = 2\pi$ due to the modulo phase operation of the counter, allowing one to write

$$\Delta\phi_{VCO}\left[n\right] = 2\pi D_{out}\left[n\right] + \left(2\pi + \phi_{e,E}\left[n\right] - \phi_{e,E}\left[n-1\right]\right) \tag{7.40}$$

Combining Equations 7.39 and 7.40,

$$D_{out}\left[n\right] = f_{FR} \cdot t_m\left[n\right] - 1 - \frac{1}{2\pi}\left(\phi_{e,E}\left[n\right] - \phi_{e,E}\left[n-1\right]\right) \tag{7.41}$$

If we define the reference count $D_{out,Ref}$ for some input time-difference reference condition, say $t_{m,Ref}$, then

$$D_{out,Ref} = f_{FR} \cdot t_{m,Ref} - 1 \tag{7.42}$$

Subtracting Equation 7.42 from Equation 7.41, one arrives at the change in the output count relative to some reference pulse width, that is,

$$\Delta D_{out}\left[n\right] = D_{out}\left[n\right] - D_{out,Ref} = f_{FR} \cdot \left(t_m\left[n\right] - t_{m,Ref}\right)$$

$$- \frac{1}{2\pi}\left(\phi_{e,E}\left[n\right] - \phi_{e,E}\left[n-1\right]\right) \tag{7.43}$$

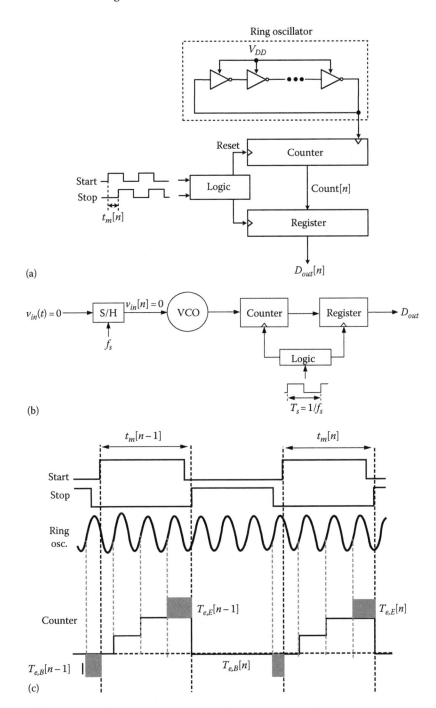

FIGURE 7.20 RO-based TDC: (a) block diagram, and (b) equivalency to the VCO-based open-loop ΔΣADC with the input set to 0, and (c) timing diagram with start/begin and stop/end phase errors highlighted.

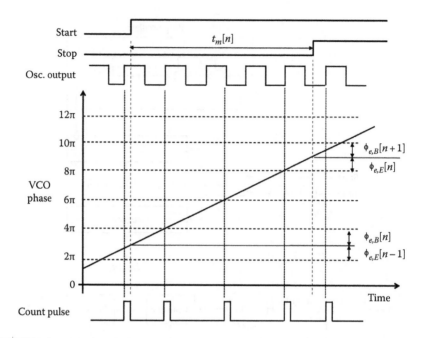

FIGURE 7.21 Timing diagram of the ring oscillator–based TDC with start/begin and stop/end phase errors highlighted.

The corresponding z-transform of the calibrated output count becomes

$$\Delta D_{OUT}\left(z\right) = K_{VCO}\cdot T_s\cdot\Delta T_m\left(z\right) - \frac{1}{2\pi}\left(1-z^{-1}\right)\Phi_{e,E}\left(z\right) \qquad (7.44)$$

where
   $\Delta T_m(z)$ represents the z-transform of the input time change
   $\Phi_{e,E}(z)$ is the z-transform of the stop or end phase error sequence, $\phi_{eEF}[n]$

Once again, we observe that the quantization error is noise shaped by the factor $(1 - z^{-1})$. As the step size of the counter/quantizer is again 2π, the PSD of the end phase error sequence in rad²/Hz is identical to that described by Equation 7.30.

The main reason for the quantization error reduction is that start/begin and stop/end phase errors are highly correlated, so that their combined sum is reduced with averaging. In practice, this is not the case with an RO-based TMΔΣ modulator. Rather, some time must be allotted to account for the time to make a decision and to read and write the data into the appropriate registers, and make sure the circuit is ready for next sampling phase. As a result, the phase of the oscillator will have changed before the start of the next sampling phase, resulting in a misalignment in the start/begin and stop/end quantization errors.

The GRO-based TMΔΣ modulator approach [44,45] enables the phase of the ring oscillator to be reset (i.e., through a gated operation) to the value it had at the end of the previous sampling instant and restored some time later so that the stop or end quantization error of the previous time sample is the same as the start/begin quantization error of the next time sample, as illustrated in Figure 7.20c. It does this by disabling the ring oscillator for a complete integer number of clock cycles from the last sampling instant.

The GRO-based TDC approach presented in [44–46] goes beyond a single ring oscillator with an enabled operation, as it uses a more complicated clock scheme. In this work, K ring oscillator TDCs are run in parallel and the output of each is combined and averaged as depicted in Figure 7.22. As will be shown shortly, the output count $D_{out}[n]$ will be identical to that for a single RO-based except that the mismatch phase errors ($\Delta\phi_{e,O}[n]$) between the various oscillators will be noise-shaped, that is,

$$D_{out}\left[n\right] = f_{FR} \cdot t_m\left[n\right] - 1 - \frac{1}{2\pi}\left(\phi_{e,E}\left[n\right] - \phi_{e,E}\left[n-1\right]\right)$$

$$+ \frac{1}{2\pi}\left(\phi_{e,O}\left[n\right] - \Delta\phi_{e,O}\left[n-1\right]\right) \tag{7.45}$$

Let us begin by assuming that K RO-based TDCs are driven by the same input start and stop signals whose time-difference is $t_m[n]$. If each RO-based TDC is driven by the same clock input, then the digital output for all K modulators would

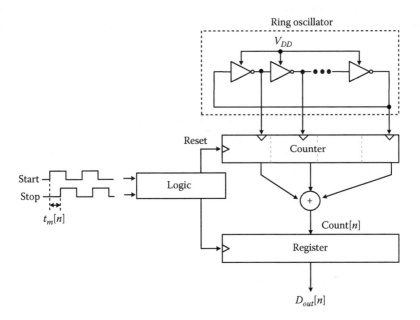

FIGURE 7.22 Block diagram of the RO-based TMΔΣ modulator.

be identical. However, because each RO-based TDC is driven with a phase-shifted clock derived from a ring oscillator, the calibrated digital output for each could be different on account of different phase quantization errors, that is,

$$\Delta D_{out,1}\left[n\right] = f_{FR} \cdot \left(t_m\left[n\right] - t_{m,Ref}\right) - \frac{1}{2\pi}\left(\phi_{e,E,1}\left[n\right] - \phi_{e,E,1}\left[n-1\right]\right)$$

$$\Delta D_{out,2}\left[n\right] = f_{FR} \cdot \left(t_m\left[n\right] - t_{m,Ref}\right) - \frac{1}{2\pi}\left(\phi_{e,E,2}\left[n\right] - \phi_{e,E,2}\left[n-1\right]\right)$$

$$\vdots$$

$$\Delta D_{out,K}\left[n\right] = f_{FR} \cdot \left(t_m\left[n\right] - t_{m,Ref}\right) - \frac{1}{2\pi}\left(\phi_{e,E,K}\left[n\right] - \phi_{e,E,K}\left[n-1\right]\right) \quad (7.46)$$

Averaging the outputs of all K TDCs results in

$$\Delta D_{out}\left[n\right] = \frac{1}{K}\sum_{i=1}^{K}\Delta D_{out,i}\left[n\right] = f_{FR} \cdot \left(t_m\left[n\right] - t_{m,Ref}\right)$$

$$-\frac{1}{K}\frac{1}{2\pi}\sum_{i=1}^{K}\left(\phi_{e,E,i}\left[n\right] - \phi_{e,E,i}\left[n-1\right]\right) \quad (7.47)$$

While these phase error terms, $\phi_{e,E,1}$, $\phi_{e,E,2}$, ..., $\phi_{e,E,N}$, are assumed to be independent and uncorrelated with the input signal $t_m[n]$, they are not uncorrelated with themselves. In fact, it is easy to show that the phase difference between adjacent stop/end phase errors is equal to the phase difference of adjacent outputs from the ring oscillator, that is,

$$\phi_{e,E,i}\left[n\right] - \phi_{e,E,i+1}\left[n\right] = \frac{2\pi}{K} \quad (7.48)$$

To illustrate this relationship, the instantaneous phase of the four clock signals derived from a four-phase ring oscillator (i.e., $K = 4$) is shown in Figure 7.23. Here we see the phase of each output shifted with respect to one another by $\pi/4$ rad. We also see that the phase difference between the two stop/end phase errors associated with osc. 1 and osc. 2 at time instant $(n-1)$ is

$$\phi_{e,E,1}\left[n-1\right] - \phi_{e,E,2}\left[n-1\right] = \frac{\pi}{4} \quad (7.49)$$

The aforementioned equation is also true for any time instant, n. Moreover, for K greater than 2, one can also observe that

$$\phi_{e,E,p}\left[n\right] = \phi_{e,E,K}\left[n\right] + \frac{2\pi}{K}(K-p) \quad p = 1,2...(K-1) \quad (7.50)$$

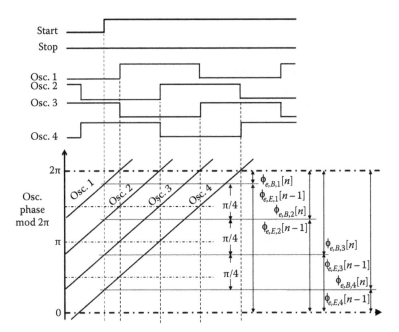

FIGURE 7.23 Illustrating the correlation of the start/begin and stop/end phase errors around the start transition for a four-phase RO-based TDC ($K = 4$). A similar plot can be drawn around the stop transition.

leading to the general result

$$\sum_{i=1}^{K}\left(\phi_{e,E,i}\left[n\right]-\phi_{e,E,i}\left[n-1\right]\right)=K\left(\phi_{e,E,4}\left[n\right]-\phi_{e,E,4}\left[n-1\right]\right) \qquad (7.51)$$

Hence, substituting Equation 7.51 back into Equation 7.47, one finds

$$\Delta D_{out}\left[n\right]=f_{FR}\cdot\left(t_{m}\left[n\right]-t_{m,Ref}\right)-\frac{1}{2\pi}\left(\phi_{e,E,4}\left[n\right]-\phi_{e,E,4}\left[n-1\right]\right) \qquad (7.52)$$

Therefore, averaging the digital outputs of K RO-based driven from a single ring oscillator with uniform phase delay across its taps provides identical performance to a single RO-based.

In practice, having a ring oscillator with all K-phases with equal delay is a poor assumption, as the individual delays are sensitive to process errors during manufacture. The effect of these mismatch phase errors will introduce an additional phase error into Equation 7.52 having the general form

$$\Delta D_{out}\left[n\right]=f_{FR}\cdot\left(t_{m}\left[n\right]-t_{m,Ref}\right)-\frac{1}{2\pi}\left(\phi_{e,E,4}\left[n\right]-\phi_{e,E,4}\left[n-1\right]\right)$$

$$+\frac{1}{2\pi}\phi_{e,O}\left[n\right] \qquad (7.53)$$

where $\phi_{e,o}[n]$ represents the combined effect of the mismatch phase offset between each clock phase as a function of the sampling index, n.

To illustrate this, consider the timing diagram for a four-phase RO-based shown in Figure 7.24 with unequal phase delays. In Figure 7.24a, the phase errors are shown around the start transition and in Figure 7.24b, a similar diagram is shown around the stop transition. The phase delay offset between clock phase 1 and clock phase 2 is denoted as $\phi_{e,O,1}$, between phase 2 and clock phase 3 is denoted as $\phi_{e,O,2}$, between phase 3 and clock phase 4 is denoted as $\phi_{e,O,3}$ and between phase 4 and clock phase 1 is denoted as $\phi_{e,O,4}$.

Consequentially, writing the output of each RO-based TDC with the phase offsets in mind, one arrives at

$$\Delta D_{out,1}\big[n\big] = f_{FR}\cdot\big(t_m\big[n\big]-t_{m,Ref}\big)-\frac{1}{2\pi}\big(\phi_{e,E,4}\big[n\big]-\phi_{e,E,4}\big[n-1\big]\big)+\frac{1}{2\pi}\big(\phi_{e,O,4}-\phi_{e,O,1}\big)$$

$$\Delta D_{out,2}\big[n\big] = f_{FR}\cdot\big(t_m\big[n\big]-t_{m,Ref}\big)-\frac{1}{2\pi}\big(\phi_{e,E,4}\big[n\big]-\phi_{e,E,4}\big[n-1\big]\big)+\frac{1}{2\pi}\big(\phi_{e,O,3}-\phi_{e,O,1}\big)$$

$$\Delta D_{out,3}\big[n\big] = f_{FR}\cdot\big(t_m\big[n\big]-t_{m,Ref}\big)-\frac{1}{2\pi}\big(\phi_{e,E,4}\big[n\big]-\phi_{e,E,4}\big[n-1\big]\big)+\frac{1}{2\pi}\big(\phi_{e,O,1}+\phi_{e,O,2}\big)$$

$$\Delta D_{out,4}\big[n\big] = f_{FR}\cdot\big(t_m\big[n\big]-t_{m,Ref}\big)-\frac{1}{2\pi}\big(\phi_{e,E,4}\big[n\big]-\phi_{e,E,4}\big[n-1\big]\big) \qquad (7.54)$$

After averaging, the digital output count becomes

$$\Delta D_{out}\big[n\big] = \frac{1}{K}\sum_{i=1}^{K}\Delta D_{out,i}\big[n\big] = f_{FR}\cdot\big(t_m\big[n\big]-t_{m,Ref}\big)-\frac{1}{2\pi}\sum_{i=1}^{K}\big(\phi_{e,E,4}\big[n\big]-\phi_{e,E,4}\big[n-1\big]\big)$$

$$+\frac{1}{2\pi}\frac{\big(\phi_{e,O,4}-\phi_{e,O,1}+\phi_{e,O,3}+\phi_{e,O,2}\big)}{4} \qquad (7.55)$$

In this particular case, the calibrated output digital code $\Delta D_{out}[n]$ now contains a term involving the phase offsets, a term that did not exist with the perfectly phase-aligned RO-based expression shown in Equation 7.47. Now for different input samples $t_m[n]$, different phase offsets will contribute to the overall phase offset term resulting in a sequence of offset phase errors, $\phi_{e,o}[n]$. A four-phase ring oscillator TDC will have a total phase error that will appear in general as

$$\phi_{e,o}\big[n\big] = h_1\big[n\big]\cdot\phi_{e,O,1}+h_2\big[n\big]\cdot\phi_{e,O,2}+h_3\big[n\big]\cdot\phi_{e,O,3}+h_4\big[n\big]\cdot\phi_{e,O,4} \qquad (7.56)$$

where $h_i[n]$ is a sequence of weighted values.

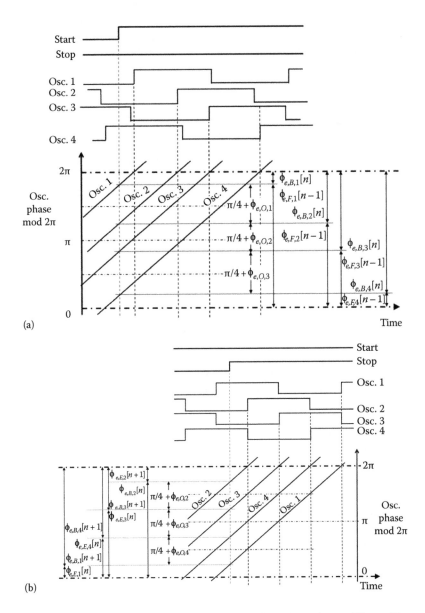

FIGURE 7.24 Accounting for mismatch errors in a four-phase ring oscillator: (a) errors around the start transition, (b) errors around the stop transition.

A number of methods have been proposed and implemented for counteracting the effects of mismatches in digital-to-analog converter circuits. These same methods are applicable to a ring oscillator–based TDCs. The principle of these offset correction methods is to manipulate the internal operation of the ring oscillator so that the sum of the total phase $\phi_{e,o}[n]$ at the output of the TDC tends to zero as n goes to infinity. From Equation 7.56, the sum of the total phase error can be expressed as

$$\sum_{n=1}^{\infty} \phi_{e,o}[n] = \left(\sum_{n=1}^{\infty} h_1[n]\right)\phi_{e,O,1} + \left(\sum_{n=1}^{\infty} h_2[n]\right)\phi_{e,O,2}$$

$$+\left(\sum_{n=1}^{\infty} h_3[n]\right)\phi_{e,O,3} + \left(\sum_{n=1}^{\infty} h_4[n]\right)\phi_{e,O,4} \quad (7.57)$$

Due to the symmetry of the clock phases, the sum of the phase offsets must sum to zero at any sampling instant, n, that is,

$$\phi_{e,O,1} + \phi_{e,O,2} + \phi_{e,O,3} + \phi_{e,O,4} = 0 \quad (7.58)$$

Therefore to achieve $\sum_{n=1}^{\infty} \phi_{e,o}[n] = 0$, the following conditions must be met

$$\sum_{n=1}^{\infty} h_1[n] = \sum_{n=1}^{\infty} h_2[n] = \sum_{n=1}^{\infty} h_3[n] = \sum_{n=1}^{\infty} h_4[n] \quad (7.59)$$

A well-known method used to achieve the aforementioned minimum total phase error due to phase mismatches is the barrel-shift algorithm that performs DEM [45]. The basic idea with this technique is to sequentially select the elements that contribute to each phase mismatch error such that over some period of time, each mismatch error has contributed an almost equal amount. The averaged output phase accumulated error made at the p-th sampling instant is then

$$\sum_{n=1}^{p+1} \phi_{e,o}[n] - \sum_{n=1}^{p} \phi_{e,o}[n] = h_1[p+1]\phi_{e,O,1} + h_2[p+1]\phi_{e,O,2}$$

$$+ h_p[p+1]\phi_{e,O,3} + h_4[p+1]\phi_{e,O,4} \quad (7.60)$$

As the term on the right-hand side consists of only a few mismatch error terms, it is clear that the mismatch errors are not accumulating. Moreover, to achieve a zero steady-state value, the terms must be all equal as p approaches infinity, that is, $h_1[p + 1] = h_2[p + 1] = h_3[p + 1] = h_4[p + 1]$.

In fact, if each mismatch offset error has a uniform PSD denoted by N_m expressed in rad/Hz, then the averaged output total phase error after averaging [45] can be shown to be given by

$$N_{m,total} = \left|1 - z^{-1}\right|^2 N_m \qquad (7.61)$$

Here we see the effect of noise-shaping on the individual mismatch offset errors.

An implementation that utilizes mismatch noise-shaping is that involving a two-stage pipeline GRO-based GRO TM $\Delta\Sigma$ modulator [44,45]. In this realization, a 47-gate multipath GRO was realized with 15-stages used in the barrel-shifting data weighting algorithm. The circuit performed two operations. One involved a single course counter driven by one phase of the ring oscillator and the other a fine residue calculation using all the taps of the ring oscillator, and some logic to deduce the edge position within a single cycle of the ring oscillator. This circuit implemented the barrel-shifting algorithm in which the mismatch-phase offset errors are noise-shaped. A block diagram of this realization is shown in Figure 7.25. This particular implementation was selected on account of its power performance advantages over a direct parallel RO-based implementation, which required multiple counters. Counters are generally expensive in terms of power consumption and silicon area requirements.

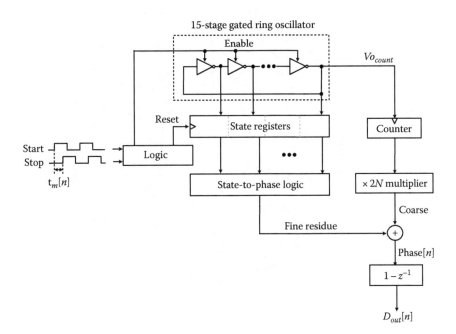

FIGURE 7.25 A two-stage pipeline GRO-based TM$\Delta\Sigma$ADC. (From Straayer, M.Z. and Perrott, M.H., *IEEE J. Solid-State Circuits*, 44(4), 1089, 2009.)

7.3.2.3 Vernier GRO-Based (VGRO) TMΔΣADC

Generating multiphase signals, for example, ring oscillator, with high resolution is at the core of many TM ΔΣ modulator approaches. The simplest method is an inverter chain, but it consumes a great deal of power and has a time resolution equal to twice the propagation delay of a single logic inverter gate. A second approach is a VCO in cascade with a phase interpolator circuit. Such a circuit has a higher resolution than an inverter chain but is sensitive to process, voltage supply, and temperature (PVT) variations as the phase interpolator operates in an open-loop fashion. A third approach is a phased-coupled VCO used in a phase-locked loop (PLL) negative feedback configuration. This approach is known to have the finest phase resolution and is insensitive to PVT effects. These three methods are well known, so we will defer the reader to visit any graduate level reference on analog design for more details.

A fourth method, which has only recently been introduced, involves a coupled ring oscillator configuration (CRO) [47]. This design involves the use of multiple rings of inverter chains and an outer ring of NMOS switches as depicted in Figure 7.26. A pseudospherical co-ordinate system is used to describe a particular inverter location in the various rings in terms of an n,m co-ordinate.

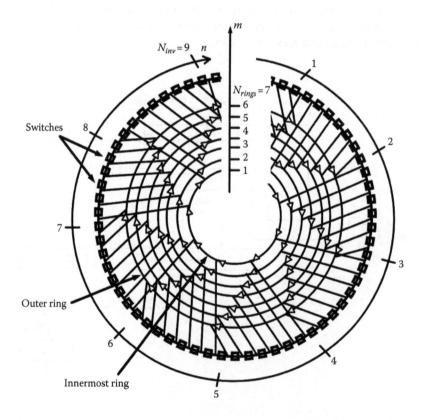

FIGURE 7.26 General implementation arrangement of a multiphase coupled oscillator configuration. (From Matsumoto, A. et al., *IEEE J. Solid-State Circuits*, 43(4), 831, 2008.)

Here we see that there are seven inner rings of inverters and that each ring contains nine inverters in cascade. The length of these rings, denoted as N_{inv} and the number of rings N_{rings}, can be made arbitrary. This design has better phase resolution than a single ring of inverter chains involving N_{inv} and an open-loop VCO with a phase interpolator but exhibits a power consumption just a little better than a single inverter chain. The basic principle of a CRO is that two different types of oscillation modes bind their phases together. For the most part, the outer-most or main ring ($m = N_{rings}$) determines the overall oscillation frequency, whereas the other rings ($m = 1, ...,$ $N_{rings} - 1$) couple with the phases of the main ring. The phase difference $\Delta\phi$ between any two inverters located in the spherical plane with co-ordinates (n_1, m_1) and (n_2, m_2) is given by the following expression:

$$\Delta\phi\left(n_2 - n_1, m_2 - m_1\right) = (m_2 - m_1)\left(\pi + \frac{2\pi}{2N_{inv}N_{rings}}\right)$$

$$+ (n_2 - n_1)\left(\pi + \frac{2\pi}{2N_{inv}}\right) \mod 2\pi \qquad (7.62)$$

In comparison, a single ring oscillator with N_{inv} inverters in cascade would have a phase difference between two inverters $(n_2 - n_1)$ apart from that given by

$$\Delta\phi\left(n_2 - n_1\right) = (n_2 - n_1)\left(\frac{2\pi}{2N_{inv}}\right) \qquad (7.63)$$

Another approach to realize a single-loop ring oscillator with a timing resolution less than a unit gate delay is through the application of a negative delay element. Consider the basic CMOS inverter circuit shown in Figure 7.27a. Here a negative delay is inserted in series with the gate of the PMOS transistor so that this transistor experiences the input signal earlier than the NMOS transistor [48]. As a result, the net delay of this gate can be made less than a conventional CMOS inverter. This is illustrated in the timing diagrams of Figure 7.27b. The top plot corresponds to the conventional timing for a single inverter circuit. The timing plot below this illustrates the timing skew introduced by the addition of a negative delay element. The following two plots are for the output signal for the conventional gate and the delay-reduced gate circuit. Different delays have been be achieved by inserting different negative timing skews [45,49].

The final method that we will describe here for increasing the timing resolution of a multiphase generator is the vernier method. The basic principle of the vernier delay line is to use two uncoupled ring oscillators, one oscillating slightly faster than the other by tuning the individual delay units to two separate delays, τ_F and τ_S, as shown in Figure 7.28a. The instantaneous phase behavior of these two oscillators is displayed modulo 2π as shown in Figure 7.28b. Here one can see how the "fast" oscillator phase behavior is phased aligned with the "slow" oscillator phase response at the very beginning, then with increasing time the "fast" oscillator phase moves

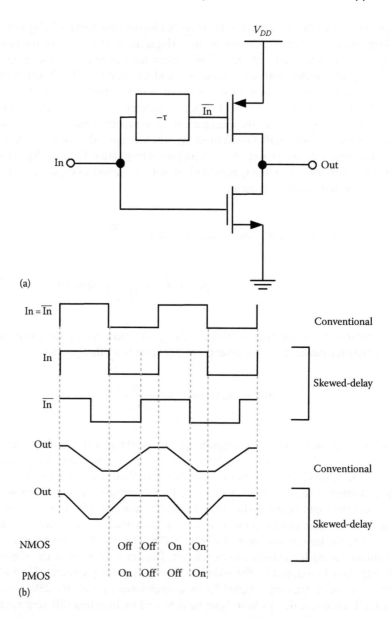

FIGURE 7.27 Negative-skewed delay cell: (a) circuit diagram and (b) timing diagram. (From Lee, S.J. et al., *IEEE J. Solid-State Circuits*, 32(2), 289, 1997.)

ahead of the "slow" oscillator until the two are again phase aligned (at the end of the time sequence). The smallest separation time or resolution between the two phase responses when they are both equal to 2π is Δt_{res} and is given by

$$\Delta t_{res} = \tau_S - \tau_F \tag{7.64}$$

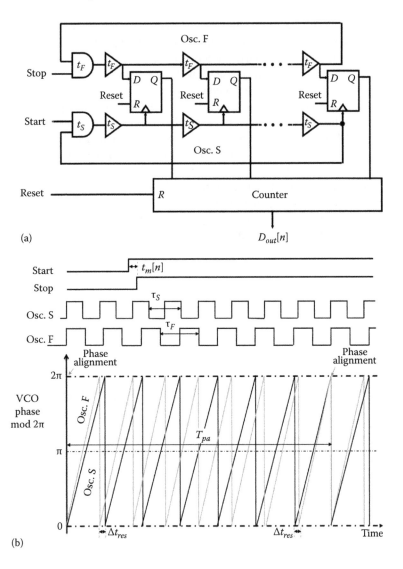

FIGURE 7.28 (a) Block diagram of a vernier GRO TDC. (b) A corresponding timing diagram of a fast and slow oscillator with a fixed time delay.

Correspondingly, the relative phase change in this time step is

$$\Delta\phi_{res} = 2\pi\left(\frac{\tau_S - \tau_F}{\tau_S}\right) \tag{7.65}$$

Clearly, the time and phase resolution can be made quite small by simply setting the delays in each ring oscillator to be very close to one another, not equal. Generally, this is done by selecting equal inverter delays, but the "slow" ring oscillator will be constructed with one additional delay element.

A flash-type TDC (i.e., one without noise-shaping) can be constructed using the vernier ring oscillator approach. The basic operation is to apply a start and stop signal to the TDC input such that the initial delay between the phase of the two oscillators equals this time-difference $t_m[n]$. As the phase of the fast oscillator rapidly cycles to catch up with the phase of the slow oscillator, a point is reached when the phase of the two oscillators is phase aligned. At this point, a number of the D-type flip-flops have been set to logic one indicating that the D input signal leads its clock input signal. A count of the number of 1s is made and the corresponding value is captured as the output $D_{out}[n]$.

One additional measurement can be made to deduce the resolution of the TDC. By measuring the time duration between adjacent phase alignment events, denoted by T_{pa}, and the total number of 1s captured by the D-type flip-flops that occurred in this time, denoted by $D_{out,ea}$, an accurate estimate of the time resolution Δt_{res} can be derived without having to use a measuring instrument with an extremely high resolution (but requires high accuracy, nonetheless), that is,

$$\Delta t_{res} = \frac{\Delta T_{pa}}{D_{out,ea}} \tag{7.66}$$

Hence, the input time difference $t_m[n]$ can then be expressed as

$$t_m\left[n\right] = \Delta t_{res} \cdot D_{out}\left[n\right] \tag{7.67}$$

Conversely, one can write the count $D_{out}[n]$ with a time quantization error included as follows

$$D_{out}\left[n\right] = \frac{1}{\Delta t_{res}} t_m\left[n\right] + \Delta t_e\left[n\right] \tag{7.68}$$

Here the time quantization error is subject to a start/begin and stop/end quantization effect, that is,

$$\Delta t_e\left[n\right] = \Delta t_{e,B}\left[n\right] + \Delta t_{e,E}\left[n\right] \tag{7.69}$$

where each time error component is bounded in magnitude by Δt_{res}. As conversion process is reset after each phase of TDC operation, there is no coupling between the start/begin and stop/end errors between sampling instants, as was seen with the other TM $\Delta\Sigma$ modulators. However, by altering the structure of the flash-type TDC such that the internal states of the two oscillators are stored, the phase difference can be read back during the next conversion cycle, thereby coupling the time quantization errors. This approach was adopted in the vernier GRO introduced in [50] resulting in a digital output with first-order noise described by

$$D_{out}\left[n\right] = \frac{1}{\Delta t_{res}} t_m\left[n\right] + \Delta t_{e,E}\left[n\right] - \Delta t_{e,E}\left[n-1\right] \tag{7.70}$$

While the vernier technique may appear to considerably improve the TDC resolution, the mismatch between the delay lines severely limits the resolution in practice. Also, a wide measurement range requires many more delay cells compared with a flash TDC with a single oscillator or delay, making it impractical in high-resolution wide-range applications. Therefore, unless a small range is allowed, vernier TDC must be combined with other circuit techniques to improve resolution without significantly increasing power and area. One such approach is based on component-invariant vernier delay line technique. Here the two delay lines are replaced by two gated oscillators, thereby eliminating the matching effort between adjacent delay line cells [51–53].

7.3.2.4 Switched-Ring Oscillator-Based TMΔΣADC

A GRO-based time-based ADC with first-order noise shaping requires the phase of the oscillator to be preserved so that the start/begin and stop/end quantization errors are coupled. However, due to various charge dynamic mechanisms, errors occur with the charge stored on the parasitic capacitors associated with the delay elements. These manifest themselves as leakages, skew, and dead-band effects [44,54,55].

To address this limitation, the switched-ring oscillator (SRO)-based TMΔΣADC was proposed in [54]. Leveraging oversampling and noise shaping, the proposed SRO-TDC achieves high resolution without the need for calibration. Ring oscillators are switched between two frequencies to achieve noise shaping of the quantization error in an open-loop manner. By decoupling the sampling clock and input carrier frequencies, the SRO-based TMΔΣADC is capable of operating at high OSRs, a feature that did not exist in any of the TDCs presented earlier.

A block diagram of the proposed approach is depicted in Figure 7.29. Here the input time difference signal is converted into a continuous-time pulse-modulated signal and applied to the control input of two voltage-controlled ring oscillators.

As the pulse-modulated signal is set between two voltage levels, the oscillation frequency of each ring oscillator is set at two different frequencies; albeit for a time duration established by the pulse width of the incoming signal, $t_m[n]$, and the other to

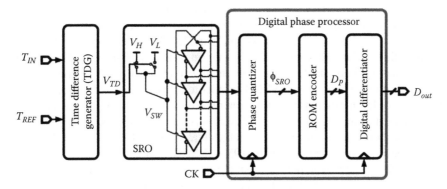

FIGURE 7.29 Block diagram of SRO-based TMΔΣADC. (From Chan, A. and Roberts, G.W., A deep sub-micron timing measurement circuit using a single-stage vernier delay line, *Proceedings of the IEEE Custom Integrated Circuits Conference*, Orlando, FL, May 2002, pp. 77–80.)

reestablish the initial phase of the next conversion cycle—similar in principle to the GRO approach but with a very different implementation (see Figure 7.20c). A timing diagram illustrating the VCO input and output behavior is shown in Figure 7.30. Note that the start/begin and stop/end phase quantization errors are arranged to be equal. The output of SRO block is fed to a phase quantizer to determine the output digital value, in much the same way that was done for the other $\Delta\Sigma$ ADCs described earlier. A ROM encoder and differentiator blocks are responsible for converting the output digital value from the quantize value to its final digital representation.

7.4 HIGH-ORDER TM$\Delta\Sigma$ MODULATORS

High-order $\Delta\Sigma$ modulators make use of greater amounts of quantization noise history to improve its overall operation. However, high-order $\Delta\Sigma$ modulators come with a higher cost in hardware complexity and silicon area footprint, loop instability, and power consumption. This section provides a brief review of several high-order TM$\Delta\Sigma$ modulator designs.

7.4.1 VCO-Based Closed-Loop TM$\Delta\Sigma$ Modulator

In Section 7.2.1, a description of an open-loop VCO-based $\Delta\Sigma$ ADC was described. One of the main drawbacks to this technique was that it was quite nonlinear. While a method of compensation was proposed based on the cancellation of even-order harmonic terms, an even better approach is to use the VCO-based ADC of Figure 7.15 as a multibit quantizer in a feedback configuration [32,37,39] as shown in Figure 7.31. Here a narrowband continuous-time loop filter with high DC-gain is used, together with a multibit DAC in the feedback path of the $\Delta\Sigma$ modulator. The loop filter is used to establish the STF and NTF of the overall $\Delta\Sigma$ modulator, as described previously in Section 7.2.1. High-order modulators can be realized by the appropriate selection of the filter order and frequency characteristics. As the VCO-based $\Delta\Sigma$ modulator is now placed in the feed-forward path of a negative feedback configuration, the nonlinearities of the quantizer are suppressed and made inconsequential.

An alternative realization is one that interchanges the sequence of the quantizer and loop filter of Figure 7.31 to that shown in Figure 7.32. Here the loop filter is realized using a digital filter. This realization is referred to as a VCO-based $\Delta\Sigma$ modulator with a tracking-loop quantizer [43]. The main goal of this work is to minimize the input signal range at input to the VCO in order to restrict the output frequencies to a very narrow frequency range, and in turn, reduce the level of distortion at its output. A simulation of the proposed approach was performed in [43] and compared to the VCO-based open-loop $\Delta\Sigma$ modulator architecture shown in Figure 7.15. In this simulation, the K_{VCO} coefficient was set to 0.95×33 MHz/V and the nonlinearity of VCO was modeled using a hyperbolic tangent function $\tanh(v)$. The simulations results are shown in Figure 7.33. The top plot illustrates the PSD for open-loop $\Delta\Sigma$ modulator architecture and the bottom plot corresponds to the PSD for the proposed closed-loop $\Delta\Sigma$ modulator architecture. The results reveal about 30 dB improvement in the SNDR with the proposed feedback approach. Experimental results have yet to confirm the validity of this approach and any unforeseen practical issues.

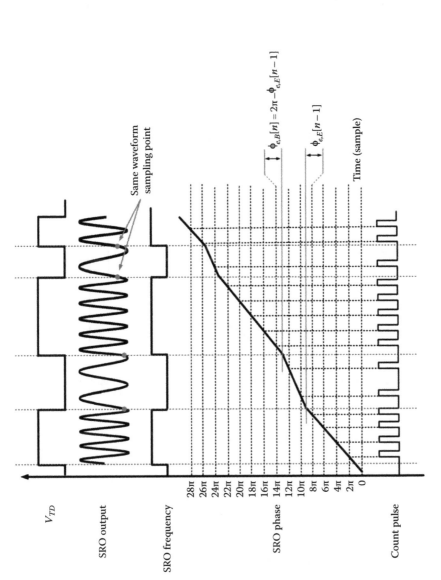

FIGURE 7.30 Illustrating the timing diagram for the SRO-based ΔΣ ADC and how the start/begin and stop/end phase errors are made equal.

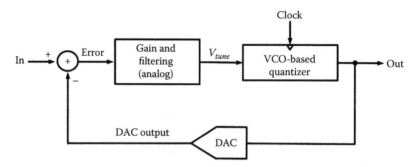

FIGURE 7.31 Block diagram of VCOΔΣADC used in a closed-loop configuration. (From Straayer, M.Z. and Perrott, M.H., *IEEE J. Solid State Circuits*, 43(4), 805, 2008; Iwata, A. et al., *IEEE Trans. Circuits Syst. II Analog Digit Signal Process.*, 46(7), 941, 1999.)

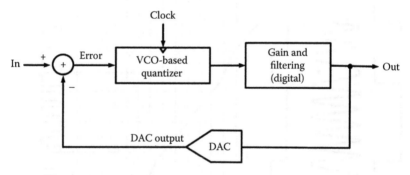

FIGURE 7.32 TM VCO-based ΔΣ with tracking-loop quantizer. (From Colodro, F. and Torralba, A., *IEEE Trans. Circuits Syst. II Expr. Brief*, 61(6), 383, 2014.)

7.4.2 TIME-MODE ΔΣADC USING DLL-LIKE STRUCTURE

A second-order TMΔΣ modulator with voltage input can be achieved by exploiting the structure of a DLL [13,36,38,56]. The general form of a DLL is shown in Figure 7.34a. Here a VCDU is tuned such that the total delay through the VCDU is equal to the period of the incoming reference clock signal. By adding a voltage summing circuit between the charge-pump and loop-filter, an input voltage can be injected into the feedback loop. In addition, a one-bit quantizer (D-type flip-flop) is added at the output of the phase detector to quantize its output. This output will be the output for the ADC. The VCDU is driven with a clock signal whose input–output phase difference will be proportional to the control voltage V_{ctrl}. In essence, this circuit acts as a voltage-to-phase converter. The resulting design is shown in Figure 7.34b.

Collectively, the VCDU, phase detector, and D-type flip-flop form a one-bit phase quantizer. The resulting single-loop configuration takes on the general form of Figure 7.31. Linearizing the system results in the equivalent z-domain block diagram

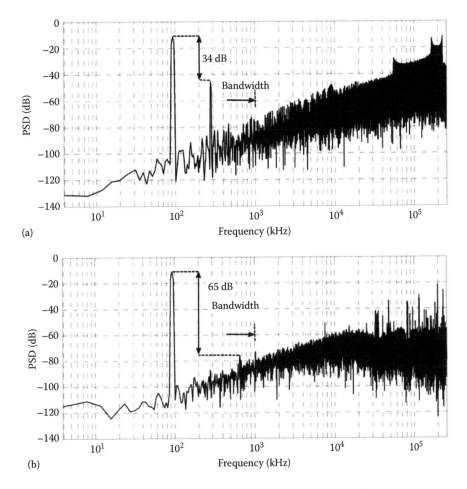

FIGURE 7.33 PSD of the VCO-based $\Delta\Sigma$ modulator with a tracking-loop quantizer: (a) nonfeedback VCO-based $\Delta\Sigma$ modulator, (b) VCO-based O$\Delta\Sigma$ modulator with feedback.

shown in Figure 7.34b. Here K_{VCDU} and K_{CP} are the gain of the VCDU and charge pump, respectively. Writing the output $D_{out}[n]$ in terms of the input $v[n]$ and phase quantization error $\phi_e[n]$, one can write in the frequency domain,

$$D_{out}(z) = \text{STF}(z)V_{in}(z) + \text{NTF}(z)\phi_e(z) \qquad (7.71)$$

where

$$\text{STF}(z) = \frac{K_{VCDU}H_{LP}(z)z^{-1}}{1 + K_{VCDU}K_{CP}H_{LP}(z)z^{-1}}$$

$$\text{NTF}(z) = \frac{1}{1 + K_{VCDU}K_{CP}H_{LP}(z)z^{-1}} \qquad (7.72)$$

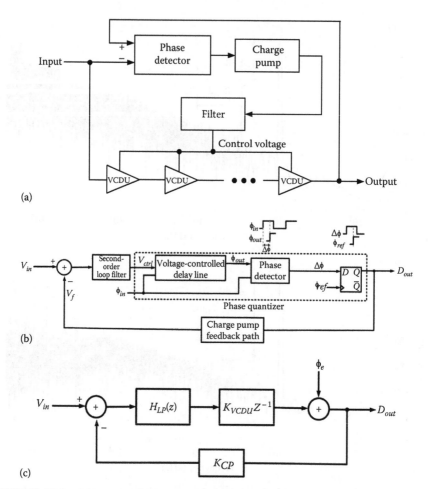

FIGURE 7.34 DLL-based ADC block diagram sharing the same mechanism as a conventional DLL: (a) Type I DLL block diagram, (b) block diagram of proposed ADC, and (c) z-domain linear equivalent representation. (From Lin, Y. and Ismail, M., *Analog Integr. Circuits Signal Process.*, 73(3), 801, 2012.)

For a second-order loop filter of the general form,

$$H_{LP}(z) = \frac{1}{\left(1 - z^{-1}\right)^2} \qquad (7.73)$$

the STF and NTF takes on the form

$$\mathrm{STF}(z) = -\frac{K_{VCDU} z^{-1}}{\left(1 - z^{-1}\right)^2 - K_{VCDU} K_{CP} z^{-1}}$$

$$\mathrm{NTF}(z) = \frac{\left(1 - z^{-1}\right)^2}{\left(1 - z^{-1}\right)^2 - K_{VCDU} K_{CP} z^{-1}} \qquad (7.74)$$

MATLAB®/Simulink® simulation reveals second-order noise shaping at the output of the time-mode $\Sigma\Delta$ADC, confirming the aforementioned theory. This design achieved 8 bits of resolution over a 10 MHz signal bandwidth [13]. Experimental results have yet to confirm the validity of this approach and any unforeseen practical issues.

7.4.3 HIGH-ORDER TM$\Delta\Sigma$ MODULATOR WITH VOLTAGE-CONTROLLED GRO (VCGRO)

To achieve a high SNDR for wideband applications, the order of the $\Delta\Sigma$ modulator must be increased. In order to achieve this, two topologies have been presented in [15] that are suitable for this application, using a single-loop and a MASH architecture. A single-loop TM$\Delta\Sigma$ modulator proposed in [32] utilizes the VCO as a quantizer to achieve third-order noise shaping. The main disadvantage of this design is that it uses voltage-domain components such as op-amp and DACs to realize the feedback structure around the quantizer. Therefore, large gain bandwidth (GBW) op-amps and extremely linear DACs are required to meet the aforementioned described system requirements.

A fully integrated time-domain high-order MASH $\Delta\Sigma$ modulator based on VCGRO has been presented in [9]. Figure 7.35 displays a block diagram of this design. The basic idea behind this approach is that the VCO in the top block is used to convert the input voltage signal $v_{in}[n]$ to the phase domain and then applied to the bottom block that forms a VCGRO-based or VCGRO TDC to digitize the quantization noise from the first modulator and a sampled version of the input voltage, $v_{in}[n]$. A digitized version of this noise is passed to the digital cancellation logic block where the quantization noise from the first modulator is canceled. An attractive feature of this structure is that it can realize a high-order NTF with a cascade of two or more VCGRO quantizers.

Based on our previous analysis, the output code count $D_{VCO}[n]$ from the VCO can be written as

$$D_{VCO}\left[n\right] = \frac{1}{2\pi}\phi_{VCO}\left[n\right] - \frac{1}{2\pi}\left(\phi_{e,VCO}\left[n\right] - \phi_{e,VCO}\left[n-1\right]\right) \tag{7.75}$$

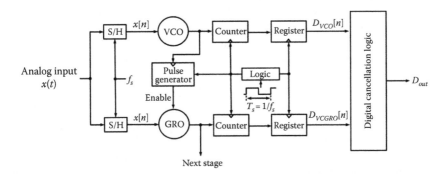

FIGURE 7.35 High-order TM$\Delta\Sigma$ modulator with MASH structure using VCO and VCGRO. (From Yu, W. et al., *IEEE Trans. Circuits Syst. I Reg. Papers*, 60(4), 856, 2013.)

and output code count from the VCGRO as

$$D_{VCGRO}\left[n\right] = \frac{1}{2\pi}\phi_{VCGRO}\left[n\right] - \frac{1}{2\pi}\left(\phi_{e,VCGRO}\left[n\right] - \phi_{e,VCGRO}\left[n-1\right]\right) \quad (7.76)$$

where $\phi_{e,VCO}[n]$ and $\phi_{e,VCGRO}[n]$ are the corresponding quantization errors at the nth sampling instant from the VCO and VCGRO. In each case, first-order noise shaping of the quantization noise is present at the outputs of each VCO. The digital cancellation block combines the two output terms such that in the z-domain

$$D_{out}\left(z\right) = z^{-1}D_{VCO}\left(z\right) - \left(1 - z^{-1}\right)D_{VCGRO}\left(z\right) \quad (7.77)$$

Based on the mathematical analysis presented in [9], together with a few approximations, the digital output $D_{out}(z)$ after cancellation can be written in terms of the input signal $v_{in}[n]$ as

$$D_{out}\left(z\right) = K_{VCO} \cdot T_s \cdot V_{in}\left(z\right) - \frac{1}{2\pi}z\left(1 - z^{-1}\right)^2 \phi_{e,VCGRO}\left(z\right) \quad (7.78)$$

where $V_{in}(z)$ and $\phi_{e,VCGRO}(z)$ are the z-transforms of input signal $v_{in}[n]$ and quantization error signal generated by the VCRGO, that is, $\phi_{e,VCGRO}[n]$. This expression highlights the claim of second-order noise shaping provided by this architecture, as the second term of Equation 7.78 contains the term $(1 - z^{-1})^2$. Experimental validation is yet to be given for this new architecture.

7.4.4 HIGH-ORDER TM$\Delta\Sigma$ MODULATOR USING A RELAXATION OSCILLATOR TECHNIQUE

Another approach to achieve high-order noise shaping based on a MASH structure was presented in [14] based on a relaxation oscillator technique. This design consists of three first-order TM$\Delta\Sigma$TDCs with a structure of a cascade of three first-order sections denoted as a 1–1–1 MASH structure (see Figure 7.9). The schematic of the first-order $\Delta\Sigma$ modulator is shown in Figure 7.36a. It includes two comparators, SR flip-flop, counter, and a circuit to convert the input time-difference interval into a charge quantity on the capacitors C. Charging and discharging the capacitors will generate a clock pulse that enables the counter through the comparator and SR flip-flop combination. The width of this pulse is proportional to the voltage difference on the capacitors.

An interesting characteristic of the relaxation oscillator-based TM$\Delta\Sigma$TDCs is that the quantization error is scrambled during successive quantization steps [14] as depicted in Figure 7.36b. Consequently, first-order noise shaping for one stage and third-order noise shaping for 1–1–1 MASH structure can be achieved.

A major performance limitation of this approach is the charge that leaks off the capacitors during their holding phase. Another issue relates to mismatches between stages.

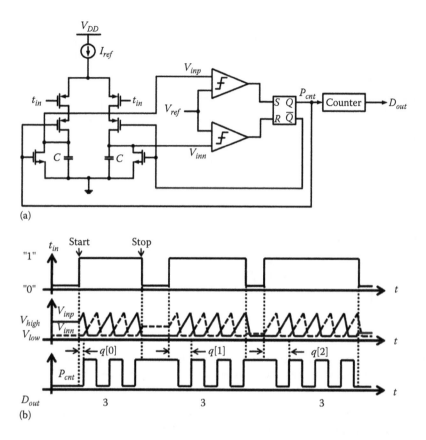

FIGURE 7.36 First-order $\Sigma\Delta$ modulator using a relaxation oscillator: (a) circuit schematic, (b) timing diagram. (From Cao, Y. et al., *IEEE J. Solid-State Circuits*, 47(9), 2093, 2012.)

7.5 MULTIBIT TM$\Delta\Sigma$ MODULATORS

Utilizing a high-order single-loop structure or a MASH technique, together with a large number of quantization levels, is instrumental in realizing a $\Delta\Sigma$ modulator with a high SNR/SNDR. Nonetheless, there are some significant design challenges that must be overcome, namely: instability, hardware complexity, and very linear multibit DACs when used in the feedback path of the modulator. In this section, several high-order TM$\Delta\Sigma$ modulator designs will be described.

7.5.1 VCDU-BASED MULTIBIT TM$\Delta\Sigma$ MODULATORS

A single-bit TM$\Delta\Sigma$ADC utilizing VCDUs as the basic building block of its realization was introduced in Section 7.2.1. One of the interesting attributes of this approach is that the resulting implementation consumes low power and has a small silicon area footprint. Figure 7.37 extends this principle to a multibit modulator by replacing the single-bit (D-type flip-flop) in Figure 7.13 with a multibit TDC to realize a multilevel quantizer [31] as shown in Figure 7.37. The actual implementation presented in

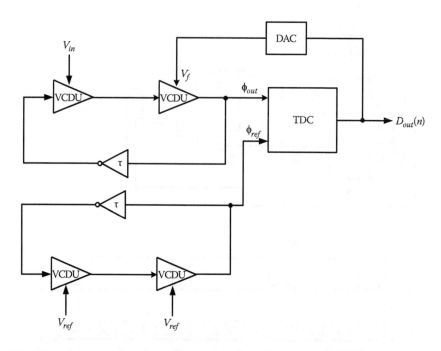

FIGURE 7.37   VCDU-based first-order, multibit TMΔΣADC. (From Taillefer, C.S., Analog-to-digital conversion via time-mode signal processing, PhD thesis, McGill University, Canada, Montreal, Quebec, Canada, 2007.)

[31] uses a 3-bit TDC and a 3-bit DAC in the feedback path. For an OSR of 76, MATLAB/Simulink simulation results show an SNR of about 76 dB, confirming the SNR improvement (i.e., 6 dB for a 1-bit increment in the quantizer resolution) over and above the 1-bit quantizer of Figure 7.13.

The main design issue for this realization is, of course, the DAC in the feedback path, but also the TDC. During dynamic operation, the TDC generates in-band spurs and must be suppressed through improved design.

7.5.2   VCDL/DCDL-Based Multibit TMΔΣ Modulators

While increasing the number of quantizer bits substantially improves the resolution of the ΔΣ ADC, it does not improve its linearity if the DAC is nonlinear. In the next proposal [13], the authors replaced the quantizer with a multibit TDC decoder circuit, and the DAC with a DTC converter circuit. This architecture is shown in Figure 7.38. Here the ΔΣ modulator consists of a single-loop configuration, with the input as a voltage sample $v_{in}[n]$, and the output a digital code, $D_{out}[n]$. In addition to the TDC-coder-DTC combination, a continuous-time loop filter is in the front-end of the block diagram.

In the feedforward path, after the loop filter block, a VCDL is used to convert the input voltage signal to a TM output signal from which the decoder converts to a corresponding digital code. Subsequently, in the feedback path, the output is passed to a digital-controlled delay line (DCDL) to convert the digital codes into delayed pulse waveforms dependent on the value of the output digital code. At the input summing

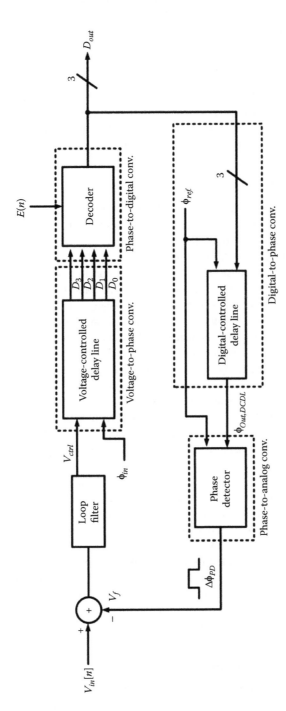

FIGURE 7.38 Multibit TMΔΣADC. (From Lin, Y. and Ismail, M., *Analog Integr. Circuits Signal Process.*, 73(3), 801, 2012.)

point, the time-difference interval of the DCDL output is subtracted from the input single $v_{in}[n]$ to create an error signal that drives the input to the loop filter. Simulation results reveal an SNDR of about 57 dB over a 10 MHz bandwidth [13].

7.6 TM $\Delta\Sigma$ DESIGN ISSUES

The performance of TM $\Delta\Sigma$ modulators is limited by four underlying factors: (1) nonlinearity of the basic delay element used in a delay line or in a ring oscillator, (2) mismatches between TM components, (3) clock jitter introduced noise, and (4) flip-flop metastability. In this section, these limitations will be described.

7.6.1 VCDU NONLINEARITY

A VCDU is often used in TMSP to convert voltage-domain signals to a corresponding time-mode signal. The main drawback of this element is that it has a limited range of linear operation, thereby limiting its overall DR of operation.

Figure 7.39a illustrates a CMOS implementation of a VCDU with a negative delay coefficient. The basic cell consists of essentially two capacitive loaded inverter circuits.

The first inverter also includes two additional transistors M_3 and M_4. Both M_3 and M_4 operate in the triode region, thereby acting as voltage-controlled resistors. The gate of M_3 is connected to the input signal v_{in} so that its resistance value can change with this level and the gate of M_4 is simply connected to V_{DD} so that its value is constant. With a specific input voltage set at the gate of M_3 and the clock input set high, M_1 turns off and M_2 turns on, thereby discharging capacitor C_W and forcing the output node to a zero state. The subsequent inverter circuit then inverts this quantity and produces a logic 1 output. Conversely, when the clock input returns to a low level, M_1 turns on and M_2 turns off, thereby charging C_W back to V_{DD}. The following inverter then puts out a logic 0.

With a periodic clock input, the output is also periodic with the identical frequency but show a slight delay with respect to the input. The propagation delay is tunable with the control voltage v_{in}. The v_{in}-input versus VCDU propagation delay transfer characteristic is shown in Figure 7.39b for a 180 nm CMOS process. While the specific delay values are unimportant here, one can see the general shape of this transfer characteristic. It has a somewhat linear region between an input voltage of 0.8 and 1.2 V, whereas for the input voltage less than 0.8 V or greater than 1.2 V, the VCDU behavior is visibly nonlinear.

This VCDU design is limited to a 0.4 V input voltage range, about 20% of the ideal headroom available from the power supply VDD, which severely limits the performance of the TM $\Delta\Sigma$ modulator. Indeed, it was shown in [ISCAS2015] that by improving the linearity of VCDUs with new circuit topologies, the dynamic range of TM $\Delta\Sigma$ modulators improves accordingly.

7.6.2 COMPONENT MISMATCHES

Mismatches between otherwise identical elements have a major impact on the linearity of TM$\Delta\Sigma$ modulators. If the delays associated with individual inverters in a ring oscillator that is used to realize a multiphase generator are mismatched, then

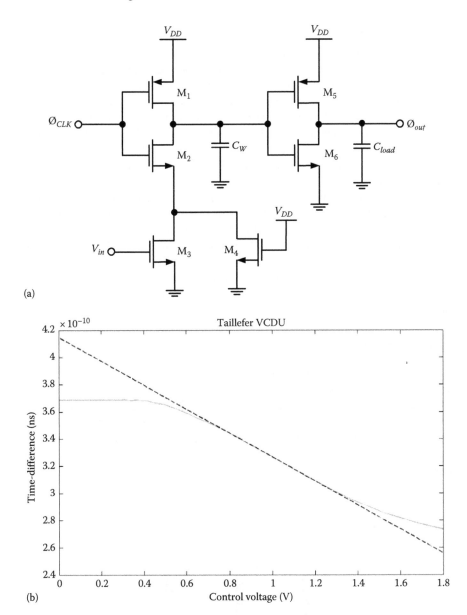

(a)

(b)

FIGURE 7.39 VCDU from [31]: (a) circuit schematic, and (b) input–output transfer characteristic.

the relative output phases will contain systematic or offset phase errors. As a consequence, in-band noise level will increase, thereby limiting the effective resolution of the TMΔΣ modulator. An analysis of these effects was provided in Section 7.3.2 related to the gate-ring oscillator approach and how data-weighted averaging could minimize these effects by noise-shaping these errors out-of-band.

7.6.3 JITTER-INDUCED NOISE

So far, we have covered issues related to nonlinearity in the transfer characteristic of a TM element such as a VCDU and mismatches between otherwise identical behaving devices. Another issue that one has to consider in the design of TM circuits is jitter-induced noise error that comes from the main reference clock. Noise associated with the clock reference generating circuit manifests itself into random variation in the placement of the clock edges as illustrated in Figure 7.40, clock jitter can be caused by electromagnetic interface (EMI), crosstalk, and wave reflections due to incorrectly terminating transmission lines, thermal noise, and/or poor power supply isolation.

Jitter is generally divided into two classifications: deterministic or random. Deterministic jitter (DJ) refers to jitter effects that are bounded in amplitude, periodic, or data dependent. Random jitter (RJ) is any jitter that does not fall into the DJ category and is fundamentally unbounded in value [57]. Jitter-induced noise effect is a fundamental limitation of ADCs and has been studied extensively [58,59]. While clock jitter is expected to also be a fundamental limitation of TM circuits, the authors are not aware of any extensive study confirming that this is indeed the case.

7.6.4 D-TYPE FLIP-FLOP DESIGN CHALLENGE

A D-type flip-flop is the most basic decision-making element of TM circuits. However, flip-flops experience a dead band effect whereby when the input time-difference signal is small in magnitude, such as when the time difference between a start and stop signal is small, the output of the flip-flop lies in an undetermined logic state, called the metastable state. Logic circuits that are reading this value cannot, as it is not a proper logic value, and instead misread the output value and can generate a logical bit error.

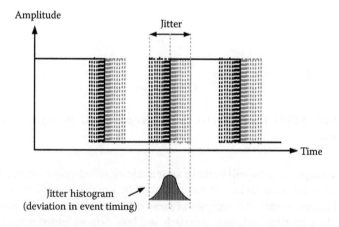

FIGURE 7.40   Jitter noise definition in time-mode $\Delta\Sigma$ modulators.

A classic method used to compensate for metastable behavior is to cascade multiple flips-flops or latches to give the front-end flip-flop more time to set its output value to the correct logic level [60]. Another method is to employ a time amplifier circuit to preamplify the small time difference prior to the decision-making flip-flop [61,62]. This approach reduces the potential for metastable-related bit errors and improves the resolution of TM$\Delta\Sigma$ modulators; however, it consumes more power and die area.

7.7   COMPARISON OF TIME-MODE (TM) VERSUS VOLTAGE-MODE (VM) $\Delta\Sigma$ MODULATORS

Over the last decade, many different voltage-mode and time-mode $\Delta\Sigma$ modulators have been implemented and their experimental results published. Many of these works can be found in the reference list at the end of this chapter. Of particular interest is how the SNDR performance of the reported $\Delta\Sigma$ modulators varies with silicon area, analog signal bandwidth, and power consumption. Scatter plots of the published works are shown Figure 7.41.

It is interesting to compare the general behavior of a TM realization with a VM realization. One can see from these three scatter plots that the SNDR performance of a TM realization is generally less than those implemented using a VM approach; however, the power and silicon area requirements are generally orders of smaller magnitude. In contrast, the analog signal bandwidth is generally much higher for a VM realization than a TM realization.

7.8   CONCLUSION

Voltage-domain $\Delta\Sigma$ modulators implemented in CMOS technologies are widely employed across the electronics industry as a main component of an ADC. However, as CMOS processes advance, MOS transistors must operate at lower voltage supply levels and this will cause major havoc on the operating characteristics of VM $\Delta\Sigma$ modulators. Time-mode $\Delta\Sigma$ modulators make use of digital-like circuits that easily scale with advances in CMOS technologies and, hence, lend themselves as a potential solution to realize ADCs in fine-lined CMOS processes.

The primary objective of this chapter was to expose the student to the principles of TM circuits for $\Delta\Sigma$ modulators more from a block diagram point-of-view rather than detail circuit perspective. While both perspectives are important, it is our belief that the block level perspective should be the student's first encounter before venturing down into the morass of transistor circuit design of $\Delta\Sigma$ modulators.

To date, numerous types of time-mode $\Delta\Sigma$ modulators have been proposed, fabricated, and tested. This includes single-bit, multibit, first-order, and higher-order type modulators. Through the application of the noise-shaping principle, both the quantization error made by a TM decision-making circuit and the systematic phase offsets associated with the component mismatches in the various timing circuits can be significantly reduced, giving way to a new generation of TM circuits that do not require any form of off-line or on-line calibration.

Results are extremely encouraging, especially in light of the present day facts that TM $\Delta\Sigma$ modulators offer low power operation and a small silicon area foot print.

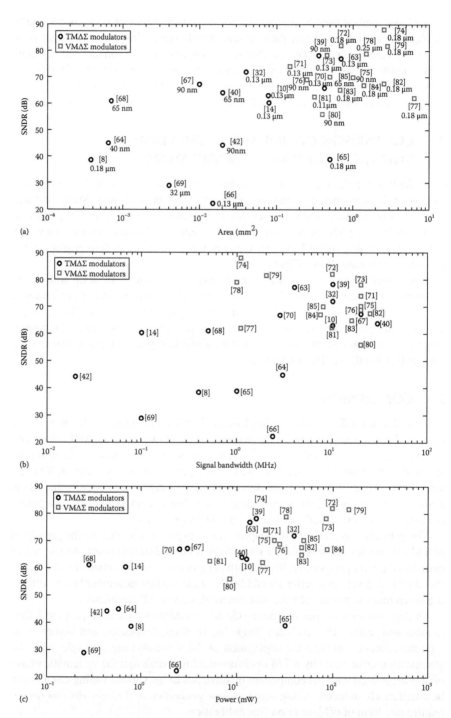

FIGURE 7.41 Historical performance comparisons: (a) SNDR vs. area, (b) SNDR vs. signal bandwidth, and (c) SNDR vs. power.

While the DR of TM circuits is not quite at the level of a VM circuit, it is the author's belief that this is just a matter of time before TM circuits reach performance levels equivalent to their voltage mode equivalents. One must recognize that the key principle of noise-shaping in TM circuits was only recently introduced and the number of people working in this area had been modest. It is our belief that this is soon to change.

REFERENCES

1. S. Henzler, *Time-to-Digital Converters*. 2010, Springer, Dordrecht, the Netherlands.
2. S.R. Norsworthy et al., *Delta-Sigma Data Converters: Theory, Design, and Simulation*. 1997, IEEE Press, New York.
3. P. Hyunsik et al., A 0.7-V 870-μW digital-audio CMOS sigma-delta modulator. *IEEE Journal of Solid-State Circuits*, 2009, **44**(4): 1078–1088.
4. E. Bilhan and F. Maloberti, A wideband sigma-delta modulator with cross-coupled two-paths. *IEEE Transactions on Circuits and Systems I: Regular Papers*, 2009, **56**(5): 886–893.
5. T. Yamamoto, M. Kasahara, and T. Matsuura, A 63 mA 112/94 dB DR IF bandpass $\Delta\Sigma$ modulator with direct feed-forward compensation and double sampling. *IEEE Journal of Solid-State Circuits*, 2008, **43**(8): 1783–1794.
6. J. Silva et al., Wideband low-distortion delta-sigma ADC topology. *Electronics Letters*, 2001, **37**(12): 737–738.
7. J. Daniels et al., A/D conversion using asynchronous delta-sigma modulation and time-to-digital conversion. *IEEE Transactions on Circuits and Systems I: Regular Papers*, 2010, **57**(9): 2404–2412.
8. C.S. Taillefer and G.W. Roberts, Delta-sigma A/D conversion via time-mode signal processing. *IEEE Transactions on Circuits and Systems I: Regular Papers*, 2009, **56**(9): 1908–1920.
9. W. Yu et al., A time-domain high-order MASH $\Delta\Sigma$ ADC using voltage-controlled gated-ring oscillator. *IEEE Transactions on Circuits and Systems I: Regular Papers*, 2013, **60**(4): 856–866.
10. K. Jaewook et al., Analysis and design of voltage-controlled oscillator based analog-to-digital converter. *IEEE Transactions on Circuits and Systems I: Regular Papers*, 2010, **57**(1): 18–30.
11. X. Zhao, H. Fang, and J. Xu, A wideband high-resolution time-interleaved sigma-delta modulator with VCO-based quantizer. *IEICE Electronics Express*, 2011, **8**(23): 1972–1977.
12. F. Yuan, Design techniques for time-mode noise-shaping analog-to-digital converters: A state-of-the-art review. *Analog Integrated Circuits and Signal Processing*, 2014, **79**(2): 191–206.
13. Y. Lin and M. Ismail, Time-based all-digital sigma–delta modulators for nanometer low voltage CMOS data converters. *Analog Integrated Circuits and Signal Processing*, Springer, New York, 2012, **73**(3): 801–808.
14. Y. Cao et al., 1-1-1 MASH $\Delta\Sigma$ time-to-digital converters with 6 ps resolution and third-order noise-shaping. *IEEE Journal of Solid-State Circuits*, 2012, **47**(9): 2093–2106.
15. R. Schreier and G.C. Temes, *Understanding Delta-Sigma Data Converters*, 1st edition. November 2004, Wiley-IEEE Press.
16. J.M. de la Rosa, Sigma-delta modulators: Tutorial overview, design guide, and state-of-the-art survey. *IEEE Transactions on Circuits and Systems I: Regular Papers*, 2011, **58**(1): 1–21.
17. M. Kozak and I. Kale, *Oversampled Delta-Sigma Modulators: Analysis, Applications, and Novel Topologies*. 2003, Kluwer Academic Publishers, Berlin, Heidelberg.

18. A.V. Oppenheim and R.W. Schafer, *Discrete-Time Signal Processing*. 2011, Pearson Education, Prentice Hall, Upper Saddle River, NJ.

19. V.D. Plassche and J. Rudy, *CMOS Integrated Analog-to-Digital and Digital-to-Analog Converters*. 2010, Springer, New York.

20. B. Razavi and B.A. Wooley, Design techniques for high-speed, high-resolution comparators. *IEEE Journal of Solid-State Circuits*, 1992, **27**(12): 1916–1926.

21. G.W. Roberts and M. Ali-Bakhshian, A brief introduction to time-to-digital and digital-to-time converters. *IEEE Transactions on Circuits and Systems II: Express Briefs*, 2010, **57**(3): 153–157.

22. H.P. Ninh, M. Miyahara, and A. Matsuzawa, A 83-dB SFDR 10-MHz bandwidth continuous-time Delta-Sigma modulator employing a one-element-shifting dynamic element matching. *IEEE International Symposium on Radio-Frequency Integration Technology (RFIT)*, Beijing, China, 2011.

23. R.T. Baird and T.S. Fiez, Stability analysis of high-order delta-sigma modulation for ADC's. *IEEE Transactions on Circuits and Systems II: Analog and Digital Signal Processing*, 1994, **41**(1): 59–62.

24. J. Lota et al., Nonlinear stability prediction of multibit delta-sigma modulators for sinusoidal inputs. *IEEE Transactions on Instrumentation and Measurement*, 2014, **63**(1): 18–26.

25. D. Macii et al., A stability criterion for high-accuracy $\Delta\Sigma$ digital resonators. *IEEE Transactions on Instrumentation and Measurement*, 2006, **55**(2): 577–583.

26. G. Manganaro and D. Leenaerts, *Advances in Analog and RF IC Design for Wireless Communication Systems*. 2013, Elsevier Science, Academic Press, San Deigo, CA.

27. H. Tao, L. Toth, and J.M. Khoury, Analysis of timing jitter in bandpass sigma-delta modulators. *IEEE Transactions on Circuits and Systems II: Analog and Digital Signal Processing*, 1999, **46**(8): 991–1001.

28. F. Gerfers and M. Ortmanns, *Continuous-Time Sigma-Delta A/D Conversion: Fundamentals, Performance Limits and Robust Implementations*. 2006, Springer-Verlag, Berlin, Heidelberg.

29. S. Lindfors et al., On the design of 2nd order multi-bit $\Delta\Sigma$ modulators. *IEEE International Symposium on Circuits and Systems, ISCAS*, Orlando, Florida, 1999.

30. T.O. Salo et al., 80-MHz bandpass $\Delta\Sigma$ modulators for multimode digital IF receivers. *IEEE Journal of Solid-State Circuits*, 2003, **38**(3): 464–474.

31. C.S. Taillefer, Analog-to-digital conversion via time-mode signal processing. PhD thesis, McGill University, Montreal, Quebec, Canada, 2007.

32. M.Z. Straayer and M.H. Perrott, A 12-Bit, 10-MHz bandwidth, continuous-time $\Delta\Sigma$ ADC with a 5-Bit, 950-MS/s VCO-based quantizer. *IEEE Journal of Solid-State Circuits*, 2008, **43**(4): 805–814.

33. J. Kim and C. SeongHwan, A time-based analog-to-digital converter using a multiphase voltage controlled oscillator. *IEEE International Symposium on Circuits and Systems, ISCAS*, Island of Kos, Greece, 2006.

34. R. Si, F. Li, and C. Zhang, A 100MHz S/s, 7 bit VCO-based ADC which is used in time interleaved ADC architectures. *Second International Conference on Consumer Electronics, Communications and Networks (CECNet)*, Yichang, China, 2012.

35. C. Tuan-Vu et al., Low-voltage, low-power, and wide-tuning-range ring-VCO for frequency $\Delta\Sigma$ modulator. *NORCHIP*, Tallinn, Estonia, 2008.

36. S. Yoder, M. Ismail, and W. Khalil, *VCO-Based Quantizers Using Frequency-to-Digital and Time-to-Digital Converters*. 2011, Springer-Verlag, New York.

37. A. Iwata et al., The architecture of delta sigma analog-to-digital converters using a voltage-controlled oscillator as a multibit quantizer. *IEEE Transactions on Circuits and Systems II: Analog and Digital Signal Processing*, 1999, **46**(7): 941–945.

38. Y.M. Tousi and E. Afshari, A miniature 2 mW 4 bit 1.2 GS/s delay-line-based ADC in 65 nm CMOS. *IEEE Journal of Solid-State Circuits*, 2011, **46**(10): 2312–2325.

39. K. Reddy et al., A 16-mW 78-dB SNDR 10-MHz BW CT $\Delta\Sigma$ ADC using residue-cancelling VCO-based quantizer. *IEEE Journal of Solid-State Circuits*, 2012, **47**(12): 2916–2927.
40. J. Daniels, W. Dehaene, and M. Steyaert, All-digital differential VCO-based A/D conversion. *Proceedings of 2010 IEEE International Symposium on Circuits and Systems (ISCAS)*, Paris, France, 2010.
41. T.K. Jang et al., A highly-digital VCO-based analog-to-digital converter using phase interpolator and digital calibration. *IEEE Transactions on Very Large Scale Integration (VLSI) Systems*, 2012, **20**(8): 1368–1372.
42. U. Wismar, D. Wisland, and P. Andreani, A 0.2 V 0.44 /spl mu W 20 kHz analog to digital/spl sigma/Δ modulator with 57 fJ/conversion FoM. *Proceedings of the 32nd European* in *Solid-State Circuits Conference (ESSCIRC)*, Montreux, Switzerland, 2006.
43. F. Colodro and A. Torralba, Linearity enhancement of VCO-based quantizers for SD modulators by means of a tracking loop. *IEEE Transactions on Circuits and Systems II: Express Briefs*, 2014, **61**(6): 383–387.
44. M. Straayer, Noise shaping techniques for analog and time to digital converters using voltage controlled oscillators. PhD thesis, MIT, Cambridge, MA, 2008, pp. 124–152.
45. M.Z. Straayer and M.H. Perrott, A multi-path gated ring oscillator TDC with first-order noise shaping. *IEEE Journal of Solid-State Circuits*, 2009, **44**(4): 1089–1098.
46. B.M. Helal et al., A low jitter 1.6 GHz multiplying DLL utilizing a scrambling time-to-digital converter and digital correlation. *IEEE Symposium on VLSI Circuits*, Kyoto, Japan, 2007.
47. A. Matsumoto et al., A design method and developments of a low-power and high-resolution multiphase generation system. *IEEE Journal of Solid-State Circuits*, 2008, **43**(4): 831–843.
48. S.J. Lee, B. Kim, and K. Lee, A novel high-speed ring oscillator for multiphase clock generation using negative skewed delay scheme. *IEEE Journal of Solid-State Circuits*, 1997, **32**(2): 289–291.
49. S.S. Mohan et al., Differential ring oscillators with multipath delay stages. *Proceedings of the IEEE Custom Integrated Circuits Conference*, San Jose, CA, 2005.
50. P. Lu, A. Liscidini, and P. Andreani, A 3.6 mW, 90 nm CMOS gated-vernier time-to-digital converter with an equivalent resolution of 3.2 ps. *IEEE Journal of Solid-State Circuits*, July 2012, **77**(7): 1626–1635.
51. A. Chan and G.W. Roberts, A deep sub-micron timing measurement circuit using a single-stage vernier delay line. *Proceedings of the IEEE Custom Integrated Circuits Conference*, Orlando, FL, May 2002, pp. 77–80.
52. G.W. Roberts and A. Chan, Timing measurement device using a component-invariant vernier delay line. US Patent #6,850,051, McGill University, Montreal, Quebec, Canada, Filed: March 26, 2002, Granted: February 1, 2005.
53. A. Chan and G.W. Roberts, Time and frequency characterization of jitter using a component-invariant vernier delay line. *IEEE Transactions on Very Large Scale Integration Systems*, January 2004, **12**: 79–95.
54. A. Elshazly et al., A noise-shaping time-to-digital converter using switched-ring oscillators-analysis, design, and measurement techniques. *IEEE Journal of Solid-State Circuits*, 2014, **49**(5): 1184–1197.
55. W. Yu, K. Kim, and S. Cho, A 148 fs rms integrated noise 4 MHz bandwidth all-digital second-order $\Delta\Sigma$ time-to-digital converter using gated switched-ring oscillator. *IEEE Conference on Custom Integrated Circuits (CICC)*, San Jose, CA, 2013.
56. R.J. Baker, *CMOS: Circuit Design, Layout, and Simulation*. 2011, Wiley-IEEE Press.
57. M.P. Li, *Jitter, Noise, and Signal Integrity at High-Speed*. 2007, Pearson Education Prentice Hall, Upper Saddle River, NJ.
58. S. Luschas and H.S. Lee, High-speed $\Delta\Sigma$ modulators with reduced timing jitter sensitivity. *IEEE Transactions on Circuits and Systems—II: Analog and Digital Signal Processing*, 2002, 49(11): 712–720.

59. K.C. Lauritzen, S.H. Talisa, and M. Peckerar, Impact of decorrelation techniques on sampling noise in radio-frequency applications. *IEEE Transactions on Instrumentation and Measurement*, 2010, **59**(9): 2272–2279.

60. J.P. Deschamps, G.D. Sutter, and E. Cantó, *Guide to FPGA Implementation of Arithmetic Functions*. Springer Science & Business Media, Berlin, Germany, April 2, 2012.

61. M. Oulmane and G.W. Roberts, Digital domain time amplification in CMOS process. *Seventh International Conference on Solid-State and Integrated Circuits Technology*, Beijing, China, 2004.

62. S.H. Chung et al., A high resolution metastability-independent two-step gated ring oscillator TDC with enhanced noise shaping. *Proceedings of 2010 IEEE International Symposium on Circuits and Systems (ISCAS)*, Paris, France, 2010.

63. S.Z. Asl et al., A 77 dB SNDR, 4 MHz MASH $\Delta\Sigma$ modulator with a second-stage multi-rate VCO-based quantizer. *IEEE Conference in Custom Integrated Circuits (CICC)*, San Jose, CA, 2011.

64. T. Konishi et al., A 40-nm 640-μm^2 45-dB opampless all-digital second-order MASH $\Delta\Sigma$ADC. *IEEE International Symposium on Circuits and Systems (ISCAS)*, Rio de Janeiro, Brazil, 2011.

65. P. Min and M.H. Perrott, A VCO-based analog-to-digital converter with second-order Sigma-Delta noise shaping. *IEEE International Symposium on Circuits and Systems (ISCAS)*, Taipei, Taiwan, 2009.

66. L. Guansheng et al., Delay-line-based analog-to-digital converters. *IEEE Transactions on Circuits and Systems II: Express Briefs*, 2009, **56**(6): 464–468.

67. P. Gao et al., Design of an intrinsically-linear double-VCO-based ADC with 2nd-order noise shaping. *Europe Conference & Exhibition in Design, Automation & Test (DATE)*, Dresden, Germany, 2012.

68. T. Konishi et al., A 61-dB SNDR 700 μm^2 second-order all-digital TDC with low-jitter frequency shift oscillators and dynamic flipflops. *Symposium on VLSI Circuits (VLSIC)*, Honolulu, Hawaii, 2012.

69. J.P. Hong et al., A 0.004 mm 250 μW $\Delta\Sigma$ TDC with time-difference accumulator and a 0.012 mm 2.5 mW bang-bang digital PLL using PRNG for low-power SoC applications. *IEEE International in Solid-State Circuits Conference Digest of Technical Papers (ISSCC)*, San Francisco, CA, 2012.

70. M. Gande et al., A 71 dB dynamic range third-order $\Delta\Sigma$TDC using charge-pump. *Symposium on VLSI Circuits (VLSIC)*, Honolulu, Hawaii, 2012.

71. G. Mitteregger et al., A 20-mW 640-MHz CMOS continuous-time $\Delta\Sigma$ADC with 20-MHz signal bandwidth, 80-dB dynamic range and 12-bit ENOB. *IEEE Journal of Solid-State Circuits*, 2006, **41**(12): 2641–2649.

72. W. Yang et al., A 100 mW 10 MHz-BW CT $\Delta\Sigma$ modulator with 87 dB DR and 91 dBc IMD. *IEEE International Conference in Solid-State Circuits (ISSCC)*, San Francisco, CA, 2008.

73. M. Park and M.H. Perrott, A 78 dB SNDR 87 mW 20 MHz bandwidth continuous-time $\Delta\Sigma$ ADC with VCO-based integrator and quantizer implemented in 0.13 μm CMOS. *IEEE Journal of Solid-State Circuits*, 2009, **44**(12): 3344–3358.

74. S.K. Gupta and F. Victor, A 64-MHz clock-rate $\Delta\Sigma$ ADC with 88-dB SNDR and −105-dB IM3 distortion at a 1.5-MHz signal frequency. *IEEE Journal of Solid-State Circuits*, 2002, **37**(12): 1653–1661.

75. P. Malla et al., A 28 mW spectrum-sensing reconfigurable 20 MHz 72 dB-SNR 70 dB-SNDR DT $\Delta\Sigma$ ADC for 802.11n/WiMAX receivers. *IEEE International Conference in Solid-State Circuits (ISSCC)*, San Francisco, CA, 2008.

76. L.J. Breems et al., A 56 mW CT quadrature cascaded $\Delta\Sigma$ modulator with 77dB DR in a near zero-IF 20 MHz band. *IEEE International in Conference Solid-State Circuits (ISSCC)*, San Francisco, CA, 2007.

77. M. Safi-Harb and G.W. Roberts, Low power delta-sigma modulator for ADSL applications in a low-voltage CMOS technology. *IEEE Transactions on Circuits and Systems I: Regular Papers*, 2005, **52**(10): 2075–2089.
78. R. Reutemann, P. Balmelli, and Q. Huang, A 33 mW 14b 2.5 M sample/s/spl sigma//spl delta/A/D converter in 0.25/spl mu/m digital CMOS. *IEEE International Conference in Solid-State Circuits*, San Francisco, CA, 2002.
79. R. Jiang and T.S. Fiez, A 14-bit delta-sigma ADC with 8× OSR and 4-MHz conversion bandwidth in a 0.18-μm CMOS process. *IEEE Journal of Solid-State Circuits*, 2004, **39**(1): 63–74.
80. Y. Ke et al., A 2.8-to-8.5 mW GSM/bluetooth/UMTS/DVB-H/WLAN fully reconfigurable CTΔΣ with 200 kHz to 20 MHz BW for 4G radios in 90 nm digital CMOS. *IEEE Symposium on VLSI Circuits (VLSIC)*, Honolulu, Hawaii, 2010.
81. K. Matsukawa et al., A fifth-order continuous-time delta-sigma modulator with single-opamp resonator. *IEEE Journal of Solid-State Circuits*, 2010, **45**(4): 697–706.
82. C.Y. Lu et al., A 25 MHz bandwidth 5th-order continuous-time low-pass sigma-delta modulator with 67.7 dB SNDR using time-domain quantization and feedback. *IEEE Journal of Solid-State Circuits*, 2010, **45**(9): 1795–1808.
83. V. Singh et al., A 16 MHz BW 75 dB DR CT ΔΣ ADC compensated for more than one cycle excess loop delay. *IEEE Journal of Solid-State Circuits*, 2012, **47**(8): 1884–1895.
84. S.D. Kulchycki et al., A 77-dB dynamic range, 7.5-MHz hybrid continuous-time/discrete-time cascaded ΔΣ modulator. *IEEE Journal of Solid-State Circuits*, 2008, **43**(4): 796–804.
85. Y.S. Shu, B.S. Song, and K. Bacrania, A 65 nm CMOS CT ΔΣ modulator with 81 dB DR and 8 MHz BW auto-tuned by pulse injection. *IEEE International Conference in Solid-State Circuits*, San Francisco, CA, 2008.

8 Fundamentals of Time-Mode Phase-Locked Loops

Fei Yuan

CONTENTS

Phase-lock loops are key building block of mixed-mode systems. They are at the heart of frequency synthesizers for data communications over wireless channels and clock and data recovery for data communications over wire channels. Charge-pump phase-locked loops, the most popular architecture of phase-locked loops, suffer from a number of drawbacks including high power consumption due to the need for charge pumps, rigid loop dynamics due to the finite frequency tuning range of the voltage-controlled oscillator (VCO) and *RC* filters, high silicon consumption due to the need for large capacitors in loop filters for filtering out transients present on the control voltage line of the VCOs, and charge-pump mismatch-induced reference spurs. A time-mode phase-locked loop is all-digital due to the replacement of its charge pump, loop filter, and VCO with a time-to-digital converter, a digital filter, and a digitally controlled oscillator (DCO), respectively. An all-digital phase-locked loop (ADPLL) is a digital system whose performance scales well with technology.

This chapter covers the fundamentals of ADPLLs. The chapter is organized as follows: Section 8.1 examines the limitations of charge-pump phase-locked loops. It is followed by the detailed examination of the phase noise of phase-locked loops in Section 8.2. The basic configuration of ADPLLs is studied in Section 8.3. An investigation of DCOs is followed. The phase noise of ADPLLs is also studied. Section 8.4 briefly examines all-digital frequency synthesizers. The chapter is summarized in Section 8.5.

8.1   LIMITATIONS OF CHARGE-PUMP PHASE-LOCKED LOOPS

Similar to analog-to-digital converters, phase-locked loops are another key building block of mixed analog-digital systems. They are at the heart of frequency synthesizers in wireless communication systems. They also play a key role in the clock and data recovery of serial data links and deskewing of parallel links [1]. Perhaps, the most widely used phase-locked loops are charge-pump phase-locked loops also known as type II phase-locked loops [2]. A charge-pump phase-locked loop typically consists of a phase-frequency detector (PFD) that detects the frequency and phase difference between a periodic reference signal and the output of a VCO and converts it to a pulse with the pulse width directly proportional to the phase difference and pulse polarity corresponding to the lead/lag information of the phases, a charge pump that maps the pulse from the preceding PFD to a constant current with the direction of the current modulated by the polarity of the phase difference, a low-pass loop filter, together with the charge pump, they function as a single-bit digital-to-analog converter that converts the output pulse of the PFD to an analog voltage whose value is proportional to the width of the pulse, and a VCO with its control voltage from the output of the loop filter and its oscillation frequency linearly proportional to the control voltage ideally. The performance of a phase-locked loop is quantified by a number of parameters with lock range, lock time, phase noise, and spurs in the lock state that are the most critical. The lock range, also known as the acquisition range, of the phase-locked loop is the maximum phase-frequency difference that the phase-locked loop can establish a lock state. Lock time is the amount of the time needed by the phase-locked loop to establish a lock state for a given phase-frequency difference. The phase noise of the phase-locked loop quantifies the spectral purity of the VCO of the phase-locked loop in the lock state. The phase noise of the phase-locked loop is affected by a number of factors such as the spectral purity of the reference and that of the local oscillator in its free-running state. The spurious tones of the phase-locked loop in the lock state are the tones in the spectrum of the oscillator that are caused by the disturbances present in the control voltage of the oscillator.

Despite their great success and wide deployment, the inherent drawbacks of charge-pump phase-locked loops become increasingly difficult to overcome in nanometer CMOS technologies. We examine them in detail.

8.1.1 SILICON CONSUMPTION

In a conventional charge-pump phase-locked loop, the loop filter is typically real-ized using an RC low-pass filter with large capacitors in order to filter out transients present on the control voltage line of the VCO. Any ripple present in the control voltage line of the VCO will be echoed with spurious tones in the spectrum of the phase-locked loop with their location set by the frequency of the ripple [3]. When the phase-locked loop is used for clock and data recovery operations in serial links, since data symbols received at the far end of the channel are contaminated with large timing jitter, a small loop bandwidth typically obtained from the low bandwidth of the loop filter is preferred in order to suppress the effect of the timing jitter of the incoming data. Although metal-insulator-metal (MIM) capacitors available in most mixed-mode CMOS, technologies offer a superior linearity; they in general suffer from a low capacitance density and are therefore costly. MOS capacitors can be used for loop filters due to their high capacitance density. The leakage current of MOS capacitors in nanometer-scale CMOS technologies, however, is significant. This inevitably affects the output voltage of the loop filter, and subsequently the level of the spurs of the oscillator [4].

8.1.2 PROGRAMMABILITY

The desire to have the full programmability of phase-locked loops in order to meet the needs of various applications demands that the loop dynamics, in particular, the loop bandwidth, of the phase-locked loops be tunable. Although a number of parameters contribute to the loop bandwidth, it is the bandwidth of the loop filter, specifically, the capacitance of the loop filter, that dictates the loop bandwidth. It is therefore highly desirable to have the loop filter realized digitally, that is, using a digital filter, such that the bandwidth, in-band gain, and stop-band attenuation of the loop filter can be programmed in order to meet different design specifications. A digitally realized loop filter also consumes much less silicon.

8.1.3 POWER CONSUMPTION

The power consumption of the charge pump of a charge-pump phase-locked loop typically constitutes a significant portion of the total power consumption of the phase-locked loop. This is because in order for the phase-locked loop to provide an adequate control voltage to the VCO so as to reduce the phase difference between the reference and the output of the local oscillator on time, the current of the charge pump must be sufficiently large. One might argue that a large control voltage of the VCO can also be obtained by lowering the capacitance of the loop filter. Lowering the capacitance of the loop filter, however, will undermine the ability of the loop filter to filter out transients present on the control voltage line of the oscillator, which will in turn give rise to more spurs in the spectrum of the phase-locked loop. Lowering the capacitance of the loop filter will also change the loop bandwidth, which will in turn affect the phase noise of the phase-locked loop. It is therefore highly desirable from a low power consumption point of view to have the charge pump eliminated.

8.1.4 SPURS

Disturbances present on the control voltage line of the oscillator manifest themselves as spurious tones in the spectrum of the phase-locked loops. To illustrate this, we follow the approach of Razavi given in [3]. Let the frequency ω of a VCO be given by

$$\omega = \omega_o + K_c v_c \tag{8.1}$$

where
   ω_o is the free-running frequency of the oscillator
   v_c is the control voltage of the oscillator
   K_c is the frequency-voltage gain of the oscillator

Note that $K_c = \Delta\omega/\Delta v_c$. If the oscillator is linear, $K_c = \Delta\omega/\Delta v_c$ is constant. If $K_c = \Delta\omega/\Delta v_c$ varies with the control voltage, the oscillator is nonlinear.

Let the output of the oscillator be $v_o(t) = V_m\cos[\phi(t)]$, where V_m is the amplitude of the output voltage of the oscillator and $\phi(t)$ is the phase of the oscillator. Utilizing $\omega = d\phi/dt$, we have

$$\phi(t) = \int_0^t \omega(\tau)d\tau + \phi(0^-). \tag{8.2}$$

Since we are only interested in the behavior of the oscillator in the steady state, it is natural to assume that the initial phase of the oscillator is zero. Note that if the frequency of the oscillator is constant, we have $\phi(t) = \omega t$. Otherwise, Equation 8.2 provides the general relation between the phase and frequency of the oscillator. The output of the oscillator is given by

$$v_o(t) = V_m \cos\left[\phi(t)\right]$$

$$= V_m \cos(\omega_o t)\cos\left[K_c\int_0^t v_c(\tau)d\tau\right]$$

$$- V_m \sin(\omega_o t)\sin\left[K_c\int_0^t v_c(\tau)d\tau\right]. \tag{8.3}$$

Let there be a sinusoidal disturbance $v_c(t) = V_c\cos(\omega_c t)$ with $V_c \ll V_m$ present on the control voltage line of the oscillator. Since

$$\int_0^t v_c(\tau)d\tau = \frac{V_c}{\omega_c}\sin(\omega_c t), \tag{8.4}$$

we have

$$v_o(t) = V_m \cos(\omega_o t) \cos\left[\frac{K_c V_c}{\omega_c} \sin(\omega_c t)\right]$$

$$- V_m \sin(\omega_o t) \sin\left[\frac{K_c V_c}{\omega_c} \sin(\omega_c t)\right]. \qquad (8.5)$$

Since V_c is sufficiently small, the following approximations hold

$$\cos\left[\frac{K_c V_c}{\omega_c} \sin(\omega_c t)\right] \approx 1 \qquad (8.6)$$

and

$$\sin\left[\frac{K_c V_c}{\omega_c} \sin(\omega_c t)\right] \approx \frac{K_c V_c}{\omega_c} \sin(\omega_c t). \qquad (8.7)$$

As a result, Equation 8.5 becomes

$$v_o(t) = V_m \cos(\omega_o t) - \frac{V_m V_c K_c}{2\omega_c}\left[\cos(\omega_o - \omega_c) - \cos(\omega_o + \omega_c)\right]. \qquad (8.8)$$

Equation 8.8 shows that a sinusoidal disturbance on the control voltage line of the oscillator will result in two tones at $\omega_o - \omega_c$ and $\omega_o + \omega_c$ in the spectrum of the oscillator. Also, the amplitude of the spurious tones is directly proportional to that of the disturbance, the gain of the oscillator, and the output voltage of the oscillator.

If the disturbance present on the control voltage line of the oscillator is a random signal, to analyze its effect over a given time interval $[0,T_s]$, we can assume that the disturbance signal will repeat periodically with period T_s. Such an approach is widely used in analysis of the effect of random signals such as noise [5]. The disturbance signal in the interval $[0,T_s]$ can be represented analytically by Fourier series

$$v_c(t) = \sum_{n=-\infty}^{\infty} \left[a_n \cos(n\omega_s t) + b_n \sin(n\omega_s t)\right], \qquad (8.9)$$

where $\omega_s = 2\pi/T_s$. As a result, the effect of the random disturbance on the spectrum of the oscillator can be analyzed in a similar way as that with a single sinusoidal tone. Utilizing the results obtained in Equation 8.8, one can show that for each of the frequency component of $v_c(t)$ at $n\omega_s$ in Equation 8.9, there are two corresponding tones at frequencies $\omega_o \pm n\omega_s$ in the spectrum of the oscillator. These spurious tones are solely due to the random disturbance present on the control voltage line of the oscillator and have nothing to do with the internal device noise of the oscillator. This observation signifies the importance of having a ripple-free control voltage of the VCO in the lock state in order to minimize the spurious tones of the oscillator.

While the spectral purity of the reference is often out of the control of designers, the disturbances present on the control voltage line of the oscillator can be minimized via better filtering. In a conventional charge-pump phase-locked loop, the charge pump and loop filter jointly function as a single-bit digital-to-analog converter that maps the output pulse of the preceding PFD to the control voltage of the downstream local oscillator. The single-bit operation of this special digital-to-analog converter inevitably gives rise to voltage ripples on the control voltage line of the oscillator, and subsequently spurious tones in the spectrum of the oscillator. Although increasing the capacitance of the loop filter will lower voltage ripples effectively, the upper bound of the capacitance of the loop filter is often dictated by the loop dynamics, which are typically specified by applications.

8.1.5 REFERENCE SPUR

Charge-pump phase-locked loops also suffer from reference spurs in the lock state, arising from the nonidealities of the charge pump such as current mismatch, pulse width mismatch, and charge injection [6,7]. In the lock state, ideally, no current from the charge pump should leak to the loop filter and the voltage of the loop filter should remain unchanged. The nonidealities of the charge pump, however, give rise to a leakage current flowing between the charge pump and the loop filter in the lock stage, corrupting the control voltage of the oscillator. This leakage current is often periodic in nature due to its close association with the reference input, which is typically periodic. As a result, periodic disturbances exist on the control voltage line of the oscillator and manifest themselves as spurious tones in the spectrum of the oscillator. These spurious tones are known as reference spurs as their frequency is closely related to that of the reference. In radio-frequency systems, reference spurs residing at the output of a frequency synthesizer will mix with the output of the preceding low-noise amplifier, which might contain signals from other channels and interferences, and generate frequency components that will fall into the wanted channel.

One commonly used technique to combat the reference spurs of a charge-pump phase-locked loop is to reduce the loop bandwidth of the phase-locked loop to well below the frequency of the reference, which is the fundamental frequency of the reference spur. This is achieved by increasing the time constant of the loop filter, however, at the cost of the increased settling time of the phase-locked loop [8]. Increasing the order of the loop filter without lowering its bandwidth though helps in reducing the reference spurs at high frequencies; the fundamental reference spur is intact. Various advanced techniques such as the random positioning of charge-pump current pulses [9], edge-interpolator [8], randomly selecting PFD [10] emerged to lower the reference spurs of phase-locked loops [11]. The essence of these techniques is to whiten the disturbances present on the control voltage line of the oscillator so as to redistribute its power over a broad range of frequency so that most will be filtered out by the loop dynamics, thereby lowering the reference spurs. These techniques, however, significantly complicate the design of phase-locked loops.

8.2 PHASE NOISE OF PHASE-LOCKED LOOPS

8.2.1 PHASE NOISE OF OSCILLATORS

The noise of VCOs critically affects the phase noise of phase-locked loops. In this section, we briefly review the mathematical treatment of the phase noise of electrical oscillators.

8.2.1.1 Leeson Model

Perhaps, the most widely cited early work on the phase noise of electrical oscillators is the empirical expression that predicts the power of the single-side-band (SSB) phase noise of an LC tank oscillator by Leeson [12]

$$L(\Delta\omega) = 10\log\left\{\frac{2FkT}{P_s}\left[1+\left(\frac{\omega_o}{2Q\Delta\omega}\right)^2\right]\left(1+\frac{\Delta\omega_f}{\Delta\omega}\right)\right\}, \quad (\text{dB}) \qquad (8.10)$$

where
  $\Delta\omega$ is the frequency displacement from the oscillation frequency ω_o of the oscillator
  Q is the quality factor of the oscillator
  F is the excess noise factor
  k is the Boltzmann's constant
  T is the absolute temperature in Kelvin
  P_s is the average power loss of the oscillator
  $\Delta\omega_f$ is the corner frequency between $1/(\Delta\omega)^2$ and $1/(\Delta\omega)^3$ regions of the phase noise spectrum of the oscillator

The flicker noise of the tail current source of the oscillator is upconverted to the vicinity of the oscillation frequency and manifests itself as $1/(\Delta\omega)^3$ region. To illustrate the upconversion of the flicker noise of the tail current source, consider the generic LC oscillator shown in Figure 8.1a. To simplify analysis, we neglect the noise generated by M1 and M2 and the thermal noise of M3, and only concentrate on the flicker noise of M3. Since M3 operates in saturation and functions as a current source with a constant current, the flicker noise generated by M3 at low frequencies is significant. We further assume that M1 and M2 operate in an *on–off* mode with period $T_o = 2\pi/\omega_o$ where ω_o is the frequency of the oscillator. The operation of M1 and M2 can therefore be depicted mathematically using the switching function $s(t)$ that is periodic with period T_o, 50% duty-cycle, and 0 ~ 1 amplitude, as shown in Figure 8.1. Since $s(t)$ is periodic, it can be represented by the Fourier series

$$s(t) = \sum_{n=-\infty}^{\infty}\left[a_n\cos(n\omega_o t) + b_n\sin(n\omega_o t)\right]. \qquad (8.11)$$

For the case shown in Figure 8.1b where M1 is *off* and M2 is *on*, C_1 is charged by V_{DD} via inductor L_1, while C_2 is both charged by V_{DD} via inductor L_2 and discharged to the ground rail via M3 simultaneously.

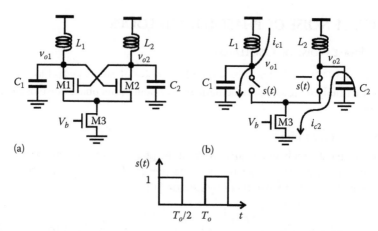

FIGURE 8.1   LC tank oscillators. (a) Schematic of LC tank oscillator. (b) LC tank oscillator with M1 off and M2 on.

In order to allow C_2 to discharge, the discharging current provided by M3 must be significantly larger than the charging current provided by V_{DD} via inductor L_2. We neglect the charging current provided by V_{DD} via inductor L_2 to simplify analysis. Let the current of M3 be $i_{DS3} = I_{DS3} + i_{nf3}$ with I_{DS3} the dc channel current and i_{nf3} the flicker noise of the channel current with its power in frequency range Δf given by

$$\overline{i_{nf3}^2} = \frac{K_f I_{DS3}}{f} \Delta f \tag{8.12}$$

where K_f is a process-dependent constant. The modulated discharge current of C_2 is given by

$$i_{c2} = s(t)\left(I_{DS3} + i_{nf3}\right). \tag{8.13}$$

Utilizing Equation 8.11, we obtain from Equation 8.13 that

$$i_{c2} = \left(I_{DS3} + i_{nf3}\right) \sum_{n=-\infty}^{\infty} \left[a_n \cos(n\omega_o t) + b_n \sin(n\omega_o t)\right]. \tag{8.14}$$

The voltage of C_2 at the end of the discharge interval is given by

$$v_{c2} = v_{c2}(0^-)$$

$$-\frac{1}{C_2} \sum_{n=-\infty}^{\infty} \int_0^{T_o/2} \left(I_{DS3} + i_{nf3}\right)\left[a_n \cos(n\omega_o t) + b_n \sin(n\omega_o t)\right] dt, \tag{8.15}$$

where $v_{c2}(0^-)$ is the initial voltage of C_2 at the onset of the discharge cycle. Equation 8.15 shows that for the flicker noise component of M3 at frequency $\Delta \omega$, i_{c2} has a corresponding component at frequency $\Delta\omega + n\omega_o$. If we assume that the quality factor of the LC tank of the oscillator is sufficiently large such that only those frequency components in the vicinity of ω_o will make their way to the output of the oscillator, then only the component of v_{c2} at frequency $\Delta\omega + \omega_o$ with $\Delta\omega \ll \omega_o$ will show up at the output of the oscillator. The rest will be diminished by the LC tank. The preceding analysis shows that the low-frequency potion of the flicker noise of M3 is upconverted to the vicinity of the oscillation frequency of the oscillator. Since the power of the flicker noise of M3 is much higher as compared with that of the thermal noise of M3 at low frequencies, the $1/(\Delta\omega)^3$ portion of the spectrum of the oscillator dominates the phase noise of the oscillator in the vicinity of ω_o, as shown in Figure 8.2.

8.2.1.2 Weigandt Model

Weigandt et al. showed that an error voltage of the output of an oscillator arising from the device noise and switching noise of the oscillator shifts the threshold-crossing time of the output voltage of the oscillator by the amount that is proportional to the voltage error and inversely proportional to the slew rate of the output of the threshold-crossing [13,14]. Because

$$\frac{dv_o}{dt} \approx \frac{v_n}{\Delta\tau},$$
(8.16)

we have

$$\overline{\Delta\tau^2} = \frac{\overline{v_n^2}}{\left(\dfrac{dv_o}{dt}\right)^2},$$
(8.17)

where

$\overline{\Delta\tau^2}$ is the mean-square value of the time displacement of threshold-crossing from the nominal threshold-crossing, that is, the power of the timing jitter

$\overline{v_n^2}$ is the power of the noise of the oscillator injected at the threshold-crossing

dv_o/dt is the slope of the output voltage of the oscillator at the threshold-crossing

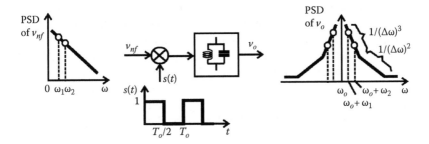

FIGURE 8.2 Upconversion of flicker noise of tail current source.

For a ring oscillator with differential-pair delay stages, because the maximum dv_o/dt is the slew rate of the differential pair delay stage of the oscillator and is given by

$$\left[\frac{dv_o}{dt}\right]_{max} = \frac{J}{C_L},$$

(8.18)

where

J is the tail current of the differential pair delay cell
C_L is the load capacitance at each of the output terminals, Equation 8.17 becomes

$$\overline{\Delta\tau^2} \approx \left(\frac{C_L}{J}\right)^2 \overline{v_n^2}.$$

(8.19)

Weigandt model establishes an explicit link between the power of the noise source of the oscillator and that of the timing jitter of the oscillator, thereby enabling designers to identify the noise sources that contribute the most to the overall timing jitter of the oscillator. Since the periodic operation nature of oscillators is not considered in Weigandt model, the unique characteristics of oscillators such as the aliasing of broadband noise and the upconversion of flicker noise are not accounted for. Further, it only takes into consideration the impact of the noise injected at the threshold-crossing. The effect of the noise injected at other time is not considered.

8.2.1.3   Razavi Model

Razavi analyzed the phase noise of oscillators using a feedback control system approach [15]. Here we briefly depict this approach. The close-loop transfer function of a linear system with a unity negative feedback, denoted by $H_c(s)$, is given by

$$H_c(s) = \frac{H_o(s)}{1 + H_o(s)},$$

(8.20)

where $H_o(s)$ is the open-loop transfer function. When the system oscillates at ω_o,

$$1 + H_o(j\omega_o) = 0$$

(8.21)

holds. Equation 8.21 is known as Barkhausen criteria. Note that Barkhausen criteria have a magnitude criterion $|H_o(j\omega)| = 1$ and a phase criterion $\angle H_o(j\omega) = -180°$. Both need to be satisfied in order to have sustained oscillation.

Consider an oscillator that oscillates at its free-running frequency ω_o. Let there be a noise signal with power spectral density $S_{in}(\omega)$ injected into the oscillator. The injection of the noise causes the oscillation frequency of the oscillator to deviate from its free-running frequency ω_o to $\omega_o + \Delta\omega$ with $\Delta\omega \ll \omega_o$ typically. The power spectral density of the output of the oscillator at $\omega_o + \Delta\omega$, denoted by $S_o(\omega_o + \Delta\omega)$, is given by

$$S_o(\omega_o + \Delta\omega) = |H_c(\omega_o + \Delta\omega)|^2 S_{in}(\omega_o + \Delta\omega).$$

(8.22)

Since $S_{in}(\omega)$ of the noise sources of MOS devices, such as the thermal and flicker noise of the channel current of MOS transistors and the thermal noise of resistors, are known, one only needs to find $H_c(\omega_o + \Delta\omega)$ in order to compute the power spectral density of the output noise of the oscillator. We notice that since $\Delta\omega \ll \omega_o$, the following first-order approximation holds

$$H_o(\omega_o + \Delta\omega) \approx H_o(\omega_o) + \left[\frac{dH_o(\omega)}{d\omega}\right]_{\omega_o} \Delta\omega. \tag{8.23}$$

Utilizing Equations 8.20 and 8.23, and noting that $H_o(\omega_o) = -1$, we have $H_o[(\omega_o + \Delta\omega)] \approx -1$. As a result

$$|H_c(\omega_o + \Delta\omega)|^2 \approx \frac{1}{\left|\dfrac{dH_o(\omega)}{d\omega}\right|^2_{\omega_o}(\Delta\omega)^2}, \tag{8.24}$$

where $|x(j\omega)|$ returns the magnitude value of $x(j\omega)$. If we let $H_o(j\omega) = A(\omega)e^{j\phi(\omega)}$ where $A(\omega)$ and $\phi(\omega)$ are the magnitude and phase of $H_o(j\omega)$, respectively, it can be shown that

$$|H_c(\omega_o + \Delta\omega)|^2 \approx \frac{1}{(\Delta\omega)^2 \sqrt{\left[\dfrac{dA(\omega)}{d\omega}\right]^2_{\omega_o} + \left[\dfrac{d\phi(\omega)}{d\omega}\right]^2_{\omega_o}}}. \tag{8.25}$$

Utilizing Herzel–Razavi's definition of the quality factor of the oscillators at ω_o [16]

$$Q(\omega_o) = \frac{\omega_o}{2}\sqrt{\left[\frac{dA(\omega)}{d\omega}\right]^2_{\omega_o} + \left[\frac{d\phi(\omega)}{d\omega}\right]^2_{\omega_o}}, \tag{8.26}$$

we arrive at

$$|H_c(\omega_o + \Delta\omega)|^2 \approx \frac{1}{4Q^2(\omega_o)}\left(\frac{\omega_o}{\Delta\omega}\right)^2. \tag{8.27}$$

The power spectral density of the output of the oscillator is given by

$$S_o(\omega_o + \Delta\omega) \approx \frac{1}{4Q^2}\left(\frac{\omega_o}{\Delta\omega}\right)^2 S_{in}(\Delta\omega). \tag{8.28}$$

If the input noise is a thermal noise source whose power spectral density is independent of frequency, Equation 8.28 shows that the power of the output noise of the oscillator will be inversely proportional to $(\Delta\omega)^2$. If the input noise is a flicker noise source whose power spectral density is inversely proportional to frequency, the power of the output noise will be inversely proportional to $(\Delta\omega)^3$. Since the power of flicker noise exceeds that of thermal noise at low frequencies, the phase noise of the output of the oscillator in the vicinity of the oscillation frequency of the oscillator is dictated by the flicker noise, whereas that at frequencies with large frequency displacements from the oscillation frequency of the oscillator is dominated by the thermal noise.

It is interesting to note that since at ω_o, $[dA(\omega)/d\omega]_{\omega=\omega_o} = 0$, we have $[dA(\omega)/d\omega]_{\omega=\omega_o+\omega_o} \approx 0$. As a result, Equation 8.26 is simplified to the widely used expression of the quality factor of oscillators

$$Q(\omega_o) \approx \frac{\omega_o}{2}\left|\frac{d\phi(\omega)}{d\omega}\right|_{\omega_o}. \tag{8.29}$$

It should be noted that the preceding model of the phase noise of oscillators does not account for the fold-over effect of thermal noise at high frequencies and the upconversion of flicker noise at low frequencies rigorously even though in Equation 8.28, we have explicitly stated the frequency of the input noise source $\Delta\omega$ and that of the output of the oscillator $\omega_o + \Delta\omega$ where the noise is measured. For a rigorous treatment, multifrequency transfer functions should be used [17]. For a linear periodically time-varying system, both transfer functions with inputs and outputs at the same frequency and aliasing transfer functions with inputs and outputs at different frequencies exist. The former quantify the relation between the inputs and outputs at the same frequency, just like a linear time-invariant system, while the latter depict the relation between the inputs and outputs at different frequencies. Readers are referred to [17] for more information on aliasing transfer functions.

For LC tank oscillators with a sufficiently high quality factor, the fold-over effect of the components of broadband noise at high frequencies to the oscillation frequency of the oscillator can be neglected. The upconversion of flicker noise to the vicinity of the oscillation frequency of the oscillator, however, cannot be neglected. In fact, it is the upconversion of the low-frequency portion of the flicker noise that contributes the most to the phase noise of the oscillator in the vicinity of the free-running frequency.

8.2.1.4 Hajimiri–Lee Model

To analyze the phase noise of arbitrary oscillators, we notice that noise injected at the peak of the output voltage of ring oscillators has the minimum impact on the phase noise, whereas that injected at the threshold-crossing of the output voltage of the oscillators has the maximum impact on the phase noise of the oscillator. To demonstrate this, consider a simple inverter ring shown in Figure 8.3. Let us first consider the threshold-crossing point. Both M1 and M2 in this case are in saturation and the inverter is essentially an inverting amplifier. The total noise generated in the channel of the transistors includes both the thermal and flicker noise of the channel current with its power given by

$$\overline{i^2}_{sat,n} = \left[4kT\gamma(g_{m1}+g_{m2}) + \frac{K_f(I_{DS1}+I_{DS2})}{f}\right]\Delta f, \tag{8.30}$$

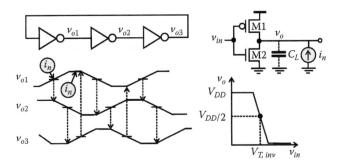

FIGURE 8.3 Ring oscillator with noise injection at threshold crossing and peak of the output voltage of the oscillator.

where γ is a process-dependent constant. The output noise voltage is given by

$$\overline{v^2}_{o,n} = (r_{o1} \| r_{o2})^2 \overline{i^2}_{sat,n}. \tag{8.31}$$

Now let us consider the case where $v_o = V_{DD}$. M1 in this case is in triode, while M2 is *off*. The noise injected to the output node contains only the thermal noise of the channel resistance of M1 with its power given by

$$\overline{i^2}_{triode,n} = \frac{4kT}{R_{on1}} \Delta f, \tag{8.32}$$

where R_{on1} is the channel resistance of M1 in triode. The output noise voltage in this case is given by

$$\overline{v^2}_{o,n} = R_{on1}^2 \overline{i^2}_{triode,n}. \tag{8.33}$$

Since $\overline{i^2}_{sat,n} > \overline{i^2}_{triode,n}$ and $r_o \gg R_{on}$, the inverter exhibits a higher output noise at the threshold-crossing than at the peak.

Based on this observation, Hajimiri and Lee introduced an impulse sensitivity function (ISF) to quantify the impulse response of oscillators [18]

$$h_\phi(t,\tau) = \frac{\Gamma(\omega_o\tau)}{q_{max}} u(t-\tau), \tag{8.34}$$

where
   $\Gamma(\omega_o\tau)$ is the ISF
   ω_o is the frequency of the oscillator
   q_{max} is the maximum charge displacement at the node where the impulse response
      $h_\phi(t,\tau)$ is measured
   $u(t-\tau)$ is the unit step function specifying the time instant at which the noise is
      injected
   τ and t are the noise launch time and response observation time, respectively [18,19]

TABLE 8.1

Phase Noise of Oscillators—Upconversion and Downconversion ($\Delta\omega \ll \omega_o$)

Noise Source	Phase Noise	Remarks
$i_o(t) = I_o\cos[(\Delta\omega)t]$	$\phi_n(t) = \dfrac{I_o c_o \sin[(\Delta\omega t)]}{2q_{max}(\Delta\omega)}$	Upconversion
$i_1(t) = I_1\cos[(\omega_o + \Delta\omega)t]$	$\phi_n(t) = \dfrac{I_1 c_1 \sin[(\Delta\omega)t]}{2q_{max}(\Delta\omega)}$	Direct transmission
$i_n(t) = I_n\cos[(n\omega_o + \Delta\omega)t]$	$\phi_n(t) = \dfrac{I_n c_n \sin[(\Delta\omega t)]}{2q_{max}(\Delta\omega)}$	Downconversion

ISF dips at the peak of the oscillator voltage and peaks at the threshold-crossing of the oscillator voltage. The periodic operation of oscillators ensures that ISF is periodic in the observation time t with its period equal to the period of the oscillators. As a result, it can be represented by Fourier series

$$\Gamma(\omega_o\tau) = \frac{c_0}{2} + \sum_{m=1}^{\infty} c_m \cos(m\omega_o\tau). \qquad (8.35)$$

The phase noise induced by the noise current $i_n(t)$ is obtained from

$$\phi_n(t) = \int_{-\infty}^{\infty} h_\phi(t,\tau)i_n(\tau)d\tau$$

$$= \frac{1}{q_{max}}\left[\frac{c_0}{2}\int_{-\infty}^{t} i_n(\tau)d\tau + \sum_{m=1}^{\infty} c_m \int_{-\infty}^{t} \cos(m\omega_o\tau)i_n(\tau)d\tau\right]. \qquad (8.36)$$

Equation 8.36 provides a theoretical foundation for the upconversion of the flicker noise and the aliasing of thermal noise of the oscillator to the noise at the oscillation frequency of the oscillator. Table 8.1 tabulates the phase noise of oscillators due to upconversion, downconversion, and direct conversion of the noise of the oscillators.

8.2.2 PHASE NOISE OF PHASE-LOCKED LOOPS

In most cases, we are only interested in the phase noise of phase-locked loops in the lock state. Since in the vicinity of the lock state of a phase-locked loop, the variation of the control voltage of the VCO of the phase-locked loop is small, the variation of the frequency of the oscillator is also small. As a result, the phase-locked loop can be treated as a linear system and methods for analyzing linear systems can be applied.

Note that this does not hold for phase-locked loops with a bang-bang phase detector. This is because bang-bang phase detectors are highly nonlinear elements. The relation between the variation of the phase of the VCO and its control voltage is given by $H_{VCO}(s) = K_{VCO}/s$. Linear phase detectors map a phase difference to a pulse with pulse width proportional to the phase difference. As a result, the transfer function of linear phase detectors is given by $H_{pd}(s) = K_{pd}$. The charge pump and loop filter map the output of the phase detector to an analog voltage. If the loop filter is implemented using a shunt capacitor only, the phase-locked loop will be a second-order system with its two poles located on the imaginary axis and will therefore be unstable. To ensure the stability of the loop, a zero is introduced by connecting a resistor in series with the capacitor of the loop filter. The transfer function of the loop filter whose input is the current from the preceding charge pump and whose output is the voltage across the RC network is given by

$$H_{LF}(s) = \frac{sRC + 1}{sC}.\tag{8.37}$$

The transfer function of the phase-locked loop shown in Figure 8.4 without considering the noise sources is given by

$$\frac{\Phi_{out}(s)}{\Phi_{in}(s)} = \frac{K_{pd}K_{VCO}R\left(s + \dfrac{1}{RC}\right)}{s^2 + sK_{pd}K_{VCO}R + \dfrac{K_{pd}K_{VCO}}{C}}.\tag{8.38}$$

The zero introduced by the resistor is evident in Equation 8.38 and is located at $s_z = -1/(RC)$. The two poles are located at

$$s_{p1,p2} = \frac{1}{2}K_{pd}K_{VCO}R\left(-1 \pm \sqrt{1 - \frac{4}{K_{pd}K_{VCO}RC}}\right).\tag{8.39}$$

It is seen that both poles are located in the left half of s-plane and the phase-locked loop is therefore stable.

To study the loop dynamics, we make use of the standard form of the characteristic equation of second-order systems

$$s^2 + 2\xi\omega_n s + \omega_n^2 = 0.\tag{8.40}$$

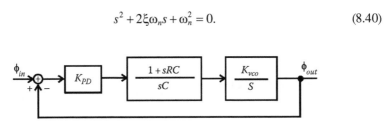

FIGURE 8.4   Linear model of phase-locked loops with a linear phase detector.

It can be shown that the *pole resonant frequency* ω_n also known as *loop bandwidth* of the phase-locked loop is given by

$$\omega_n = \sqrt{\frac{K_{PD}K_{VCO}}{C}}. \tag{8.41}$$

The *damping factor* ξ of the phase-locked loop is given by

$$\xi = \frac{1}{2}R\sqrt{K_{PD}K_{VCO}C}. \tag{8.42}$$

Equations 8.41 and 8.42 reveal that the damping factor ξ can be tuned by varying R without affecting the loop bandwidth ω_n. In other words, ξ and ω_n can be adjusted individually. Depending upon the value of ξ, the response of the phase-locked loop to a step input can be critically damped if $\xi = 1$, overdamped if $\xi > 1$, or underdamped if $0 < \xi < 1$. If the response is overdamped, the system has two distinct real poles. If the response is critically damped, the system has two identical real poles. If the response is underdamped, the system has a pair of complex conjugate poles. The rising time of the response of the phase-locked loop differs in these three cases. An underdamped response will have a shorter rise time, while an overdamped response will have a longer rise time. It should, however, be noted that a shorter rise time does not infer that the system will have a shorter lock time. This is because an underdamped response is an oscillating response with a exponentially decaying amplitude, which typically has a longer settling time, and subsequently a longer lock time.

To derive the transfer functions from the noise sources to the output of the phase-locked loop, we add noise sources N_{in}, N_{pd}, N_{LP}, and N_{VCO}, denoting the noise from the input, the phase detector, loop filter, and the oscillator, respectively, to the block diagram of the phase-locked loop, as shown in Figure 8.5. It can be shown that

$$\text{NTF}_{in}(s) = \sqrt{\omega_n}\left(\frac{1+\dfrac{s}{\omega_z}}{s^2 + 2\xi\omega_n s + \omega_n^2}\right), \tag{8.43}$$

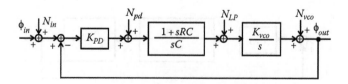

FIGURE 8.5   Linear model of phase-locked loops with noise sources considered.

$$\text{NTF}_{pd}(s) = \frac{K_{VCO}}{C} \left| \frac{1 + \dfrac{s}{\omega_{z1}}}{s^2 + 2\xi\omega_n s + \omega_n^2} \right|, \tag{8.44}$$

$$\text{NTF}_{LF}(s) = \frac{k_{VCO}s}{s^2 + 2\xi\omega_n s + \omega_n^2}, \tag{8.45}$$

and

$$\text{NTF}_{VCO}(s) = \frac{s^2}{s^2 + 2\xi\omega_n s + \omega_n^2}, \tag{8.46}$$

where
  $\omega_z = 1/(RC)$ is the frequency of the zero
  $\omega_{z1} = 1/(K_{VCO}RC)$

It is seen from the preceding results that the transfer function from the noise source of the oscillator to the output of the phase-locked loop has a high pass characteristic, that from the noise of the charge pump and loop filter to the output of the phase-locked loop has a band pass characteristic, and that from the noise of the input and the phase detector to the output of the phase-locked loop has a low pass characteristic.

Let us focus on the transfer function from the noise of the oscillator and that from the input to the output of the phase-locked loop only. We observe that an increase in ω_n will lead to a reduction in the noise transfer function from the oscillator to the output of the phase-locked loop. On the other hand, it will increase the noise transfer function from the input to the output of the phase-locked loop. These observations reveal that in order to suppress the noise generated by the oscillator, the loop bandwidth ω_n of the phase-locked loop should be maximized. The minimization of the effect of the noise injected at the input, on the other hand, requires that the loop bandwidth ω_n be minimized. If a phase-locked loop is used for clock and data recovery in serial links where the input is a received data stream at the far end of the channel with a large timing jitter typically, the loop bandwidth should be minimized in order to suppress the effect of the jitter of the incoming data stream. If a phase-locked loop is part of a frequency synthesizer with its input from a crystal oscillator, the loop bandwidth should be maximized in order to suppress the effect of the noise from the local oscillator.

As mentioned earlier that in addition to phase noise, the behavior of a charge-pump phase-lock loop in the lock state is also subject to the effect of reference spurs, arising from the nonidealities of the charge pump. Reference spurs are located at the frequencies that are displaced from the oscillation frequency of the oscillator by the reference frequency and their harmonics. The loop bandwidth should be set to well below the frequency of the reference so that the reference spurs can be adequately attenuated by the loop dynamics.

8.3 ALL-DIGITAL PHASE-LOCKED LOOPS

As the function of a linear phase detector is to generate a pulse whose width is directly proportional to the phase difference between the input of the phase-locked loop and the output of the VCO of the phase-locked loop, the resultant time variable can be digitized by a TDC, as shown in Figure 8.6 [20]. The digital output of the TDC can then be fed to a digital loop filter whose output controls a DCO. As a result, an ADPLL can be formed. In an ADPLL, not only the power-greedy charge pump and silicon-consuming loop filter are removed, the loop bandwidth can also be made fully programmable so as to meet the need of various applications. The removal of the charge pump also eliminates the source of reference spurs. ADPLLs also exhibit a reduced lock-in time [21]. DCOs also exhibit a much larger frequency tuning range and greatly improved linearity [22]. The fact that an ADPLL is realized using digital components allows it to benefit fully from technology scaling.

8.3.1 DIGITALLY CONTROLLED OSCILLATORS

A DCO is an oscillator whose frequency is controlled by a digital word. The work of Dunning et al. is perhaps one of the earliest studies of ADPLLs [23]. Dunning's ADPLL consists of a digitally controlled current-starved ring oscillator, a frequency acquisition block, and a phase acquisition block. The charging and discharging currents of the current-starved delay stages of the oscillator are 4-bit binary-coded so that the oscillator has a total of 16 different charging and discharging currents, and subsequently 16 different oscillation frequencies, as shown in Figure 8.7. The frequency acquisition block performs frequency comparison and acquisition, while the phase acquisition block performs phase comparison and acquisition after frequency acquisition is completed.

DCO constructed using the cascading structure shown in Figure 8.8 is another popular architecture of DCOs [24]. The DCO supports both the coarse tuning and fine-tuning of the oscillation frequency. The coarse tuning adjusts the frequency of the oscillator by varying the number of the delay stages of the oscillator, while the fine-tuning varies the frequency of the oscillator by changing the delay of the fine-tuning stage of the oscillator. If we assume that the number of the delay stages of the oscillator is N and the per-stage delay is $\tau_{c,d}$ while the tuning range of the fine-tuning delay stage is given by $\tau_L \leq \tau_{f,d} \leq \tau_H$, the largest frequency discontinuity will take place when the delay of the fine-tuning stage is set to either τ_L or τ_H.

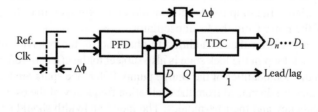

FIGURE 8.6 TDC-based phase-frequency detector. (From Kratyuk, V. et al., *IEEE Trans. Circuits Syst.*, 54(7), 247, 2007.)

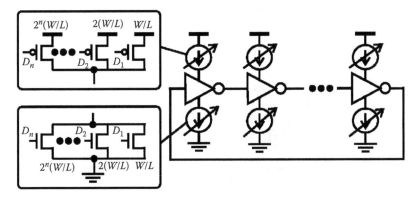

FIGURE 8.7 Digitally controlled current-starved oscillators. (From Dunning, J. et al., *IEEE J. Solid-State Circuits*, 30(4), 412, 1995.)

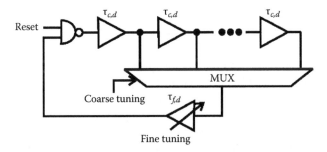

FIGURE 8.8 Digitally controlled oscillators with cascading configuration. (From Chung, C. et al., *IEEE Trans. Circuits Syst. II*, 58(3), 149, 2011.)

For example, if the number of the coarse delay stage is N and the delay of the fine tuning stage is set to τ_L before frequency adjustment, while the number of the coarse delay stage is $N + 1$ and the delay of the fine tuning stage is set to τ_H, the amount of the total loop delay is therefore given by $N\tau_{c,d} + (\tau_H - \tau_L)$. Similarly, if the number of the coarse delay stage is N and the delay of the fine tuning stage is set to τ_H before frequency adjustment, while the number of the coarse delay stage is $N - 1$ and the delay of the fine tuning stage is set to τ_L, the amount of the total loop delay is also $N\tau_{c,d} + (\tau_H - \tau_L)$. In both cases, the amount of total loop delay is the maximum and so is the total amount of frequency variation. To avoid a sudden change in the oscillation frequency of the oscillator due to digital frequency tuning, frequency overlaps between adjacent frequency tuning bands are needed. This, however, gives rise to the nonmonotonic frequency tuning characteristics of the oscillator [24].

Perhaps the most widely cited study on ADPLLs is the work conducted at Texas Instruments led by R. Staszewski where a Bluetooth radio utilizing an all-digital frequency synthesizer was designed in a 130 nm CMOS technology [25] and a single-chip GSM/EDGE transceiver utilizing an ADPLL was realized in a 90 nm CMOS technology [26]. The DCO used in these designs is an LC tank oscillator with a digitally switched varactor array, as shown in Figure 8.9.

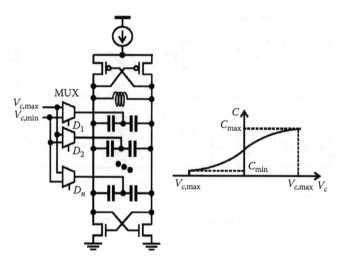

FIGURE 8.9 Digitally controlled LC oscillators with switched varactor banks. (From Staszewski, R. et al., *IEEE J. Solid-State Circuits*, 39(12), 2278, 2004.)

Unlike a conventional VCO whose frequency is varied continuously, the frequency of a DCO can only be varied in a stepwise constant manner with a frequency step Δ_f. As a result, a frequency quantization error e_f lower-bound by $-0.5\Delta_f$ and upper-bound by $0.5\Delta_f$ exists. Similar to the probability density function of the quantization error of analog-to-digital converters, the frequency quantization error of the DCO is distributed uniformly in $[-0.5\Delta_f, 0.5\Delta_f]$ with its probability density function given by

$$p_f = \begin{cases} 1/\Delta_f, & \text{if } -0.5\Delta_f \leq e_f \leq 0.5\Delta_f \\ 0, & \text{otherwise.} \end{cases} \tag{8.47}$$

Equation 8.47 ensures that

$$\int_{-\infty}^{\infty} p_f de_f = \int_{-0.5\Delta_f}^{0.5\Delta_f} \frac{1}{\Delta_f} de = 1. \tag{8.48}$$

The power of the frequency quantization error, denoted by P_f, is obtained from [25]

$$P_f = \int_{-0.5\Delta_f}^{0.5\Delta_f} e_f^2 p_f de_f = \frac{\Delta_f^2}{12}. \tag{8.49}$$

To obtain the power spectral density of the frequency quantization noise $S(f)$, we notice that $S(f)$ is constant from dc to the Nyquist frequency, that is, half of the reference frequency f_{REF}, of the phase-locked loop in which the oscillator resides. Since the power of the frequency quantization error is the total power of the frequency quantization noise over $[0, f_{REF}/2]$, we have

$$\int_0^{f_{REF}/2} S(f)df = \frac{\Delta_f^2}{12},$$

(8.50)

from which we obtain the power spectral density of the frequency quantization error of the DCO [22,27]

$$S(f) = \frac{\Delta_f^2}{12} \frac{1}{f_{REF}}.$$

(8.51)

Since the transfer function of oscillators is $2\pi/s$, the power spectral density of the quantization noise at the output of the oscillator is therefore given by

$$S_{DCO}(\omega) = \left(\frac{2\pi}{\omega}\right)^2 \frac{\Delta_f^2}{12} \frac{1}{f_{REF}}.$$

(8.52)

8.3.2   PHASE NOISE OF ALL-DIGITAL PHASE-LOCKED LOOPS

It was shown earlier that in a conventional voltage-mode phase-locked loop, noise sources in every block of the phase-locked loop contribute to the phase noise of the phase-locked loop. In a typical ADPLL, only the noise of the DCO and that of the TDC-based PFD are considered [26].

Let us first examine the noise of DCOs. If the DCO is an LC oscillator with a digitally switched varactor bank such as the one shown in Figure 8.9, both the thermal noise of the switching MOS transistors and the flicker noise of the head current source of the oscillator contribute to the phase noise of the oscillator. If the oscillator is a current-starved ring oscillator with digitally controlled charging and discharging currents such as the one shown in Figure 8.7, the thermal noise of the transistors of the oscillator constitutes the phase noise of the oscillator. In addition to these physical noise sources of the oscillator, DCOs are also subject to the effect of frequency quantization noise with its power spectral density given by Equation 8.51. The frequency quantization noise is white and its power spectral density is directly proportional to the step size of frequency adjustment and inversely proportional to the reference frequency.

The second is the quantization noise of the TDC PFD. If we assume that the time resolution of the TDC is Δ_t, then the quantization error of the TDC is lower-bound by $-0.5\Delta_t$ and upper-bound by Δ_t, that is, $\Delta_t \le e_t \le \Delta_t$ where e_t is the time quantization error of the TDC. The quantization error is distributed uniformly in $[-0.5\Delta_t, 0.5\Delta_t]$ with its probability density function given by

$$p_t = \begin{cases} 1/\Delta_t, & \text{if} -0.5\Delta_t \le e_t \le 0.5\Delta_t \\ 0, & \text{otherwise.} \end{cases}$$

(8.53)

Equation 8.53 ensures that

$$\int_{-\infty}^{\infty} p_t de_t = \int_{-0.5\Delta_t}^{0.5\Delta_t} \frac{1}{\Delta_t} de_t = 1. \tag{8.54}$$

The power of the quantization error, denoted by P_t, is obtained from [25]

$$P_t = \int_{-0.5\Delta_t}^{0.5\Delta_t} e_t^2 p_t de_t = \frac{\Delta_t^2}{12}. \tag{8.55}$$

Since the phase error corresponding to the time quantization error e_t is obtained from

$$\Delta_\phi = 2\pi \left(\frac{e_t}{T_o} \right), \tag{8.56}$$

where T_o is the period of the oscillator, we therefore arrive at the power of the phase error

$$P_\phi = \frac{(2\pi)^2}{12} \frac{\Delta_t^2}{T_o^2}. \tag{8.57}$$

To obtain the power spectral density of the quantization noise $S_t(f)$ of the TDC, we assume that $S_t(f)$ is constant from dc to half of the reference frequency f_{REF} of the phase-locked loop. As a result,

$$\int_0^{f_{REF}/2} S_t(f) df = \frac{(2\pi)^2}{12} \frac{\Delta_t^2}{T_o^2}, \tag{8.58}$$

from which we obtain the double-sided power spectral density of the quantization error of the TDC

$$S_t(f) = \frac{(2\pi)^2}{12} \frac{\Delta_t^2}{T_o^2} \frac{1}{f_{REF}}. \tag{8.59}$$

Equation 8.59 shows that the power spectral density of the quantization noise of the TDC is directly proportional to the resolution of the TDC and inversely proportional to the frequency of the reference.

Having obtained the power spectral density of the quantization noise of the TDC and that of the DCO, we now investigate the phase noise of the ADPLL. Figure 8.10 shows the simplified block diagram of an ADPLL with the noise from the TDC and DCO shown. It was shown previously that the noise transfer function

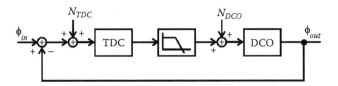

FIGURE 8.10 Phase noise of ADPLLs.

from the PFD to the output of a phase-locked loop has a low-pass characteristic, while that from the DCO to the output of the phase-locked loop has a high-pass characteristic. For frequency synthesizers, although the input is typically from a crystal oscillator with superior phase noise, the quantization noise of the TDC phase detector enters the phase-locked loop in a similar way as the input of the phase-locked loop does. Since the loop bandwidth of the phase-locked loop of the synthesizer is typically optimized to minimize the noise from the oscillator, a large amount of the quantization noise of the TDC phase detector will propagate to the output of the phase-locked loop with little attenuation. It is therefore highly desirable to lower the quantization noise of the TDC so as to improve the phase noise of the phase-locked loop. If a noise-shaping TDC such as a GRO-based TDC is used to quantify the pulse from the preceding phase-frequency detector, the reduced quantization noise of the TDC will improve the phase noise of the phase-locked loop [28].

8.4 ALL-DIGITAL FREQUENCY SYNTHESIZERS

An ADPLL can be migrated to an all-digital frequency synthesizer by adding a frequency divider in the feedback path of the loop. If the divider is an integer, the synthesizer is an integer-N frequency synthesizer. In the lock state, the frequency of the local oscillator f_{VCO} is N times that of the reference frequency f_{REF}. As a result, the minimum frequency adjustment step that the frequency synthesizer can provide is f_{REF}. Although the lower the reference frequency, the better the frequency resolution, the loop bandwidth must also be lowered accordingly in order to minimize reference spurs. Lowering the loop bandwidth, on the other hand, will not only increase the settling time of the loop, it will also undermine the ability of the loop to suppress the noise of the oscillator. In order to obtain a fine frequency resolution without sacrificing the settling time and phase noise performance, a technique known as digiphase [29] or fractional-N [30] can be used. Fractional division is obtained by periodically modulating the control bits of a dual-modulus divider such that the modulus of the frequency divider is toggled between two integers N and $N + 1$. The frequency of the oscillator alternates between the frequencies of the two lock states corresponding to moduli N and $N + 1$, as shown in Figure 8.11. For example, to achieve an $N + 1/4$ division ratio or a fractional modulo of 4, a divider-by-$(N + 1)$ is performed after every three divide-by-N operations. The carry of the accumulator of the divider controller is therefore the sequence of...000100010001. The oscillator will stabilize

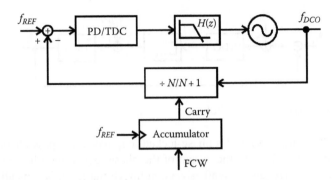

FIGURE 8.11 All-digital fractional-N frequency synthesizer.

at the frequency that is the time average of the two lock frequencies corresponding to moduli N and $N + 1$. By varying the percentage of the amount of the time that the frequency divider spends at each of the two division ratios, the frequency of the locked oscillator is given by

$$f_{REF} = \left(N + \frac{k}{M} \right) f_{VCO}, \qquad (8.60)$$

where
  M is the fractional modulus
  k assumes any number between 0 and M

If the toggling between the two division ratios is periodic, such as the example given previously where a divide-by- $(N + 1)$ operation occurs every three divide-by- N operations, periodic transient tones will appear at the output of the frequency divider. These periodic transient tones will make their way to the control voltage of the oscillator and generate spurious tones called fractional spurs in the spectrum of the oscillator. Note fractional spurs differ fundamentally from reference spurs as the former are due to the periodicity of the output of the frequency divider due to periodic toggling between the dual moduli, whereas the latter are due to the periodic tones at the output of the charge pump caused by the nonidealities of the charge pump. Since fractional spurs originate from the periodicity of the output of the frequency divider, one effective way to reduce fractional spurs is to whiten the output of the frequency divider so that its power is evenly distributed over a broad frequency range rather than concentrated at a few frequencies. Figure 8.12 shows the block diagram of such a design [31]. The output of the pseudorandom generator is compared with the frequency control word via comparators, and the output is used to select division moduli N and $N + 1$. Although periodic tones are removed, the white noise injected by the pseudorandom number generator and comparators will worsen the phase noise of the synthesizer.

To reduce the in-band noise of the preceding all-digital fractional-N frequency synthesizer with dithering injection, Riley et al. showed that a $\Delta\Sigma$ modulator with a dc input can be employed to generate the random control bits used to select the

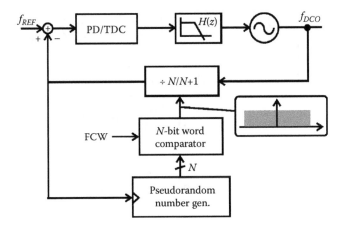

FIGURE 8.12   All-digital fractional-N frequency synthesizer with dithering injection.

modulus of the frequency divider [32]. Note that the output of a $\Delta\Sigma$ modulator with a dc input is a random sequence of 1s and 0s that does not contain any fixed tone and only contains quantization noise. The noise-shaping characteristics of the $\Delta\Sigma$ modulator ensure a low level in-band quantization noise achieved by moving the excessive quantization noise to higher frequencies outside the loop bandwidth, which can then be removed effectively by the loop dynamics. The level of the in-band quantization noise can be changed by varying the order and oversampling ratio of the modulator. Typically, the order of the $\Delta\Sigma$ modulator should be 3 or higher [33].

Figure 8.13 shows the typical configuration of an all-digital fractional-N frequency synthesizer with the frequency divider modulated by a $\Delta\Sigma$ modulator [28,34]. If the loop bandwidth of the synthesizer is overly large, the excessive quantization noise displaced to high frequencies from low frequencies will be inadequately attenuated by the loop dynamics, thereby deteriorating the in-band phase noise of the synthesizer. The loop bandwidth of the synthesizer therefore needs to be properly chosen in order to achieve a good balance between phase noise performance and settling time. A large loop bandwidth certainly benefits fast settling and minimizing the effect of the noise of the oscillator, while a small loop bandwidth is favored for minimizing the effect of the quantization noise of the frequency divider and the contribution of the noise from the TDC phase detector.

All-digital fractional-N frequency synthesizers with $\Delta\Sigma$ modulator frequency divider control are perhaps the most popular frequency synthesizers with superior performance. The need for a high-order $\Delta\Sigma$ modulator greatly complicates the design. As the performance of voltage-mode $\Delta\Sigma$ modulators scales poorly with technology, an alternative to this architecture is needed. The all-digital frequency synthesizer shown in Figure 8.14 (without the integrator in the forward path) eliminates fractional spurs inherent to conventional fractional-N frequency synthesizers without $\Delta\Sigma$ modulated modulus control [25,26]. The TDC is placed in the feedback path rather than in the forward path as in conventional all-digital fractional-N synthesizers. Both the frequency divider and $\Delta\Sigma$ modulator, which are essential part of the all-digital frequency synthesizer shown Figure 8.13, are removed. This frequency

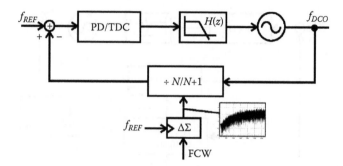

FIGURE 8.13   All-digital fractional-N frequency synthesizer with $\Delta\Sigma$ modulator frequency divider control.

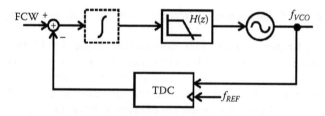

FIGURE 8.14   All-digital fractional-N frequency synthesizer.

synthesizer is known as divider-less frequency synthesizer and is significantly simpler as compared with that in Figure 8.13. Further, it is purely all-digital as all blocks of the synthesizer are digital. An integrator is added in the forward path to convert the frequency error into a phase information, which corresponds to the phase difference between the output of the oscillator and the References 34–36. The TDC computes the ratio of f_{DCO} to f_{REF}. The output of the TDC is compared with the frequency control word (FCW). The result is converted to a phase variable using an integrator. It is then used to control the digital loop filter whose output adjusts the frequency of the DCO. The absence of the frequency divider and the associated dithering block not only simplifies the design, it also eliminates the source of fractional spurs.

The nonidealities of the TDC such as quantization error and nonlinearity will undoubtedly contribute to the phase noise of the synthesizer due to its location in the feedback path. TDCs with a fine resolution and a good linearity are critical.

8.5 SUMMARY

This chapter started with a close examination of the drawbacks of charge-pump phase-locked loops. These drawbacks include the high silicon of loop filter, the high power consumption of charge pumps, the poor programmability of loop filters, and the existence of reference spurs caused by the nonidealities of charge pumps. It was followed by the detailed examination of the modeling of the phase noise of oscillators. We showed that the upconversion of the flicker noise of bias current sources is the dominant source of phase noise in the vicinity of the oscillation frequency of the oscillators. A detailed study of the phase noise of phase-locked loops was presented. The distinct noise transfer characteristics of noise from the input and the oscillator to the output of the phase-locked loops impose different design constraints on the phase-locked loops for frequency synthesizers and those for clock and data recovery. The basic architectures of DCOs, namely, digitally tuned current-starved ring oscillators, digitally tuned ring oscillators with cascading configurations, and digitally tuned LC oscillators, were studied. It was followed with the detailed study of the frequency quantization error of DCOs and the time quantization error of TDC-based phase detectors, and the phase noise of ADPLLs. The chapter ended with the brief examination of the architecture of all-digital fractional-N frequency synthesizers with an emphasis on fractional spurs.

REFERENCES

1. T. Wang and F. Yuan. A new current-mode incremental signaling scheme with applications to Gb/s parallel links. *IEEE Transactions on Circuits and Systems I*, 54(2):255–267, February 2007.
2. B. Razavi. *Design of Integrated Circuits for Optical Communications*. McGraw-Hill, Boston, MA, 2003.
3. B. Razavi. *RF Microelectronics*, 2nd edn. Prentice Hall, Englewood Cliffs, NJ, 2011.
4. C. Hung and S. Liu. A leakage-compensated PLL in 65-nm CMOS technology. *IEEE Transactions on Circuits and Systems II*, 56(7):525–529, 2009.
5. A. van der Ziel. *Noise, Sources, Characterization, Measurement*, 1st edn. Prentice Hall, Englewood Cliffs, NJ, 1970.
6. W. Rhee. Design of high-performance CMOS charge pumps in phase-locked loops. In *Proceedings of IEEE International Symposium on Circuits and Systems*, Orlando, FL, 1999, pp. 545–548.
7. F. Yuan. *CMOS Current-Mode Circuits for Data Communications*. Springer, New York, 2006.
8. J. Choi, W. Kim, and K. Lim. A spur suppression technique using an edge-interpolator for a charge-pump PLL. *IEEE Transactions on VLSI Systems*, 20(5):969–973, May 2012.
9. C. Thambidurai and N. Krishnapura. Spur reduction in wideband PLLs by random positioning of charge pump current pulses. In *Proceedings of IEEE International Symposium on Circuits and Systems*, Paris, 2010, pp. 3397–3400.
10. T. Liao, J. Su, and C. Hung. Spur-reduction frequency synthesizer exploiting randomly selected PFD. *IEEE Transactions on VLSI Systems*, 21(3):589–592, March 2013.
11. K. Wang, A. Swaminathan, and I. Galton. Spurious tone suppression techniques applied to a wide-bandwidth 2.4 GHz fractional-N PLL. *IEEE Journal of Solid-State Circuits*, 43(12):2787–2797, December 2008.
12. D. Leeson. A simple model of feedback oscillator noise spectrum. *Proceedings of IEEE*, 54(2):329–330, February 1966.
13. T. Weigandt, B. Kim, and P. Grey. Analysis of timing jitter in ring oscillators. In *Proceedings of IEEE International Symposium on Circuits and Systems*, London, 1994, pp. 27–30.
14. T. Weigandt. Low-phase-noise, low-timing-jitter design techniques for delay cell based VCOs and frequency synthesizer. PhD dissertation, University of California, Berkeley, CA, 1998.
15. B. Razavi. A study of phase noise in CMOS oscillators. *IEEE Journal of Solid-State Circuits*, 31(3):331–343, March 1996.

16. F. Herzal and B. Razavi. A study of oscillator jitter due to supply and substrate noise. *IEEE Transactions on Circuits and Systems II*, 46(1):56–62, January 1999.

17. F. Yuan and A. Opal. *Computer Methods for Mixed-Mode Switching Circuits*. Kluwer Academic Publishers, Boston, MA, 2004.

18. A. Hajimiri, S. Limotyrakis, and T. Lee. Jitter and phase noise in ring oscillators. *IEEE Journal of Solid-State Circuits*, 34(6):790–804, June 1999.

19. T. Lee and A. Hajimiri. Oscillator phase noise: A tutorial. *IEEE Journal of Solid-State Circuits*, 35(3):326–336, March 2000.

20. V. Kratyuk, P. Hanumolu, U. Moon, and K. Mayaram. A design procedure for all-digital phase-locked loops based on a charge-pump phase-locked-loop analogy. *IEEE Transactions on Circuits and Systems II*, 54(7):247–251, March 2007.

21. G. Yu, Y. Wang, H. Yang, and H. Wang. Fast-locking all-digital phase-locked loop with digitally controlled oscillator tuning word estimating and presetting. *IET Circuits Devices and Systems*, 4(3):207–217, 2010.

22. R. Staszewski, C. Hung, M. Lee, and D. Leipold. A digitally controlled oscillator in a 90 nm digital CMOS process for mobile phones. *IEEE Journal of Solid-State Circuits*, 40(11):2203–2211, November 2005.

23. J. Dunning, G. Garcia, J. Lundberg, and E. Nuckolls. An all-digital phase-locked loop with 50-cycle lock time suitable for high-performance microprocessors. *IEEE Journal of Solid-State Circuits*, 30(4):412–422, April 1995.

24. C. Chung, C. Ko, and S. Shen. Built-in self-calibration circuit for monotonic digitally controlled oscillator design in 65-nm CMOS technology. *IEEE Transactions on Circuits and Systems II*, 58(3):149–153, March 2011.

25. R. Staszewski, K. Muhammad, D. Leipold, C. Hung, Y. Ho, J. Wallberg, C. Fernando et al. All-digital TX frequency synthesizer and discrete-time receiver for bluetooth radio in 130-nm CMOS. *IEEE Journal of Solid-State Circuits*, 39(12):2278–2291, December 2004.

26. R. Staszewski, J. Wallberg, S. Rezeq, C. Hung, O. Eliezer, S. Vemulapalli, C. Fernando et al. All-digital PLL and transmitter for mobile phones. *IEEE Transactions on Nuclear Science*, 40(12):2469–2482, December 2005.

27. P. Madoglio, M. Zanuso, S. Lavantino, C. Samori, and A. Lacaita. Quantization effects in all-digital phase-locked loops. *IEEE Transactions on Circuits and Systems II*, 54(12):1120–1124, December 2007.

28. C. Hsu, M. Straayer, and M. Perrott. A low-noise wide-BW 3.6-GHz digital $\Delta\Sigma$ fractional-N frequency synthesizer with a noise-shaping time-to-digital converter and quantization noise cancellation. *IEEE Journal of Solid-State Circuits*, 43(12):2776–2786, December 2008.

29. G. Gillette. The digiphase synthesizer. In *Proceedings of 23rd Annual Frequency Control Symposium*, Fort Monmouth, NJ, 1969, pp. 25–29.

30. J. Gibbs and R. Temple. Frequency domain yields its data to phase-locked synthesizer. *Electronics*, 27:107–113, April 1978.

31. V. Reinhardt. Spur reduction technique in direct digital synthesizer. In *Proceedings of 47th Frequency Control Symposium*, October 1993, pp. 230–241.

32. T. Riley, M. Copeland, and T. Kwanieski. Delta-sigma modulation in fractional-N frequency synthesis. *IEEE Journal of Solid-State Circuits*, 28(5):553–559, 1991.

33. S. Meninger and M. Perrott. A fractional-n frequency synthesizer architecture utilizing a match compensated PFD/DAC structure for reduced quantization-induced phase noise. *IEEE Journal of Solid-State Circuits*, 50(11):839–849, November 2003.

34. E. Temporiti, C. Wu, D. Baldi, R. Tonietto, and F. Svelto. A 3 GHz fractional all-digital PLL with a 1.8 MHz bandwidth implementing spur reduction techniques. *IEEE Journal of Solid-State Circuits*, 44(3):824–834, 2009.

35. M. Lee, M. Heidari, and A. Abidi. A low-noise wideband digital phase-locked loop based on a coarse-fine time-to-digital converter with subpicosecond resolution. *IEEE Journal of Solid-State Circuits*, 44(10):2808–2816, October 2009.

36. E. Temporiti, C. Wu, D. Baldi, R. M. Cusmai, and F. Svelto. A 3.5 GHz wideband ADPLL with fractional spur suppression through TDC dithering and feed-forward compensation. *IEEE Journal of Solid-State Circuits*, 45(12):2723–2736, 2010.

9 Time-Mode Circuit Concepts and Their Transition to All-Digital Synthesizable Circuits

Moataz Abdelfattah and Gordon W. Roberts

CONTENTS

This chapter outlines the general concepts related to time-mode signal processing (TMSP) and some of its state-of-the-art applications. These provide a very good alternative to conventional techniques, which suffer from problems such as linearity and accuracy limitations among others. The ultimate goal of this chapter is to arrive at all-digital time-domain circuits that can be synthesized using existing digital CAD tools and to make the design process fully automated in contrast to its conventional analog counterpart.

9.1   INTRODUCTION

Over the past few decades, advancements in CMOS technology have favored the performance and integration densities of digital circuits, while analog circuits suffer from the adverse effects of channel length reduction [1]. Advancements in CMOS technology focus on improving digital aspects such as switching speeds and design density. Apart from the lack of accurate analog models, long-channel devices are expensive to use, as they consume unreasonable levels of silicon areas and question the reason for using an expensive fine-line CMOS technology in the first place.

Furthermore, and to reduce cost and increase productivity, analog and mixed-signal circuits are integrated with digital circuits. Analog circuits are thus affected by the reduced feature size of the devices, which forces lower operating voltages, which may not be optimal for analog operation and result in smaller voltage swings and worse linearity in addition to other problems such as current leakage and coupling noise between digital and analog sections of the system.

A conventional data-processing system to perform the processing of analog information [2] is illustrated in Figure 9.1. Most of the processing is usually performed in the digital domain; consequently, a front-end analog-to-digital converter (ADC)

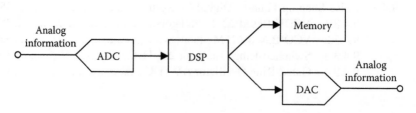

FIGURE 9.1   Conventional data-processing system.

is used to convert the input analog information into digital bits and continue the processing in digital domain. TMSP is a candidate for replacing traditional signal-processing techniques as a means to counteract the challenges imposed by conventional processing techniques used in ADC design.

TMSP is an approach that will enable analog signal-processing techniques to be realized in advanced CMOS processes, such as those that utilize FinFET device technologies. Moreover, such circuits can be captured by existing digital design tools and methodologies, thereby improving the overall productivity of analog design.

The fundamental building block of TMSP is a "modified" digital inverter circuit. As a result, TMSP circuits have the capacity be synthesized and realized using existing computer-aided design (CAD) tools and methodologies for digital circuits. This includes those used for design-for-manufacturability, that is, test and yield learning and optimization. TMSP uses time-differences to represent information; however, as time is not an actual physical variable, another variable must be used to represent this information. Recently, this variable was chosen to be the electric charge on a capacitor. This approach is different from previous approaches whereby various time-to-voltage or voltage-to-time conversion operations were included within each block. The conversion from the voltage (or current) domain to the time domain and back leads to signal loss and linearity issues.

The idea behind TMSP is based on considering time as the variable under processing instead of conventional analog variables such as current or voltage. Any positive, negative, or zero analog quantity can correspond to a positive, negative, or zero time-mode variable, respectively. A time-mode variable is defined as the time (or phase) difference between two digital rising edges. Figure 9.2 illustrates the correspondence between voltage domain and time-difference signals. The attractiveness of TMSP is based on the digital nature of the signals representing a time-mode variable; hence, the circuits involved in TMSP consist of a digital CMOS construction, while the variable under processing is still an analog quantity. Therefore, in TMSP, analog accuracy and digital design can be combined together.

As will be explained later on, most of the achievements reported by researchers and performed in TMSP are restricted to the implementation of fast ADCs or delta-sigma modulators. This limits their application to the front-end ADC operation or the back-end DAC operation of Figure 9.1. TMSP has never been considered as a direct general processing tool for analog information.

Processing signals in the time-domain is advantageous particularly, because ADC design is getting more and more challenging. As the resolution of the ADC increases, the silicon area and design cost increase exponentially. Considering TMSP as an alternative, postprocessing can be done in the digital domain using analog (time-mode) variables without the actual need for an ADC. The system will still keep all the benefits of a digital design. Two disadvantages for TMSP would be that the circuit outputs have timing limitations and that there is an inherent trade-off between dynamic range and bandwidth of time-mode circuits, which can be overcome as well.

In the rest of this chapter, the basics of TMSP design will be discussed starting with the definition of a "time-mode variable" and the TMSP ideology. Section 9.3 describes the basic building blocks needed to build TMSP circuits such as the "voltage-controlled delay unit" (VCDU) or the "switched-delay unit" (SDU) analogous

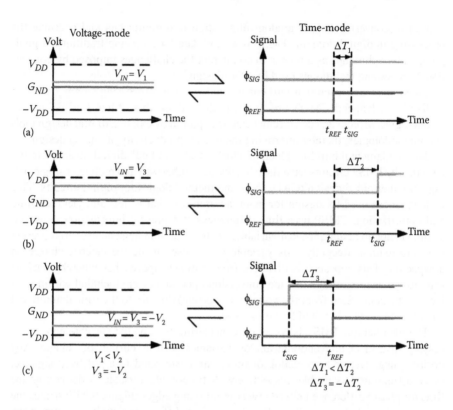

FIGURE 9.2 Definitions of various time differences as compared to voltage-domain signals: (a) a small positive voltage and time difference, (b) a large positive voltage and time difference, and (c) negative voltage and time difference.

to inverters in the digital domain, for example. These building blocks are used to implement more complex operations such as time-latching, multiplication/division, integration, etc. Then we move on to signal-processing applications such as data conversion and filtering, and explain the techniques devised to implement these functions in the time domain. Also, a method to convert between the various domains must exist to ensure continuity with a DSP-based paradigm, which leads to the design of time-to-digital converters (TDCs) and digital-to-time converters (DTCs). Phase-locked loops (PLLs), types of filters (FIR, IIR, etc.) and data conversion techniques (Counter-type, Flash, Vernier-delay, etc.) are discussed in detail. Finally, some recent advancements and emerging applications such as quadrature phase shift keying (QPSK) modulation and new VCDU architectures are introduced, which opens the door to a wide possibility of countless future applications.

9.2 TM CONCEPTS

TMSP can be defined as the detection, storage, and manipulation of analog information using time-mode variables. Clearly, this is a sampled data mechanism that provides a means to implement analog signal-processing functions in any technology

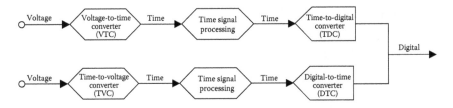

FIGURE 9.3 TMSP ideology.

using the most basic element available, that is, propagation delay [2]. In CMOS technology, the lower limit would be the delay of a static inverter, while theoretically, there is no upper limit. This is not the case for voltage-based analog design.

To relate time-mode processing to conventional voltage processing, the block diagram shown in Figure 9.3 can be utilized. Since most information comes from sensors in the form of a time-varying voltage, a Voltage-to-Time Converter (VTC) is employed to convert the input signal into a time-mode variable. The time signal is then processed by various circuits, resulting in a time-difference output. Finally, the processed time signal is transformed into a digital representation using a TDC. By adopting TMSP for the processing of analog time-mode variables, there will be no more need for digital quantization and multibit digital computations. The process can be reversed using the complementary blocks known as DTC as well as time-to-voltage converters (TVC), also shown in Figure 9.3.

Although the emphasis of this chapter is to develop the TMSP method as a mostly digital technique, some analog circuits related to this topic are also included here to provide a general overview of the many TMSP techniques that have been proposed.

9.2.1 TM Variable and Operations

A time-mode variable, ΔT, can be defined in one of the two ways [2] similar to a voltage/current variable. It can be the differential time or phase difference between rising edges of two step-like digital signals where one of the signals is considered as the reference (or ground) and the other one as the signal (e.g., Figure 9.2). The second alternative to define ΔT is the duration of a single digital pulse. While the first definition is bipolar, this definition is unipolar, that is, can be used to implement only positive quantities.

Operations in the time domain can be either asynchronous or synchronized to a common reference signal, which we call the clock. Some synchronous operations can be problematic because of the possibility of noncausality, for example, subtraction. This can usually be circumvented by forcing an extra unit delay in the signal path, or converting an otherwise synchronous design to an asynchronous one.

We can define basic arithmetic operations in the time domain as is done in the voltage domain. These include time addition, time subtraction, and scalar multiplication and division. However, one of the most important operations is integration, since it is an integral part in many signal-processing algorithms. Another important operation is "memory" or time-difference variable storage for later retrieval. This can also serve as a delay cell to solve the noncausality issue of synchronous TM systems. Other operations include comparison, delay, etc.

9.3   TM BUILDING BLOCKS

In this section, different building blocks essential to understand and implement basic functions in TMSP will be reviewed [3,4].

9.3.1   ANALOG APPROACHES

The transition from analog circuits to synthesizable all-digital TM circuits goes through intermediate transitions whereby the circuits can be considered as "part analog" and "part digital." In this subsection, such circuits are described briefly.

9.3.1.1   Voltage-Controlled Delay Unit

A VCDU linearly delays an input time event (i.e., a rising or falling edge) with respect to a sampled input voltage $V_{IN}[n]$ such that the rising edge is delayed by an amount that is proportional to this input voltage level. Figure 9.4 illustrates both the symbol and its corresponding timing diagram. The VCDU consists of a digital input and output port, denoted as ϕ_{IN} and ϕ_{OUT}, and a voltage-controlled input port $V_{IN}[n]$. For an input time waveform that has a low-to-high transition at time t_{IN} and a sampled input voltage $V_{IN}[n]$, the output digital or step signal has a low-to-high transition some time later at $t_{OUT} = t_{IN} + G_\phi V_{IN}[n]$, as illustrated in Figure 9.4b. In practice, an actual circuit implementation may introduce a time offset, which suggests the general behavior of a VCDU is best described in general terms as

$$T_{Delay}\left[n\right] = t_{OUT} - t_{IN} = G_\phi V_{IN}[n] + b_\varnothing \tag{9.1}$$

where G_ϕ and b_ϕ are arbitrary gain and offset coefficients.

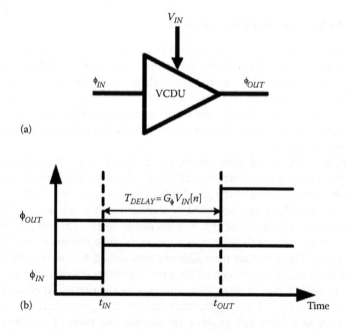

(a)

(b)

FIGURE 9.4   A VCDU: (a) symbol and (b) timing diagram.

To have a one-to-one linear time-invariant (LTI) relation between the input voltage and the input–output delay, the constant term b_ϕ should be removed. This can be done by adopting a two VCDU arrangement as shown in Figure 9.6. Here the same digital input drives the input to each VCDU but with different control voltage levels, $V_{IN}[n]$ and V_{REF}. The time difference between the two low-to-high output transitions can then be described as

$$T_{DELAY}\left[n\right]=t_{OUT,\,SIG}-t_{OUT,\,REF}=G_\varnothing(V_{IN}\left[n\right]-V_{REF}) \qquad (9.2)$$

Here one arrives at a VTC with no time offset. This of course assumes that the two VCDUs are well matched.

9.3.1.2 Voltage-to-Time Adder

A voltage-to-time adder can be realized using almost the same configuration as the offset-free two-VCDU arrangement of Figure 9.6. By applying two separate digital inputs, say $\phi_{IN,SIG}$ and $\phi_{IN,REF}$, with edge transitions of $t_{IN,SIG}$ and $t_{IN,REF}$, respectively, to the inputs of two matched VCDUs driven by control voltages $V_{IN}[n]$ and V_{REF} as shown in Figure 9.5 leads to the output time-difference equation

$$\Delta T_O\left[n\right]=t_{OUT,\,SIG}-t_{OUT,\,REF}=t_{IN,\,SIG}-t_{IN,\,REF}+G_\varnothing(V_{IN}\left[n\right]-V_{REF}) \qquad (9.3)$$

Here one sees that the output time difference $\Delta T_O[n]$ is equal to the time difference of the two input digital signals plus a scaled factor of the input voltage difference, hence, forming a dual-mode adder circuit.

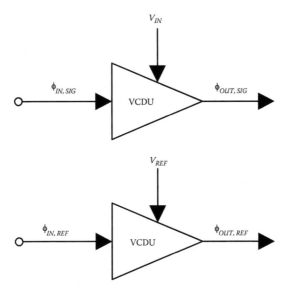

FIGURE 9.5 Voltage-to-time adder realization from two matched VCDUs.

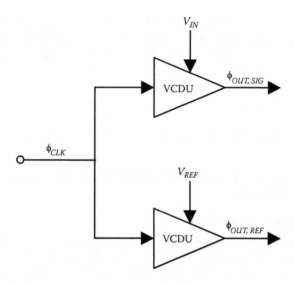

FIGURE 9.6   A two VCDU arrangement with no time offset.

9.3.1.3 Voltage-to-Time Integrator

By connecting the output of the dual-mode adder circuit of Figure 9.6 back to its input through an inverter block as shown in Figure 9.7, a voltage-to-time integrator results. One can show, after some algebra, that the output time difference between the two digital output signals ϕ_{OUT} and ϕ_{REF} at any sampling instant is the sum of the previous output time difference and a scale version of the present instant input voltage difference, that is,

$$\Delta T_O[n] = \Delta T_O[n-1] + G_{\varnothing}(V_{IN}[n] - V_{REF}) \qquad (9.4)$$

This is effectively a circuit that performs a discrete-time integration operation on the input voltage difference. One may recognize that the resulting circuit shown in Figure 9.7 resembles a ring oscillator in differential form. One can therefore say that the difference in the control voltage of each ring oscillator is integrated onto the phase difference between the two oscillators.

It would be desirable for many signal-processing algorithms to cascade two integrators so that a higher-order integration function can be realized. However, since a time-difference variable is not a physical stationary quantity, the summation of two time-difference variables cannot be done without first transforming them into some intermediate physical quantity such as an electrical charge, voltage, or current variable.

9.3.1.4 Gated-Ring Oscillator

In many time-mode applications, a ring oscillator that can be turned on or off through an external control is used as shown in Figure 9.8. During the hold phase, the state of

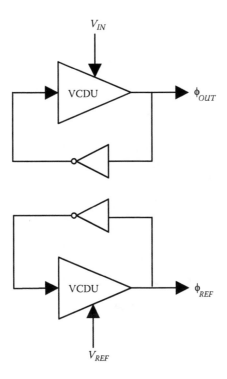

FIGURE 9.7   Voltage-to-time integrator using two matched VCDU blocks with inverters in its feedback path. This realization is equivalent to a differential ring oscillator.

the internal transistor, that is, charge on various transistor gate, is meant to be held constant for an indefinite time period. Of course, charge leakage effects will cause these state values to change and introduce errors into the phase of the next oscillation phase, limiting the hold phase to some short-time value.

9.3.1.5   Time-Mode Quantizers

Any signal-processing algorithm would be expected to make decisions as time progresses. Below are two types of decision-making elements will be described: single-bit and multibit quantizers.

A single-bit quantizer compares the edge position of one digital signal against the edge position of another. If one leads the other, an output logic high level is set; otherwise, a logic low level is set as depicted by the symbol shown in Figure 9.9a. A very simple implementation that can realize this operation is a D-type flip-flop. If we assume two digital signals ϕ_{IN} and ϕ_{REF} are applied to the D and clock inputs of a D-type flip-flop as shown in Figure 9.9b, then the D-type flip-flop will go to a logic high level as ϕ_{IN} leads ϕ_{REF}. Of course, the reverse situation where the edge transition of ϕ_{IN} lags that of ϕ_{REF} would result in an output logic low level.

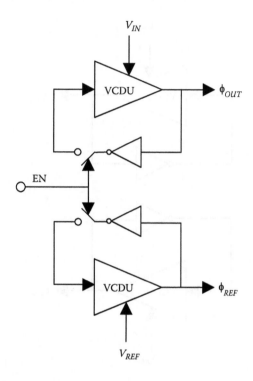

FIGURE 9.8   GRO implementation.

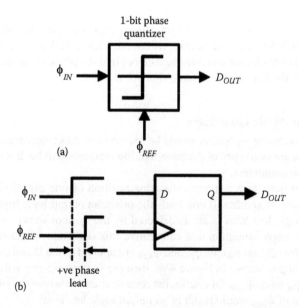

FIGURE 9.9   Single-bit quantizer: (a) symbol, and (b) single D-type flip-flop implementation.

A multibit quantizer compares the edge position of one digital signal ϕ_{IN} against the edge position of a reference signal ϕ_{REF} but generates a multibit digital output. In effect, an m-bit quantizer divides the input time-difference range into 2^m equal partitions and identifies which region the input time-difference lies. A symbol of the multibit quantizer is shown in Figure 9.10a. One very elegant implementation is shown in Figure 9.10b. It consists of a cascade of 2^m VCDU stages with individual delay T_{VCDU}, which together form a delay line. In addition, each tap along the delay line is connected to the clock input of a D-type flip-flop. The D-input of all the flip-flops is driven by the input digital signal ϕ_{IN} and is compared to the various delayed reference signals. As each flip-flop will see a rising input at its clock input T_{VCDU} seconds after the previous one, the reference signal will climb toward the input signal with a step size of T_{VCDU} seconds, eventually overtaking it; we shall refer to this instant as the switchover point. All the flip-flops trigger before the switchover point that would be set logically low and all others would be set logically high. Simply adding up all the zeros and passing this result as the output would indicate the regions that the switchover point occurs. This process is illustrated by the timing diagram shown in Figure 9.10c.

It should be noted that time amplifiers can be used at the front-end of a quantizer. This serves the purpose of setting the input signal to the full-scale range prior to the quantization step. This maximizes the signal-to-quantization noise ratio of the overall conversion process [5,6].

9.3.2 ALL-DIGITAL TM BUILDING BLOCKS

The ultimate goal of this work is to make the design of asynchronous TM circuits feasible using readily available digital structures used by designers everywhere. To this end, the time-latch design [7] introduced in this section can be used in more complex TM-mode circuits such as time adders/subtractors, time multiplication/division-by-a-constant circuits. In addition, time-to-time integration can be implemented without the need for digital oscillators circuits as in reference 8. This eliminates the timing conditions related to the oscillation period and poses fewer restrictions on the circuit's dynamic range requirements. The design of first- and second-order integrators will be briefly explained.

9.3.2.1 Switched Delay Unit

VCDUs are frequently used in TMSP to introduce controlled delays using an analog control voltage. For the most part, such delays are realized by varying the time constant of a charging/discharging process. Another way to look at it would be to fix the time constant but vary the charging/discharging time through the use of an on-off switch. This eliminates the need for an analog control voltage.

Performing a role similar to that of the digital inverter, an SDU can be considered as one of the basic building blocks of time-mode systems [4]. Figure 9.11 shows the schematic circuit of the SDU and a timing diagram. The SDU has two digital inputs, Φ_{IN} and SW, and one digital output Φ_{OUT} and two stages of inversion. The front-end inverter stage consisting of transistors M_1–M_4 is loaded with additional capacitance C to control its charging/discharging rate. The inverter closest to the

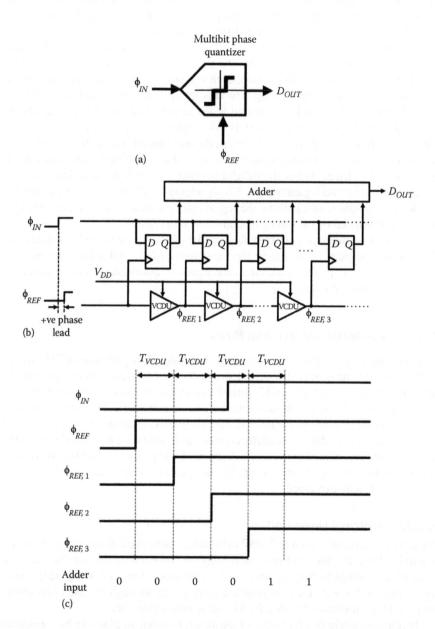

FIGURE 9.10 Multibit quantizer: (a) symbol, (b) one particular implementation, and (c) timing diagram.

output consisting of transistors M_5 and M_6 is used to buffer the charging/discharging node from any loading effect at the output. Initially, C is charged to V_{DD} and the SW signal is set high. On the arrival of a rising edge at Φ_{IN}, C starts to discharge through M_2–M_4. If the SW signal is held high, the output of the inverter Φ_{OUT} is triggered from low to high after some time interval, which we denote as T_{SDU}. One can show

$$T_{SDU} = \frac{C(V_{DD} - V_{TH,INV})}{I} \tag{9.5}$$

where

I is the discharging current

$V_{TH,INV}$ is the threshold voltage of the second inverter involving M_5 and M_6 as
described in [7]

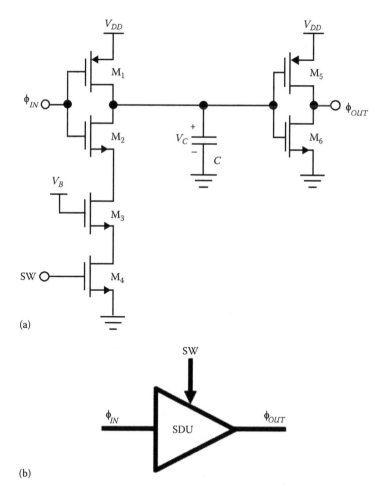

(a)

(b)

FIGURE 9.11   An SDU: (a) schematic and (b) symbol diagram.   *(Continued)*

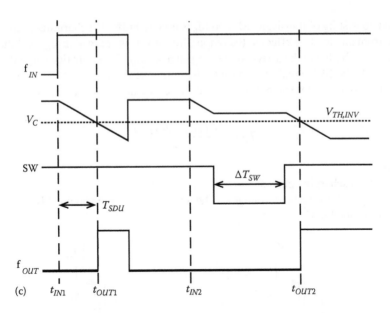

FIGURE 9.11 (Continued) An SDU: (c) timing diagram.

Now, if the SW signal experiences a low pulse with ΔT_{SW} duration during the discharge process before the output is triggered, as shown in the timing diagram of Figure 9.11b, the total delay from input to output rising edges becomes

$$T_{OUT} = T_{SDU} + T_{SW} \tag{9.6}$$

A differential configuration (not shown here) can enhance the unit's performance by reducing the errors due to charge injection, leakage currents, and other circuit-related effects.

9.3.2.2 Time-Mode Read/Write Memory Cell (TLatch)

The SDU can be incorporated into a circuit that can memorize a TM variable and allow for its retrieval some later time, essentially forming a read/write time-memory cell. Figure 9.12 illustrates the schematic or logic diagram of this memory structure. It consists of two inputs, $\phi_{IN,SIG}$ and $\phi_{IN,REF}$, which carry the input TM variable ΔT_{IN} and two output signals, $\phi_{OUT,SIG}$ and $\phi_{OUT,REF}$, which carry the output TM variable ΔT_{OUT}. In addition, there are two control digital inputs acting as the read and write signals (R and W, respectively).

As with other memories, it operates over three phases: write, idle, and read. In the write phase, the input signal and the reference pass through two delay cells. After the arrival of both edges at the input, the propagation delays of both paths will be set to infinity via the internal SW control lines to postpone the occurrence of the rising edges at the output of the memory cell until the read phase is initiated. As long as the delay lines are set in their open state (i.e., infinite delay), the memory cell is in its idle state and keeps the stored TM

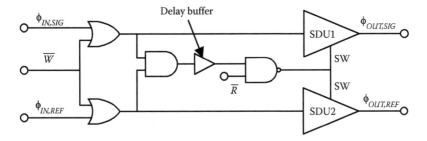

FIGURE 9.12 Schematic/logic diagram of a read/writable time-memory cell.

variable, as charge stored on the internal capacitor C of each SDU. During the read phase, the delay in each path will be set back to its nominal value to deliver the rising edges at the output of the memory cell, resulting in input–output relationship

$$\Delta T_{OUT}[n] = \Delta T_{IN}[n-1] \tag{9.7}$$

The relative timing of these various signals is captured in the timing diagram shown in Figure 9.13. The signals $V_{C,SDU1}$ and $V_{C,SDU2}$ in this figure represent the voltages across the capacitors inside SDU1 and SDU2, respectively. Here, we see that the initial TM input signal is written into the TM memory and retrieved some time later as an output signal.

A memory cell with an add function [4], which we shall refer to as a TLatch, can be created by simply attaching two memory cells in parallel but with different read signals (R_{SIG} and R_{REF}) as shown in Figure 9.14. Both the schematic diagram and the symbol are shown. One can show that time difference between the low-to-high transitions of the two read signals ΔT_R simply adds to the value stored in the TLatch, that is,

$$\Delta T_{OUT}[n] = \Delta T_{IN}[n-1] + \Delta T_R[n-1] \tag{9.8}$$

A timing diagram illustrating the operation of the TLatch is shown in Figure 9.15.

9.3.3 TM Multipliers and Dividers

Scalar operations such as multiplication or division by a constant form the basis of complicated circuit functions and can be used to implement complex analog or digital circuits such as filters.

Multiplying a TM quantity by a factor of 2 can easily be realized by adding a TM variable twice. A circuit that performs this function involving two TLatch circuits is shown in Figure 9.16. It works by basically storing the input time-mode value in a TLatch and adding this value to itself in a second TLatch. The output will be available asynchronously after a delay of approximately $2T_{SDU}$, that is,

$$\Delta T_{OUT}[n] = 2 \cdot \Delta T_{IN}[n-1] \tag{9.9}$$

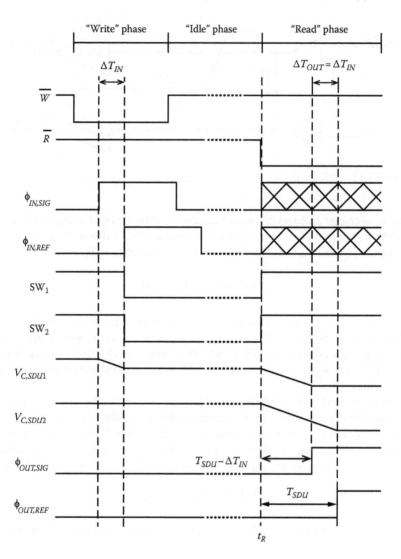

FIGURE 9.13 Timing diagram of the TM memory cell. $V_{C,SDU1}$ and $V_{C,SDU2}$ are the voltages across the internal load capacitors C of each SDU.

Clearly, this can be extended to multiplication by higher powers of 2 by simply cascading several of these units together.

The division-by-2 operation, however, is not as straightforward. One could possibly derive several complex ways to implement it accurately using DLL-based or Vernier delay methods, but the goal is to find simple feed-forward structures analogous to those found in the digital domain. Figure 9.17 shows one such structure. The idea is to simply store the input TM value in two separate SDUs and once the data is ready, the two SDU outputs are shorted together using digital transmission gates to share the

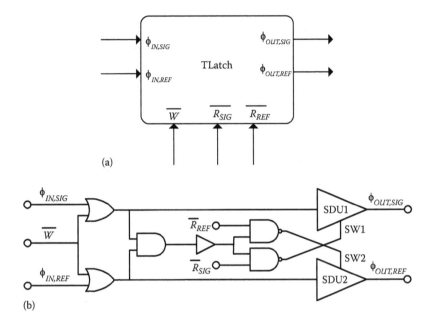

(a)

(b)

FIGURE 9.14 A TLatch circuit: (a) symbol and (b) schematic/logic diagram.

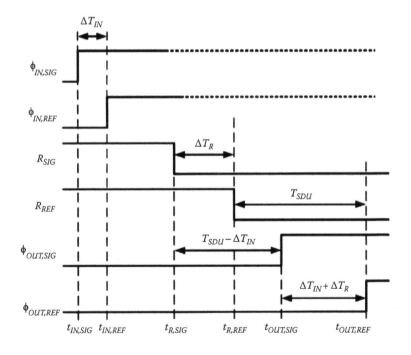

FIGURE 9.15 Timing diagram of the TLatch memory with adder function cell of Figure 9.14.

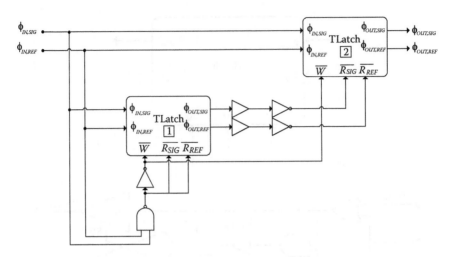

FIGURE 9.16 A schematic/logic diagram of a TM multiply-by-2 circuit.

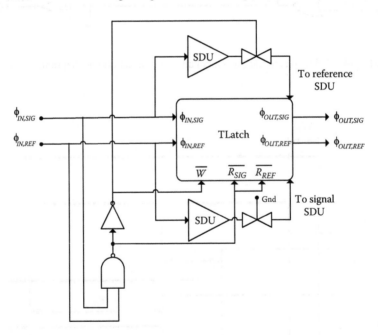

FIGURE 9.17 A schematic/logic diagram of a divide-by-2 TM circuit. The TLatch was slightly modified to include logic signals that manipulate the charge stored in its internal SDU capacitors.

charge stored across the SDU capacitors and then the output can be read out after an intrinsic delay of approximately T_{SDU} seconds with the input–output relation given by

$$\Delta T_{OUT}[n] = \frac{1}{2}\Delta T_{IN}[n-1] \tag{9.10}$$

Division by higher powers of 2 is feasible by simply cascading these circuits. Combinations of the multiplication and division circuits can be used to virtually realize all coefficients of a larger system via binary quantization, that is, in powers of 2.

9.3.4 TIME-MODE INTEGRATORS

Analog circuits require differentiation/integration functions as basic signal-processing tools. Specifically, integrators are widely used in data converters, filters, and PLLs to provide certain transfer characteristics and perform functions such as feedback error control and error correction in the corresponding system loops.

A first-order TM integrator can be constructed using two TLatch circuits as shown in Figure 9.18a. One TLatch circuit is used for storing the input TM variable, while the other is used to combine this value to the previous output value. Unlike other designs [8], this design operates asynchronously and does not require the use of any oscillators. This circuit performs the integration operation described by

$$\Delta T_{OUT}(n) = \Delta T_{IN}(n-1) + \Delta T_{OUT}(n-1) \tag{9.11}$$

Clearly, this is a first-order delayed TM integrator with both a time-difference input and output. The design of a first-order delay-free TM integrator is not yet known, but, in most practical situations, there is usually a way to work around this limitation (more on this in Section 9.5). To illustrate its step behavior, Figure 9.18b illustrates the SPICE simulation of this circuit subject to a 10 ns step input signal. As is evident, the results behave according to Equation 9.11.

Higher-order integration functions are often used to realize high-performance $\Sigma\Delta$ data converters and high-order frequency-selective filters. A second-order integrator can be realized using a set of TLatch blocks and some logic as shown in Figure 9.19a. The input–output behavior can be described by the second-order difference equation:

$$\Delta T_{OUT}(n) = \Delta T_{IN}(n-1) + 2\Delta T_{OUT}(n-1) - \Delta T_{OUT}(n-2) \tag{9.12}$$

The circuit operates as follows: The TM input at time instant $[n-1]$ is first stored in TLatch 1. Next, the value stored in TLatch 3, which corresponds to the TM output at instant $[n-1]$, is added to the value stored in TLatch 4, which corresponds to the output at instant $[n-2]$. Then, the result is added to the value stored in TLatch 1. Now, the output from TLatch 1, which is actually the required output at instant $[n]$, is simultaneously stored in TLatch 2 and 3 in place of the previous output at instant $[n-1]$. At the same time, the TM output at instant $[n-1]$ is stored in TLatch 5 and transferred to TLatch 4 to replace the previous output at instant $[n-2]$. The process is repeated when a new TM input arrives.

To illustrate its step behavior, Figure 9.19b illustrates the SPICE simulation of this circuit subject to a 10 ns step input signal. As is evident, the results behave according to Equation 9.12.

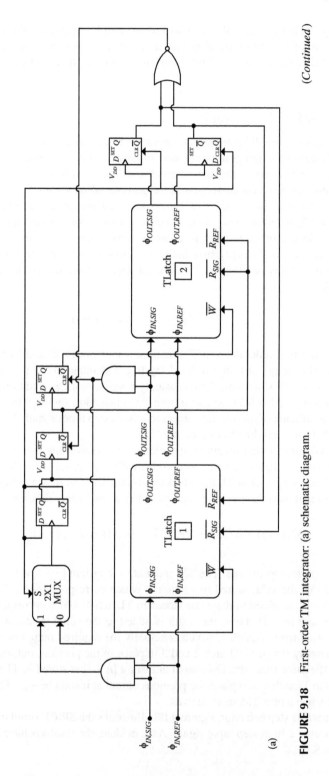

(a)

FIGURE 9.18 First-order TM integrator: (a) schematic diagram.

(Continued)

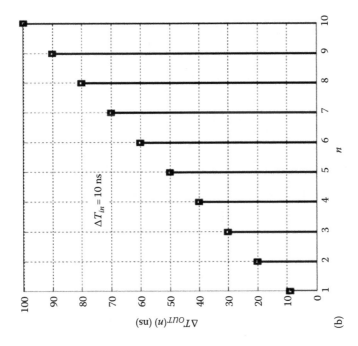

FIGURE 9.18 (Continued) First-order TM integrator: (b) simulation results to a step input.

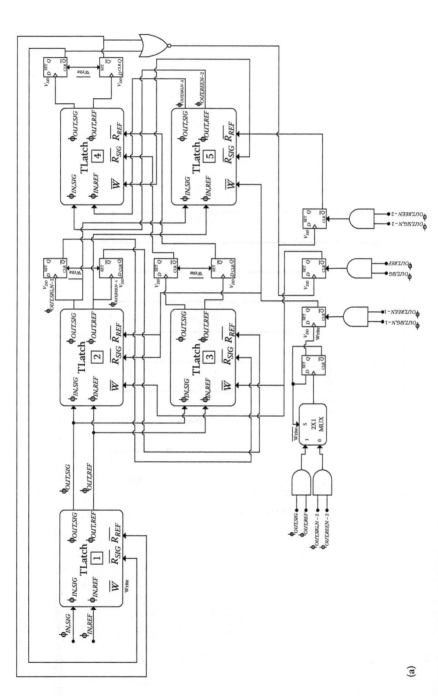

(a)

FIGURE 9.19 Second-order TM integrator: (a) schematic diagram.

(Continued)

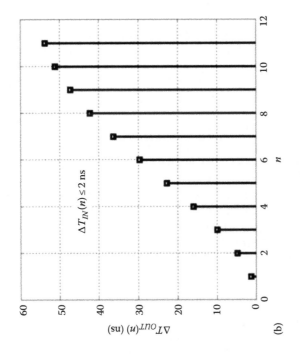

FIGURE 9.19 (*Continued*) Second-order TM integrator: (b) simulation results to a step input.

9.4   DATA CONVERTERS

TMSP provides some interesting technical advantages for implementing data converters such as small silicon footprint, low power consumption, and the ability to implement analog signal-processing functions (ASP) in advanced CMOS processes using digital-like circuits. In this section, the TM building blocks presented in the previous section will be used to develop various types of data converter realizations. This will include VCO-based $\Delta\Sigma$ modulators for analog-to-digital conversion, and DTC and TDC.

9.4.1   NYQUIST-RATE DIGITAL-TO-TIME CONVERTERS

A (DTC) takes a digital input and converts it into an output TM variable. The time difference between the rising edge of a digital output signal and a clock reference will be used to represent the analog equivalent of the input digital code.

A delay-locked loop (DLL) when combined with a many-to-one multiplexer (MUX) forms a DTC. A DLL is quite similar to a PLL; it is a servomechanism in which the total delay through a path of VCDUs, as opposed to the frequency of a VCO, is automatically adjusted so that it equals the period of the incoming clock signal. Figure 9.20 illustrates the basic idea behind this DTC. Here a clock signal is applied to the input of a DLL consisting of a stage of N (= 8) VCDUs in cascade, a phase detector, and a loop filter. The delayed output signal of the DLL is compared to the clock input, and the phase difference modulo 2π is extracted using a phase detector. The phase detector output is used to control the input voltage of the VCDUs in the delay chain through a negative feedback loop. Once the loop is locked, the total phase through the VCDU chain is equal to the period of the incoming clock signal, say T_{CLK}. Correspondingly, each VCDU will therefore introduce a delay of T_{CLK}/N, or in this particular case, $T_{CLK}/8$. By digitally selecting which of the eight-phase-delayed outputs should be passed to the output using a many-to-one MUX, a DTC results.

9.4.2   NYQUIST-RATE TIME-TO-DIGITAL CONVERTERS

Nyquist-rate TDCs have been around for a long time and are widely employed in various applications such as automatic test equipment, time-of-flight measurements, and for measuring the timing jitter associated with clock signals. Contrary to a DTC, a TDC is used to convert a time interval, say between two rising edges associated with a digital signal, into a binary-formatted integer value. The following is a brief description of the various TDCs that have been discussed in the open literature. As one will see, many of the topologies are quite similar to those used to construct ADCs in the conventional voltage/current domain [9].

9.4.2.1   Single Counter

A very simple TDC is one that counts the number of clock cycles that completes within the time interval between the rising edges of the start and stop signals as measured by a counter driven by a high-frequency reference clock as depicted by the block diagram shown in Figure 9.21. The method is very simple to implement;

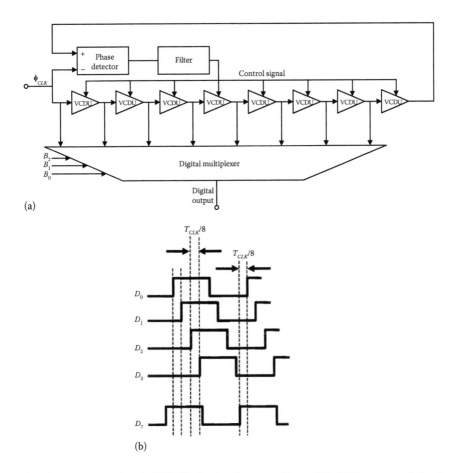

(a)

(b)

FIGURE 9.20 DLL-based DTC: (a) circuit schematic, (b) possible DTC outputs relative to the clock input.

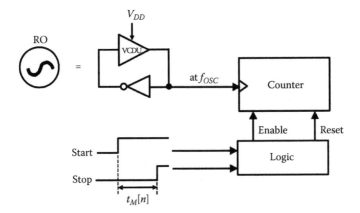

FIGURE 9.21 Counter-based TDC.

however, the resolution attained by this method is largely dependent on the clock frequency f_{OSC} of the ring oscillator and it can be no better than a single clock period. The output digital value $D_{OUT}[n]$ can be described as

$$D_{OUT}\left[n\right]=\frac{1}{T_{OSC}}\left(t_{M}\left[n\right]-T_{E}\left[n\right]\right) \tag{9.13}$$

where
  $t_M[n]$ denotes the input time difference between the start and stop signals
  T_{OSC} is the oscillation period of the ring oscillator
  $T_E[n]$ represents the sampling time quantization error in seconds

9.4.2.2   Flash TDC

Flash TDCs are analogous to flash ADCs for voltage-amplitude encoding. Flash TDCs are the fastest type of TDCs and are well suited for use in on-chip timing measurement systems, because they are capable of performing a measurement on every clock cycle. One such realization is shown in Figure 9.22. This kind of TDC compares an input signal edge ϕ_{SIG} with respect to various reference edges all displaced in time derived from a single reference signal ϕ_{REF}. The basic operation of this TDC is equivalent to that described for the multibit quantizer of Section 9.3.1. Here a delay line is implemented as a cascade of VCDUs, but they are phased-locked using a DLL acting on the input clock reference ϕ_{REF}. The DLL mitigates against temperature, supply voltage, and process variations (PVT) that the delay line will be subject to during normal operation.

The time difference between the input rising edge and set of delayed reference signals is compared with the flip-flop and the corresponding output code is either counted or simply applied to a digital decoder circuit that performs the count through a truth table operation. Figure 9.22 illustrates the latter case. The output digital code can be described as

$$D_{OUT}\left[n\right]=\frac{1}{\Delta t_{RES}}\left(t_{M}\left[n\right]+T_{E}\left[n\right]\right) \tag{9.14}$$

where Δt_{RES} is the difference in the oscillation period of the two ring oscillators, $T_{OSC,1}=1/f_{OSC,1}$ and $T_{OSC,2}=1/f_{OSC,2}$.

The drawback to this implementation is that the temporal resolution can be no higher than the delay through a single gate in the semiconductor technology used. To achieve subgate temporal resolution, the flash converter can be constructed with a Vernier delay line such as that shown in Figure 9.23. To reduce PVT effects on the matching between the two delay lines, a DLL configuration can be used to fix the delay of each delay line.

9.4.2.3   Component-Invariant Vernier Oscillator TDC

The advantage of a Vernier class of TDCs is that it eliminates the effects caused by component variation in the delay lines of Vernier delay flash TDCs. It is slower than a flash TDC, since it takes many cycles to complete a single measurement. A Vernier oscillator TDC is composed of two ring oscillators producing plesiochronous square waves to

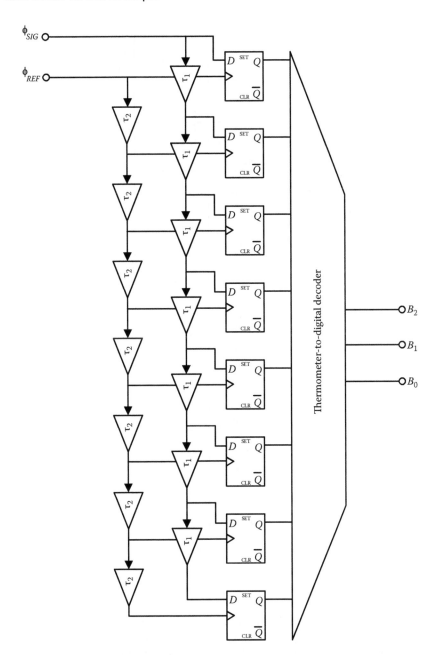

FIGURE 9.22 Vernier-based TDC.

quantize a time interval based on a very small frequency difference between the two oscillators. Phase locking the delay elements using a DLL can correct for any process variations and temperature effects in the VCDU delays. The phase detector and the counter measure the time difference between these two rising edges. The corresponding digital output code $D_{OUT}[n]$ is identical to that described by Equation 9.14.

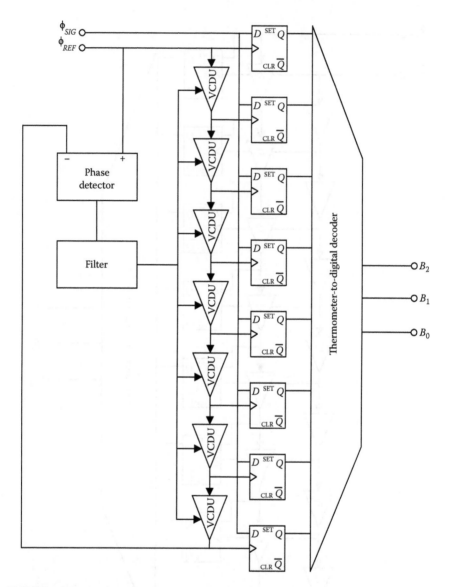

FIGURE 9.23 Flash-based TDC.

Due to its small size and relatively high temporal resolution, the Vernier oscillator is amenable for use in on-chip test systems. In addition, its main advantage is the use of the oscillators reduces the matching requirements on the delay buffers used to quantize a time interval (Figure 9.24) [10–13].

9.4.3 Noise-Shaping Time-to-Digital Converters

Various architectures have been proposed recently to implement TDCs. As mentioned earlier, a flash TDC is the simplest, but its resolution is technology limited by

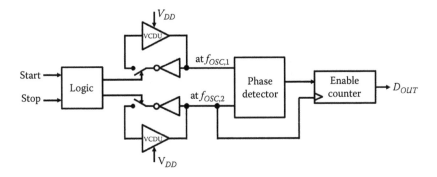

FIGURE 9.24 Component-invariant Vernier TDC.

the minimum gate delay. Vernier and pipelined TDCs can overcome this limitation, but they still achieve only moderate resolution. To improve resolution, noise-shaping TDCs using delta-sigma modulation can be used.

9.4.3.1 VCDU-Based ΔΣ Modulators

The block diagram of a single-loop, first-order ΔΣ modulator is shown Figure 9.25a. It consists of an integrator, quantizer (ADC) and a DAC in the feedback path. All three elements have a TM equivalent. The TM integrator can be seen in Figure 9.7, a single-bit quantizer was shown in Figure 9.9, and a one-bit DAC can be realized using a single VCDU shown in Figure 9.4. The addition is realized by simply cascading separate VCDU. Replacing each block with their TM equivalents leads to the ΔΣ modulator shown in Figure 9.25b. Following the analysis in [1], the digital output $D_{OUT}[n]$ is given by

$$D_{OUT}\left[n\right] = v_{IN}\left[n-1\right] + \frac{T_E\left[n\right] - T_E\left[n-1\right]}{G_\phi} \tag{9.15}$$

where
  $v_{IN}[n]$ is the input sampled voltage signal
  G_ϕ is the voltage-to-time-delay gain coefficient related to the input–output behavior of the VCDU

Here $T_E[n]$ represents the quantization error introduced by the quantizer at the nth sampling instant.
  A fully differential implementation of this ΔΣ modulator can also be realized by feeding back both the output and its complement back to VCDUs placed in series with each TM integrator circuit as shown in Figure 9.25c. Extension to multibit ΔΣ modulator is relatively straightforward by replacing the quantizer with the multibit quantizer shown in Figure 9.10, or one similar, and by using a voltage-mode multibit DAC [2].

9.4.3.2 VCO-Based ΔΣ Modulators

A block diagram of a ΔΣ modulator with a feedforward structure is illustrated in Figure 9.26a. This structure illustrates the simplest approach to constructing a

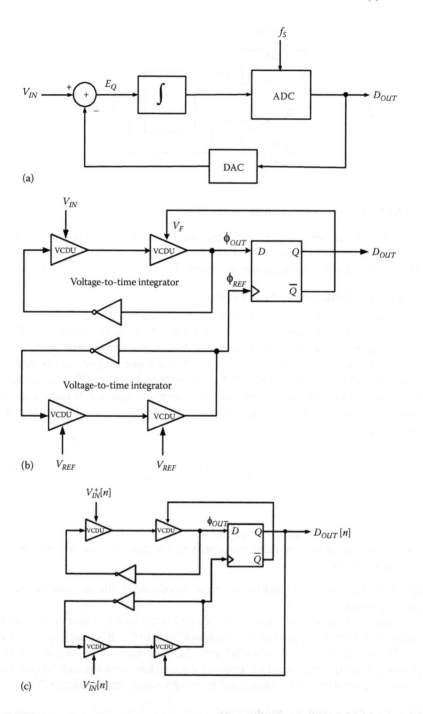

FIGURE 9.25 A single-loop, first-order TM VCDU-based ΔΣ modulator: (a) block diagram of a first-order ΔΣ modulator, (b) single-ended implementation, and (c) fully differential implementation.

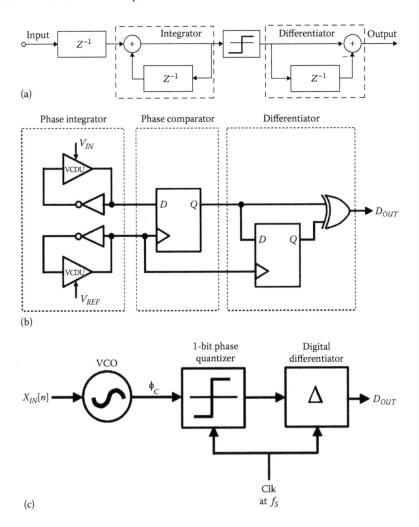

FIGURE 9.26 VCO-based $\Delta\Sigma$ modulator: (a) block diagram with individual transfer functions, (b) VCDU implementation, and (c) oscillator-based perspective in generalized block diagram form.

first-order $\Delta\Sigma$ modulator [14]. After the unit delay, the input signal is forwarded to an integrator, and is then followed by a quantizer. The output of the quantizer is then passed through a differentiator. Since the quantization error introduced by the quantizer is not integrated, it will be differentiated while the input signal passes unchanged, as it is both integrated and differentiated. Assuming a linear additive error model for the quantizer, the output of the feedforward $\Delta\Sigma$ modulators can be written as

$$D_{OUT}[n] = x_{IN}[n-1] - \left(e[n] - e[n-1]\right) \qquad (9.16)$$

Taking the z-transform leads to

$$D_{OUT}(z) = z^{-1}X(z) - (1 - z^{-1})E(z) \qquad (9.17)$$

where $X_{IN}(z)$, $D_{OUT}(z)$, and $E(z)$ represent the z-transform on the input $x_{IN}[n]$, output $D_{OUT}[n]$, and quantization error signal $e[n]$, respectively. It is common-place to refer to the coefficient in front of the $X_{IN}(z)$ term as the signal transfer function $STF(z)$ and the coefficient in front of the $E(z)$ term as noise transfer function $NTF(z)$. The form of the $NTF(z)$ suggests that a quantization error that is uniform across all frequencies will contribute a noise component that has a high-pass shape at the output and that most of the noise component will be pushed into higher frequencies away from DC. By limiting the input to a bandwidth of frequencies that are located close to DC, the resulting signal-to-noise ratio (SNR) over this bandwidth can be made quite large. From a noise perspective, this is precisely the role that any $\Delta\Sigma$ modulator is expected to perform.

From a TMSP perspective, the integrator analog block of Figure 9.26a can be implemented using a voltage-to-time integrator (Figure 9.7): the 1-bit quantizer (Figure 9.9) and the digital differentiator is easily accomplished with another D-type flip-flop implementing the delay and a XOR gate performing the modulo-2 subtraction resulting in the circuit schematic shown in Figure 9.26b. As the voltage-to-time integrator effectively forms a voltage-controlled ring oscillator (VCO), the realization of Figure 9.26b is commonly referred to as a *VCO-based $\Delta\Sigma$ modulator*. A more generalized view of the VCO-based modulator is shown in block diagram form in Figure 9.26c.

The digital output from the VCO-based $\Delta\Sigma$ modulator can be described as

$$D_{OUT}[n] = K_{VCO} \cdot T_S \cdot x[n] - \frac{1}{2\pi}\left(\phi_{e,E}[n] - \phi_{e,E}[n-1]\right) \qquad (9.18)$$

where

$D_{OUT}[n]$ is the nth time difference between the present count value and the previous count value

$x_{IN}[n]$ is the input voltage sample at time n

K_{VCO} is the VCO phase gain coefficient (rad/V)

$\phi_{e,E}[n]$ is the back-end (stop) quantization phase error

As the phase error contribution at any time instant, n, is dependent on the difference of present and past errors, the phase error is said to be noise-shaped. Further details related to this development can be found in Chapter 7.

The main drawback of a feedforward or VCO-based $\Delta\Sigma$ modulator is that it lacks a feedback path around the analog components, such as the integrator and the quantizer. Nonlinearities and any other nonidealities will simply be combined with the incoming signal, thereby limiting its overall performance. Also due to nonuniform sampling, there will be strong harmonic components or idle tones at the output of the modulator, which necessitates adopting calibration techniques to compensate for this accumulated nonlinearity. Nonetheless, the performance of $\Delta\Sigma$ modulator can be made quite high with a simple modification as described next.

9.4.3.3   Switched-Ring Oscillator $\Delta\Sigma$ TDC

The switched-ring oscillator (SRO) $\Delta\Sigma$ modulator is a more recent development [15] where a time-difference generator (TDG) is placed in front of a VCO-based $\Delta\Sigma$ modulator realized with a multibit quantizer as shown in Figure 9.27a. The TDG detects the time difference $t_M[n]$ between the input signal and a reference signal as shown in Figure 9.27b. The frequency of these signals is assumed to be f_C ($=1/T_C$). The generated time-difference pulse is a continuous-time pulse-width modulated signal that represents the input and has the same rate as the input signal. The TDG output is then fed to the VCO and is used to switch the VCO between two frequencies, f_H and f_L. The VCO output phase is then quantized and fed to a digital differentiator to generate the final modulator output. Both the quantization and

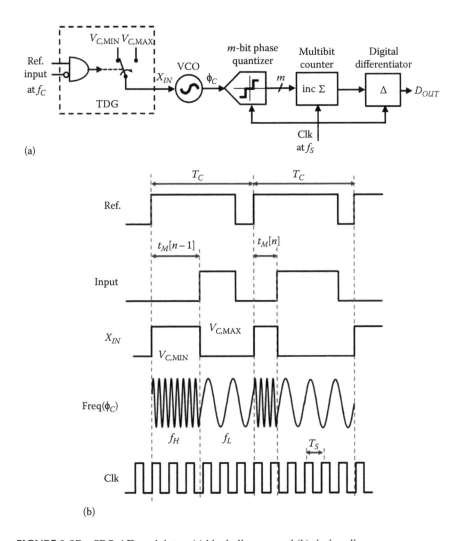

(a)

(b)

FIGURE 9.27   SRO $\Delta\Sigma$ modulator: (a) block diagram and (b) timing diagram.

differentiation operation are performed with a separate sampling clock operating at frequency f_S (=1/T_S). This sampling rate is generally operating at a much higher rate than the input signal rate f_C.

Mathematically, the digital output from the SRO $\Delta\Sigma$ modulator can be described by

$$D_{OUT}\left[n\right] = K_{VCO} \cdot T_S \cdot \left(\frac{t_M\left[n\right]}{T_C}\right) - \frac{1}{2\pi}\left(\phi_{e,E}\left[n\right] - \phi_{e,E}\left[n-1\right]\right) \qquad (9.19)$$

This expression reveals the key attribute of the SRO $\Delta\Sigma$ modulator; specifically, the input signal rate f_C is separate from the sampling rate f_S and that the quantization error is noise-shaped. Having the sampling rate separate from the input signal rate enables one to set a high oversampling ratio, thereby improving the overall SNR performance of the $\Delta\Sigma$ modulator. Further details related to this development can be found in Chapter 7.

9.4.3.4   Gated-Ring Oscillator $\Delta\Sigma$ TDC

The counter-based Nyquist-rate TDC described in the previous subsection can be modified slightly in order to noise-shape the quantization noise created by the counting processes. Through the application of a GRO, a multibit quantizer, and a differentiator, a $\Delta\Sigma$ TDC [16] can be created as shown in Figure 9.28a. Such a realization has come to be known as a GRO $\Delta\Sigma$ modulator.

The main objective of the GRO is to hold the state of the oscillator constant so that the phase of the oscillator can return to the exact same value it had when the stop edge arrived to start the next count cycle. By doing so, the quantization error made at the end of the previous count $T_{e,E}[n-1]$ plus the error made at the start of the next count cycle $T_{e,B}[n]$ will be equal to the time delay between any two adjacent taps of the ring oscillator, that is, $T_{OSC}/2^m$. Thus, the past and present errors become highly correlated and will introduce a noise-shaping factor in the digital output count value, that is,

$$D_{OUT}\left[n\right] = \frac{t_M\left[n\right]}{T_{OSC}} - \frac{1}{2^m}\frac{\left(T_{e,E}\left[n\right] - T_{e,E}\left[n-1\right]\right)}{T_{OSC}}$$

$$= \frac{t_M\left[n\right]}{T_{OSC}} - \frac{1}{2^m}\frac{1}{2\pi}\left(\phi_{e,E}\left[n\right] - \phi_{e,E}\left[n-1\right]\right) \qquad (9.20)$$

where
  $D_{OUT}[n]$ is the nth time difference between the present count value and the previous count value
  $t_M[n]$ is the time-difference between the start and stop edge transitions
  T_{OSC} is the period of the ring oscillator
  2^m is the number of phase taps associated with the multiphase ring oscillator

As the latter term involving the quantization errors is dependent on the difference of past and present errors, the phase error will again be noise-shaped.

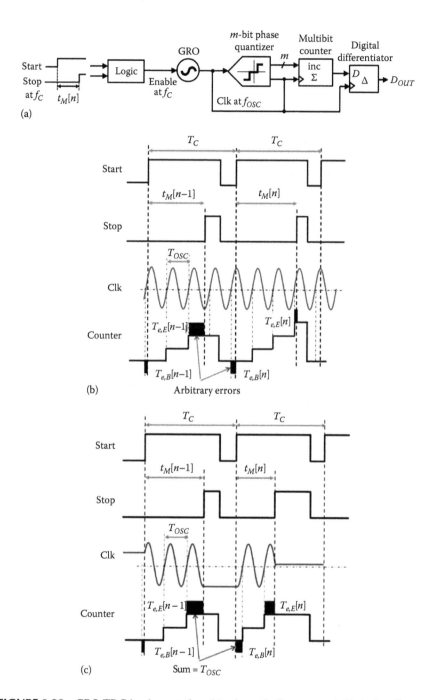

FIGURE 9.28 GRO TDC implementation: (a) schematic diagram, and (b) timing diagram illustrating the arbitrary quantization errors (shaded area) that occurs around the start (begin) and stop (end) edge transitions, and (c) timing diagram of GRO with noise-shaping, the start and stop errors are correlated.

To better visualize the operation of the GRO $\Delta\Sigma$ modulator, Figure 9.28b illustrates the timing behavior of a free-running ring oscillator. At the arrival of the rising edge of the stop signal, the counter is stopped and the result is captured and stored. However, the oscillator continues to change its phase, resulting in a different timing error at the start of the next count phase. Hence, the timing errors are uncorrelated and contribute equally to the total timing error. These timing errors would be identical to that produced by the Nyquist-rate single counter-based TDC described earlier. In contrast, the timing diagram of the GRO $\Delta\Sigma$ modulator is shown in Figure 9.28c. At the arrival of the rising edge of the stop signal, the counter is stopped and the count value is captured and stored. At the same time, the oscillator is disabled, causing its output to be held constant. At the start of the next count phase, the oscillator is enabled and its output continues from the exact same point it had at the end of the previous count phase. This results in the start and stop quantization errors to be highly correlated, as described in Equation 9.20.

The GRO $\Delta\Sigma$ modulator approach report in [16] goes beyond just noise-shaping the quantization errors, as it also demonstrated a method in which to noise-shape the mismatch phase errors associated with each stage of the ring oscillator. Further details related to this development can be found in Chapter 7 of this book.

9.5   TIME-MODE FREQUENCY-SELECTIVE FILTERING

Frequency-selective filtering in the time-domain can be particularly advantageous as a means to realize arbitrary system functions. Moreover, TM filters can be effective at reducing jitter in digital clock signals and can also be used to implement antialiasing functions for time-domain signals among other applications analogous to those in the voltage domain. Filtering operations can be classified on a grand scale into finite impulse response (FIR) and infinite impulse response (IIR) filtering. In terms of hardware, IIR systems require both the delay and summation operations of the FIR structure as well as an additional integrator element with localized feedback.

9.5.1   FIR

In terms of impulse response, FIR filters are usually the easiest to implement. An FIR filter is represented by a simple, nonrecursive difference equation that defines the output $y[n]$ of the filter in terms of its input $x[n]$, as follows:

$$y\left[n\right] = b_O x\left[n\right] + b_1 x\left[n-1\right] + \rightleftharpoons + b_N x\left[n-N\right]$$ (9.21)

where b_i, $i = 0, 1, 2, ..., N$ represent the filter coefficients and N is the order of the filter. This equation is usually implemented using a set of taps corresponding to delayed versions of the input $x[n]$.

A TM FIR example taken from [17] is shown in Figure 9.29. The structure is based on the VCDU circuit presented earlier in this chapter, except that in this case, there are multiple branches simultaneously connected to one of the comparator terminals. To implement the filter, the inputs to the inverters Figure 9.29 are fed with

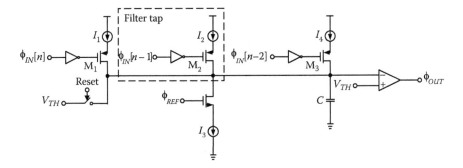

FIGURE 9.29   A second-order FIR TM filter. (Example taken from Ravinuthula, V. et al., *Int. J. Circ. Theory Appl.*, 37, 631, 2009.)

different delayed versions of the input signal $\phi_{IN}[n]$ derived from a delay line configuration. For each tap, the input is delayed exactly by one sample compared to the tap before. Therefore, if M_1 is switched on for the duration of the pulse of Φ_{IN} at sampling instance n, then M_2 and M_3 will be fed by the corresponding pulses at instances $[n-1]$ and $[n-2]$, respectively. This results in the following difference equation relating ΔT_{IN} and ΔT_{OUT} as follows:

$$\Delta T_{OUT}\left[n\right] = \frac{I_1}{I_4}\Delta T_{IN}\left[n\right] + \frac{I_2}{I_4}\Delta T_{IN}\left[n-1\right] + \frac{I_3}{I_4}\Delta T_{IN}\left[n-2\right] \qquad (9.22)$$

This corresponds to a second-order TM-FIR filter function.

To implement higher-order filters, more taps are added using additional VCDUs such that the total number of taps is equal to the filter order N.

9.5.2   IIR

Time-mode IIR filtering can be realized in much the same way as for FIR filters [18]; however, to better illustrate the principle behind this method, consider the single stage weighted-delay TM circuit shown in Figure 9.30. In Figure 9.30a, the block diagram representation is shown. Here the weighted-delay cell consists of two main blocks: a VCDU and a TVC. Figure 9.30b illustrates one possible implementation of the TVC consisting of some logic and a switched-capacitor arrangement used to hold the charge corresponding to the magnitude of the TM input signal [19]. The exact value of the capacitors plays no role in the time-to-voltage conversion process, hence is insensitive to process variations. The difference equation describing the input–output relationship of this cell is

$$\Delta T_{OUT}\left[n\right] = K_1\Delta T_{IN}\left[n-1\right] \qquad (9.23)$$

where K_1 represents the equivalent input–output conversion gain largely established by the gain of the VCDU cell.

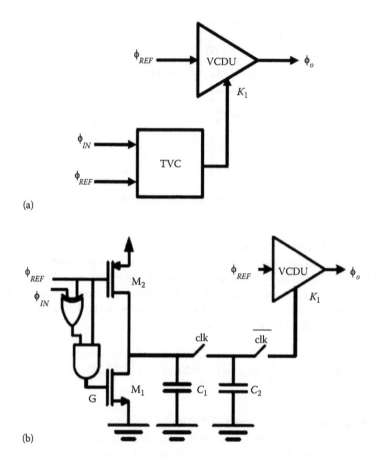

FIGURE 9.30   A weighted-delay TM circuit: (a) block diagram representation and (b) VCDU–TVC implementation.

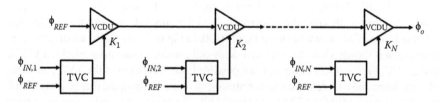

FIGURE 9.31   A block diagram of a weighted-delay summer/subtractor TM circuit.

Another important building block required to implement IIR filters is summation. Cascading weighted-delayed cells together as shown in Figure 9.31 results in the following input–output difference equation,

$$\Delta T_{OUT}\big[n\big] = K_1 \Delta T_{IN,1}\big[n-1\big] + K_2 \Delta T_{IN,2}\big[n-1\big] + \rightleftharpoons + K_N \Delta T_{IN,NB}\big[n-1\big] \quad (9.24)$$

which clearly illustrates the summation operation. Through the application of a fully differential structure or a negative VCDU gain coefficient, both a summation or subtraction operation can be realized with basically the same structure as that shown in Figure 9.31. It is interesting to note the structure of Figure 9.31 can be used to realize an FIR filter as long as the input to each TVC is a delayed version of the input signal, Φ_{IN}.

In terms of realizing an IIR filter, consider the difference equation for a generalized first-order filter function shown:

$$\Delta T_{OUT}\left[n\right] = K_1\Delta T_{OUT}\left[n-1\right] + K_2\Delta T_{IN}\left[n-1\right] \tag{9.25}$$

This equation can be realized using the adder/summation circuit shown in Figure 9.32. This approach can be used to generalize the principle of IIR filters to higher-order realizations. However, as described in [18], this approach is limited to Nth-order realizations that are constructed exclusively from delay integrators. Rather, more recently, a more general approach was introduced that can map *any* digital-based approach to a time-mode realization.

In reference 8, the simulation of the internal workings of a fifth-order doubly terminated LC ladder network implementing an elliptic transfer function using TM circuits was demonstrated. An LC-ladder-based design was selected, as it is well known to have excellent component sensitivity properties. Using the digital filter synthesis method described in [20], a block diagram of the digital circuit in the z-domain can be created as shown in Figure 9.33. Every loop contains a cascade of a delay-free integrator and a delayed integrator, resulting in no delay-free loops. Coefficients α_0–α_{12} can be related to the original LC ladder component values through the bilinear transformation technique, as described in [20].

The presence of the delay-free integrators is problematic for TMSP, so the transformation illustrated in Figure 9.34 was employed. This transformation manipulates internal loops such that it eliminates the need for delay-free integrators and replaces them by second-order integrator functions. Such integrators were described in Section 9.3.4. As a result of these transformations, the circuit now contains first- and second-order integrators, as well as multipliers implementing the alpha coefficients. In this realization, these coefficients are approximated by a binary representation that is assembled by a combination of multiply-by-2 and divide-by-2 circuits. For more details, we refer our readers to [8] for more details.

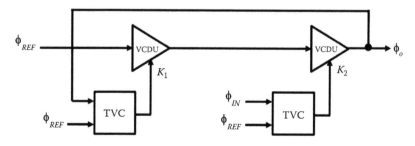

FIGURE 9.32 A first-order IIR filter realization.

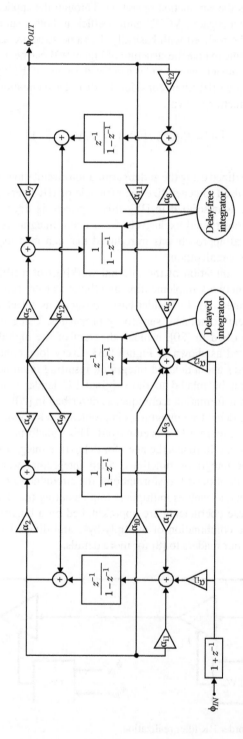

FIGURE 9.33   A digital implementation of a fifth-order elliptic filter transfer function.

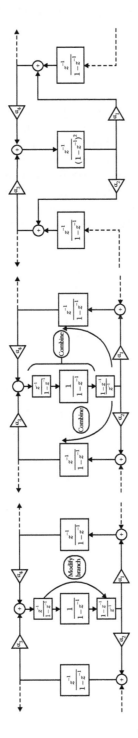

FIGURE 9.34 Removing delay-free loops using a transformation technique.

9.6   MORE APPLICATIONS AND RECENT ADVANCEMENTS

This section describes several new developments related to TMSP.

9.6.1   QPSK MODULATION

A recent paper [21] proposes a TMSP-based QPSK modulator design. One important difference between the proposed architecture and the conventional one shown in Figure 9.35 is that this design uses a digital clock input rather than an analog carrier. Unlike the conventional design in which the 2-bit modulating signal rides on the sine wave carrier as a phase-shift keying signal, the proposed design proposes that the same 2-bit modulating signal is carried by the clock (carrier) signal in terms of the delay of its positive clock edge with respect to the reference clock.

The architecture of the TMSP design is depicted in Figure 9.36. In the proposed design, the VCDU component acts as the main building block that generates the required phase shift of the carrier signal. Two identical instances of a VCDU is used, tuned to create 90° phase shifts, along with one inverter and a 4-to-1 MUX driven by a 2 bit digital modulating code I, Q where $\{I, Q\} \in \{1, 0\}$. While VCDU#1 receives its input clock in the same phase of the digital carrier, the other VCDU receives an inverted version of the carrier. Therefore, they generate 90° and 270° phase difference at their respective outputs with respect to the carrier. Among the

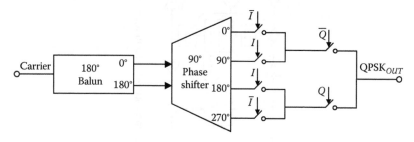

FIGURE 9.35   Conventional QPSK modulator.

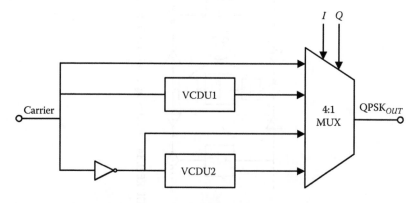

FIGURE 9.36   A TM QPSK modulator implementation.

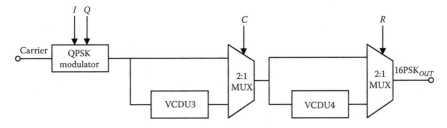

FIGURE 9.37 A TM 16-PSK implementation.

phase differences {0, 90, 180, and 270}, 0 and 180 are obtained directly from the digital carrier and its inverted version; they are fed to the input of the MUX. A TMSP demodulator can then be used to extract the modulating signal. The basic principle of this modulator can be extended to higher-order PSK designs as shown in Figure 9.37.

9.6.2 ALL DIGITAL PHASE LOCKED LOOP

As mentioned earlier, a PLL is a key block used for both upconversion and downconversion of radio signals that can also be regarded as a TM filter. It has been traditionally based on a charge-pump structure, which is not easily amenable to scaled CMOS integration and suffers from high level of reference spurs and opposes the all-digital focus of this chapter. A digitally controlled oscillator (DCO), which deliberately avoids any analog tuning voltage controls, was proposed and demonstrated for Bluetooth and GSM applications. This allows a fully digital implementation of its loop control circuitry as was first proposed in [22]. To explain how the ADPLL works, a new term called frequency command word (FCW) is introduced, which is defined as

$$FCW = \frac{F_{OUT}}{F_{REF}}$$

The operation of the block is dedicated to the comparison of the desired FCW together with the real ratio between the output and the reference frequencies. As shown in Figure 9.38, a TDC is adopted to measure the phase difference between the rising edges of the reference and output signals. The normalized version of this

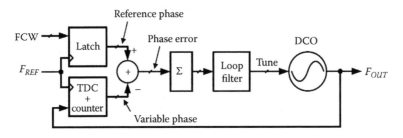

FIGURE 9.38 All-digital PLL.

phase difference will be compared to FCW = 1. For greater values of FCW, the ratio of the frequencies includes integer and fractional parts. The integer part will be easily calculated as the number of the output rising edges during one period of the reference clock and the fractional part is measured using the TDC. By detecting the difference between the measured frequency ratio and the intended FCW, a DCO is utilized for frequency modification of the signal at the output of the ADPLL and in a negative feedback configuration.

In the ADPLL, however, the digitally controlled oscillator (DCO) and TDC quantize the time and frequency tuning functions, respectively, which can lead to spurious tones and phase noise increase. As such, finite TDC resolution can distort data modulation and spectral mask at near integer-N channels, while finite DCO step size can add far-out spurs and phase noise.

Also, a major underreported issue is an injection pulling of the DCO due to harmonics of the digital activity at closely spaced frequencies, which can also create spurs. In [23], the authors have addressed all these problems and RF performance matching that of the best-in-class traditional approaches.

9.7   CONCLUSIONS

TM signal processing concepts were discussed together with their basic building blocks such as delay, addition, multiplications, and integration, to name just a few. Such operations can be used to realize more complex analog processing functions such as filtering and data conversion, and many other functions are expected in the very near future. Two-way conversion between the time and digital domains is a necessary transitional step. Several designs for TDC and DTC were described. This includes Nyquist-rate and $\Delta\Sigma$ data converters. Filtering operations based on FIR and IIR classification were also discussed. The transition to all-digital synthesizable TMSP is underway and the implementation example of an all-digital higher-order elliptic filter in TMSP asserts this fact. The authors expect major advances in TMSP in the next few years.

REFERENCES

1. C. Taillefer and G. W. Roberts, Delta-sigma analog-to-digital conversion via time-mode signal processing, *IEEE Transactions on Circuits and Systems I*, 56(9), 1908–1920, September 2009.
2. C. Taillefer, Analog-to-digital conversion via time-mode signal processing, PhD thesis, McGill University, Montreal, Quebec, Canada, September 2007.
3. M. Ali-Bakhshian, Digital processing of analog information adopting time-mode signal processing, PhD thesis, McGill University, Montreal, Quebec, Canada, August 2012.
4. M. Ali-Bakhshian and G. W. Roberts, Digital storage, addition and subtraction of time-mode variables, *Electronic Letters*, 47(16), 910–911, August 2011.
5. M. Oulmane and G. W. Roberts, CMOS digital time amplifiers for high resolution timing measurement, *Analog Signal Processing Journal*, 43(3), 269–280, June 2005.
6. G. W. Roberts, M. Safi-Harb, and M. Oulmane, A novel technique for characterizing rise/fall times for high speed digital circuits and analog slew rates. U.S. Patent #7,474,974, January 6, 2009.

7. M. Ali-Bakhshian and G. W. Roberts, A digital implementation of a dual-path time-to-time integrator, *IEEE Transactions on Circuits and Systems I*, 59(11), 2578–2591, November 2012.

8. M. Abdelfattah, G. W. Roberts, and V. P. Chodavarapu, All-digital time-mode elliptic filters based on the operational simulation of LC ladders, *2014 IEEE International Symposium on Circuits and Systems (ISCAS)*, Melbourne, Victoria, June 2014, pp. 2125–2128.

9. G. W. Roberts and M. Ali-Bakhshian, A brief introduction to time-to-digital and digital-to-time converters, *IEEE Transactions on Circuits and Systems II*, 57(3), 153–157, January 2010.

10. A. Chan and G. W. Roberts, A synthesizable, fast and high-resolution timing measurement device using a component-invariant Vernier delay line, *Proceedings of the IEEE International Test Conference*, Baltimore, Maryland, October 2001, pp. 858–867.

11. A. Chan and G. W. Roberts, Time and frequency characterization of jitter using a component-invariant Vernier delay line, *IEEE Transactions on Very Large Scale Integration Systems*, 12(1), 79–95, January 2004.

12. G. W. Roberts and A. Chan, Timing measurement device using a component-invariant Vernier delay line, U.S. Patent #6,850,051, February 1, 2005.

13. P. Dudek, S. Szczepanski, and J. V. Hatfield, A high-resolution CMOS time-to-digital converter utilizing a Vernier delay line, *IEEE Journal of Solid-State Circuits*, 35(2), 240–247, February 2000.

14. M. Hovin, A. Olsen, T. S. Lande, and C. Toumazou, Delta-sigma modulators using frequency-modulated intermediate values, *IEEE Journal of Solid-State Circuits*, 32(1), 13–22, January 1997.

15. A. Elshazly, S. Rao, B. Young, and P. K. Hanumolu, A noise-shaping time-to-digital converter using switched-ring oscillators—Analysis, design, and measurement techniques, *IEEE Journal of Solid-State Circuits*, 49(5), 1184–1197, May 2014.

16. M. Z. Straayer and M. H. Perrott, A 12-bit, 10-MHz bandwidth, continuous-time $\Delta\Sigma$ ADC with a 5-bit, 950-MS/s VCO-based quantizer, *IEEE Journal of Solid-State Circuits*, 43(4), 805–814, 2008.

17. V. Ravinuthula, V. Garg, J. G. Harris, and J. A. B. Fortes, Time-mode circuits for analog computation, *International Journal of Circuit Theory and Applications*, 37, 631–659, June 2009.

18. M. Guttman and G. W. Roberts, Sampled-data IIR filtering using time-mode signal processing circuits, *IEEE International Symposium on Circuits and Systems*, Taipei, Taiwan, May 2009.

19. A. Ameri and G. W. Roberts, Time-mode reconstruction IIR filters for $\Sigma\Delta$ phase modulation applications, *Proceedings of the 21st IEEE/ACM Great Lakes Symposium on VLSI*, Lausanne, Switzerland, May 2011.

20. L. E. Turner and B. K. Ramesh, Low sensitivity digital ladder filters with elliptic magnitude response, *IEEE Transactions on Circuits and Systems*, CAS-33, 697–706, 1986.

21. S. Saha, B. Kar, and S. Sur-Kolay, A novel architecture for QPSK modulation based on time-mode signal processing, *18th International Symposium on VLSI Design and Test*, Peelamedu, Coimbatore, July 2014, pp. 1–6.

22. R. B. Staszewski, D. Leipold, K. Muhammad, and P. T. Balsara, Digitally controlled oscillator (DCO)-based architecture for RF frequency synthesis in a deep-submicrometer CMOS process, *IEEE Transactions on Circuits and Systems II*, 50(11), 815–828, November 2003.

23. R. B. Staszewski, K. Waheed, S. Vemulapalli, F. Dulger, J. Wallberg, C. M. Hung, and O. Eliezer, Spur-free all-digital PLL in 65 nm for mobile phones, *IEEE International Solid-State Circuits Conference*, San Francisco, CA, February 2011, pp. 52–54.

10 Time-Mode Temperature Sensors

Fei Yuan

CONTENTS

Integrated temperature sensors are of a great importance in applications such as medical implants, smart sensors for environment monitoring, and on-chip temperature monitoring of very-large-scale integration (VLSI) systems, to name a few. Traditionally, integrated temperature sensors are realized using a voltage-mode proportional to absolute temperature (PTAT) circuit whose output voltage is a linear function of temperature, a temperature reference circuit whose output voltage is independent of temperature and can be adjusted digitally, and a voltage comparator that compares the output voltage of the PTAT circuit and that of the temperature [1]. The continuous shrinking of the supply voltage due to technology scaling greatly reduces the voltage resolution of voltage-mode temperature sensors, making the improvement of temperature measurement resolution increasingly difficult. Time-mode approaches for temperature measurement where temperature is represented by the width of a pulse whose width is proportional to temperature, on the other hand, offer the desired immunity from the effect of technology scaling, making them particularly attractive for integrated temperature sensors.

This chapter studies time-mode integrated temperature sensors. Section 10.1 studies relaxation oscillator temperature sensors, while Section 10.2 investigates ring oscillator-based temperature sensors. Temperature sensors that utilize time-to-digital converters are studied in Section 10.3. Digital set point temperature sensors are investigated in Section 10.4. Section 10.5 compares the performance of some recently reported time-mode temperature sensors. The chapter is summarized in Section 10.6.

10.1 RELAXATION OSCILLATOR TEMPERATURE SENSORS

When the control voltage of a voltage-controlled oscillator varies with temperature, the frequency of the oscillator will be a function of temperature. A one-to-one mapping between temperature and the frequency of the oscillator can be established. If the relation is linear, for a given temperature, one can obtain the digital representation of the temperature by counting the number of the oscillation cycles of the oscillator within a known time interval. A notable advantage of oscillator-based temperature sensors is their simple configurations, and subsequently low-power consumption.

Relaxation oscillators offer the intrinsic advantage of a low-frequency sensitivity to PVT effects. This is because the frequency of these oscillators is solely determined by the charging and discharging currents of passive capacitors and predefined threshold voltages with which the voltages of the capacitors compare. If the charging and discharging currents and the reference voltage of the comparators are independent of PVT effect, the oscillation frequency of these oscillators will be independent of PVT effect. This differs fundamentally from generic inverter-based ring oscillators whose charging and discharging currents and threshold voltages are the strong functions of the supply voltage and temperature.

Figure 10.1 shows the simplified schematic of the relaxation oscillator temperature sensor proposed in [2]. The frequency of the PTAT oscillator varies with temperature in a linear fashion, while that of the timing oscillator is independent of temperature. The frequency of the timing oscillator is purposely set to be lower than that of the PTAT oscillator such that a unique relation between the number of the oscillation cycles of the PTAT oscillator per oscillation period of the timing oscillator and temperature can be established. The number of the oscillation cycles of the PTAT oscillator per oscillation period of the timing oscillator is the digital representation of temperature.

The timing oscillator is a relaxation oscillator with its frequency set by the reference voltage V_{REF}, charging current I, and relaxation capacitor C. The PTAT oscillator is also a relaxation oscillator with a PTAT charging current. The operation

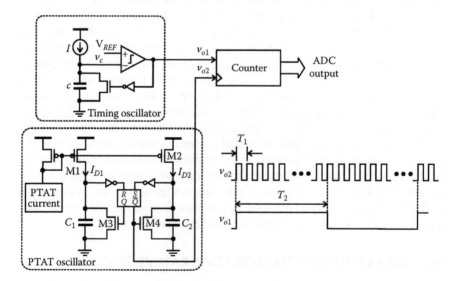

FIGURE 10.1   Relaxation oscillator temperature sensor. (From Zhou, S and Wu, N. A novel ultra low power temperature sensor for UHF RFID tag chip. In *Proceedings of IEEE Asian Solid-State Circuits Conference*, November 2007, pp. 464–467.)

of the PTAT oscillator is briefly depicted as follows: Assume $v_{c1} = 0$ and $v_{c2} = 1$ initially. In this case, M3 is *off* and M4 is *on*. C_1 is now charged and v_{c1} rises with time with the rising rate set by the PTAT current and the value of C_1. C_2 at the same time is discharged through the path provided by M4 with a discharge time constant $R_{ds4}C_2$ where R_{ds4} is the *on* resistance of M4. When v_{c1} climbs to the threshold voltage of the inverter, it will trigger the RS register and change the logic state of its output. As a result, M3 will turn on and M4 will turn off. C_1 will be discharged through M3 and C_2 will be charged by the PTAT current. This process repeats and a sustained oscillation with its frequency set by the PTAT current is established.

Let us now turn our attention to the operation of the relaxation oscillator temperature sensor. Assume that initially v_c of the timing oscillator is zero and $v_{o1} = 1$. The counter is reset and v_c rises with time. The counter records the number of the oscillation cycles of the output voltage of the PTAT oscillator. Since the control current of the oscillator is PTAT, the frequency of the PTAT oscillator is proportional to temperature. When v_c exceeds V_{REF}, $v_{o1} = 0$ and the counter will stop recording and the content of the counter will yield the digital representation of temperature.

A number of issues need to be considered to ensure the proper operation of the relaxation oscillator temperature sensor. First, the frequency of the timing oscillator must be independent of both temperature and supply voltage fluctuation. This requires that both V_{REF} and I be independent of both temperature and supply voltage fluctuation. They should therefore be provided by temperature-compensated Reference 3. Second, the accuracy of the temperature sensor is set by the smallest frequency variation that can be detected. For a given oscillation frequency of the timing oscillator, the higher the frequency of the PTAT oscillator, the better is the resolution. This, however, is at the cost of the increased power consumption of the PTAT oscillator. Finally, the conversion time of the temperature sensor is set by the oscillation period of the timing oscillator. For a fixed frequency of the PTAT oscillator, although the higher the frequency of the timing oscillator, the shorter the conversion time, the resolution of the sensor will drop as well.

10.2   RING OSCILLATOR TEMPERATURE SENSORS

Ring oscillators can also be used as temperature sensors to take the advantages of their simple configuration and low power consumption. The simplified schematic of the ring oscillator-based temperature sensor proposed by Kim et al. is shown in Figure 10.2 [4]. The frequency of the ring oscillator with a PTAT biasing current, that is, the PTAT oscillator, is directly proportional to temperature, while that of the ring oscillator with a constant biasing current, that is, the timing oscillator, is independent of temperature. The highlighted portion of the PTAT oscillator generates a PTAT biasing current and that of the timing oscillator generates a temperature-independent biasing current. For details on how to generate PTAT currents and temperature-independent currents, please refer to (Yuan, 2010) where a comprehensive coverage of the temperature-dependent characteristics of semiconductor devices and an exhaustive compilation of recently reported reference circuits are provided.

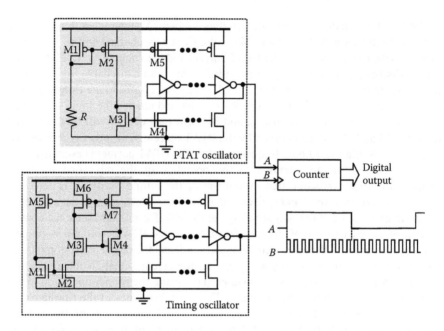

FIGURE 10.2   Ring oscillator-based temperature sensor. (From Kim, C. et al., CMOS temperature sensor with ring oscillator for mobile DRAM self-refresh control, *Proceedings of IEEE International Symposium on Circuits and Systems*, May 2008, pp. 3094–3097.)

The frequency of the PTAT oscillator is lower than that of the timing oscillator so that the number of the oscillation cycles of the timing oscillator in each oscillation cycle of the PTAT oscillator, which is temperature dependent, gives the digital representation of temperature. It is interesting to note that in the relaxation oscillator temperature sensor studied earlier, the frequency of the timing oscillator is higher as compared with that of the PTAT oscillator, while in the ring oscillator–based temperature sensor, the frequency of the timing oscillator is made lower as compared with that of the PTAT oscillator.

The oscillator-based temperature sensor proposed by Park et al. utilizes the temperature dependence of the frequency of a generic ring oscillator with devices operating in both weak inversion and strong inversion regions [5]. Figure 10.3 shows the simplified schematic of the temperature sensor. The timing signal v_{in2} is provided by an external timing oscillator. Digitally controlled capacitor arrays are used as the load of the delay stages. The full-swing operation of the inverting stages requires that the transistors of the inverting stages operate through cut-off, weak inversion, and strong inversion. The temperature dependence of the channel current of MOSFETs originates from the temperature dependence of the mobility of minority charge carriers and that of the threshold voltage of MOSFETs. Recall that the channel current of a MOSFET in strong inversion is given by

$$I_{DS,sat} = \frac{1}{2}\mu C_{ox}\frac{W}{L}(V_{GS} - V_T)^2, \tag{10.1}$$

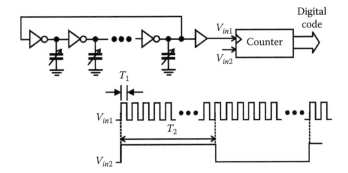

FIGURE 10.3 Ring oscillator–based temperature sensor. (From Park, S. et al., A 95 nW ring oscillator-based temperature sensor for RFID tags in 0.13 μm CMOS, in *Proceedings of IEEE International Symposium on Circuits and Systems*, May 2009, pp. 1153–1156.)

when the transistor is in saturation (preciously in pinch-off) and

$$I_{DS,tri} = \mu C_{ox} \frac{W}{L} \left[(V_{GS} - V_T) V_{DS} - \frac{V_{DS}^2}{2} \right] \tag{10.2}$$

when the transistor is in the triode region. When the device is in the weak inversion region, its channel current is given by

$$I_{DS,sub} \approx I_{D0} e^{\frac{V_{GS} - V_T}{nV_t}}, \tag{10.3}$$

where

$$I_{D0} = 2n\mu C_{ox} S V_t^2, \tag{10.4}$$

$$n = 1 + \frac{C_{ox}}{C_{js}}, \tag{10.5}$$

$S = W/L$ is the aspect ratio
C_{js} is the capacitance of the depletion region under the channel
$V_t = kT/q$ is the thermal voltage
q is the charge of an electron
$V_{DS} \gg V_t$ is assumed

Often $n = 1.5$ is used [6]. The normalized temperature coefficient of the channel current, denoted by TCC, in strong inversion and that in weak inversion are obtained from

$$TCC_{sat} = \frac{1}{I_{DS,sat}} \frac{\partial I_{DS,sat}}{\partial T} = \frac{1}{\mu} \frac{\partial \mu}{\partial T} - \frac{2}{V_{GS} - V_T} \frac{\partial V_T}{\partial T}, \tag{10.6}$$

$$\begin{aligned}
\mathrm{TCC}_{tri} &= \frac{1}{I_{DS,tri}} \frac{\partial I_{D,tri}}{\partial T} \\
&\approx \frac{1}{\mu} \frac{\partial \mu}{\partial T} - \frac{V_{DS}}{\left(V_{GS} - V_T\right)V_{DS} - \dfrac{V_{DS}^2}{2}} \frac{\partial V_T}{\partial T},
\end{aligned} \qquad (10.7)$$

$$\begin{aligned}
\mathrm{TCC}_{sub} &= \frac{1}{I_{DS,sub}} \frac{\partial I_{D,sub}}{\partial T} \\
&= \frac{1}{\mu} \frac{\partial \mu}{\partial T} + \frac{2}{T} - \frac{1}{nV_t}\left(\frac{\partial V_T}{\partial T} - \frac{V_T}{T}\right).
\end{aligned} \qquad (10.8)$$

The channel current of MOSFETs in weak inversion exhibits a much larger thermal sensitivity as compared with that in strong inversion. The temperature dependence of the mobility of the minority charge carriers and that of the threshold voltage result in the temperature dependence of the channel current, and subsequently the temperature dependence of the frequency of the ring oscillator.

10.3   TDC TEMPERATURE SENSORS

Oscillator-based temperature sensors not only require two oscillators, specifically a timing oscillator and a PTAT oscillator to function properly; the oscillation frequency of these two oscillators must also differ largely in order to yield a good temperature resolution. Both result in a high level of power consumption. Clearly, if one wants to lower the power consumption, oscillators need to be removed. TDC-based temperature sensors first represent temperature by a pulse whose width is directly proportional to temperature. The width of the temperature-dependent pulse is then digitized using a TDC. Although there are many ways to perform time-to-digital conversion, those that consume the least amount of power are the ideal candidates for low-power temperature sensors. Figure 10.4 shows an example of such a temperature sensor. The temperature-dependent pulse is digitized using a cyclic time-to-digital converter [7,8].

The temperature sensor employs two delay lines of the same length. Similar to oscillator temperature sensors, the delay of one of the delay lines is proportional to temperature. This delay line is termed the PTAT delay line. The delay of the other delay line is independent of temperature and the delay line is identified as the reference delay line. The propagation delay of the delay stage of the PTAT delay line is a function of temperature due to the temperature-dependent characteristics of both the mobility of minority charge carriers and the threshold voltage of MOSFETs, as to be seen shortly.

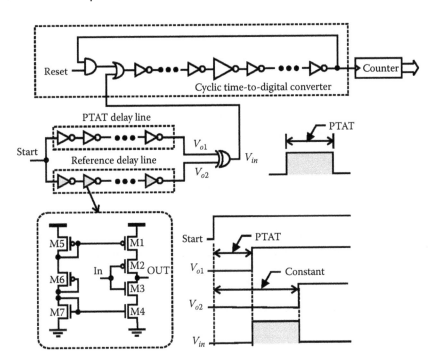

FIGURE 10.4 TDC temperature sensor. (From Chen, P. et al., A time-to-digital-converter-based CMOS smart temperature sensor. *IEEE Journal of Solid-State Circuits*, 40(8):1642–1648, August 2005; Chen. P. et al., A precise cyclic CMOS time-to-digital converter with low thermal sensitivity. *IEEE Transactions on Nuclear Science*, 52(4):834–838, August 2005.)

The propagation delay of the delay stage of the reference delay line, on the other hand, is made independent of temperature. The temperature-independent delay is achieved by using the special delay stages with its schematic shown in the figure. The delay cell is a current-starved inverter with the charging and discharging currents controlled by transistors M5–M7. The mobility of minority charge carriers varies with temperature as per Tsividis, 1999.

$$\mu(T) = \mu(T_o)\left(\frac{T}{T_o}\right)^m, \tag{10.9}$$

where
   $\mu(T)$ and $\mu(T_o)$ are the mobility of minority charge carriers at temperature T and reference temperature T_o, respectively
   $m \approx -1.2 \sim -2.0$
   T_o is often set to room temperature (300 K) as the performance of integrated circuits is typically optimized at room temperature

The threshold voltage of MOS transistors is given by [10]

$$V_T = V_{To} + \gamma \left(\sqrt{V_{SB} + 2\phi_F} - \sqrt{2\phi_F} \right), \tag{10.10}$$

where

V_{SB} is the source-bulk (substrate) voltage
ϕ_F is Fermi potential calculated from

$$\phi_F = V_t \ln \left(\frac{N_A}{n_i} \right) \tag{10.11}$$

where

N_A the doping of the substrate
n_i is the concentration of intrinsic charge carriers in silicon ($n_i \approx 1.5 \times 10^{10}/\text{cm}^3$ at $300°K$ [11])
$\phi_F \approx 0.3$ V for a typical p-type silicon substrate [10]

γ is the body effect constant calculated from

$$\gamma = \frac{\sqrt{2q\epsilon_{si}N_A}}{C_{ox}}, \tag{10.12}$$

with

$$C_{ox} = \frac{\epsilon_{ox}}{t_{ox}} \tag{10.13}$$

the gate capacitance per unit area, $\epsilon_{ox} = 3.5 \times 10^{-13}$ F/cm the dielectric constant of oxide, $\epsilon_{si} = 1.05 \times 10^{-12}$ F/cm the dielectric constant of silicon, and t_{ox} the thickness of the gate oxide. V_{To}, the threshold voltage of MOSFETs when $V_{SB} = 0$, is a function of a number of parameters including Fermi potential [9–11]. Since ϕ_F is a strong function of temperature, the threshold voltage also varies with temperature. $V_T(T)$ decreases with temperature almost linearly and is typically depicted using the following first-order model [9]

$$V_T(T) = V_T(T_o) + \alpha_{V_T}(T - T_o), \tag{10.14}$$

where

$V_T(T)$ and $V_T(T_o)$ are the threshold voltages at T and T_o, respectively
α_{V_T} is the temperature coefficient of the threshold voltage

The typical value of α_{V_T} is in the range of −0.5 mV/°C to −4 mV/°C, with −2.4 mV/°C the most frequently used [11]. Making use of Equations 10.9 and 10.14, we arrive at

$$I_{SD5} = \frac{1}{2}\mu_p(T_o)\left(\frac{T}{T_o}\right)^m C_{ox}\left(\frac{W}{L}\right)_6 \left[V_{SG5} - V_T(T_o) - \alpha_{V_T}(T - T_o) \right]^2. \tag{10.15}$$

Since we want I_{SD5} to be independent of temperature, the first-order constraint

$$\frac{\partial I_{SD5}}{\partial T} = 0 \tag{10.16}$$

is imposed. As a result,

$$V_{SG5} = V_T(T_o) + \alpha_{V_T}(T - T_o) + \frac{2\alpha_{V_T}T}{m}. \tag{10.17}$$

The channel current in this case becomes

$$I_{SD5} = \frac{1}{2}\mu_p(T_o)\left(\frac{T}{T_o}\right)^m C_{ox}\left(\frac{W}{L}\right)_6\left(\frac{2\alpha_{V_T}T}{m}\right)^2. \tag{10.18}$$

The channel current of transistor M7 is given by

$$I_{DS7} \approx \frac{1}{2}\mu_n C_{ox}S(V_{GS7} - V_T)^2. \tag{10.19}$$

Since both the mobility of electrons given by Equation 10.9 and the threshold voltage of MOSFETs given by Equation 10.14 are the functions of temperature and both have negative temperature coefficients, it is possible to make I_{DS} independent of temperature by properly choosing V_{GS}. To illustrate this point, we differentiate Equation 10.19 with respect to temperature T and impose

$$\frac{\partial I_{DS7}}{\partial T} = 0, \tag{10.20}$$

$$\frac{\partial V_{GS7}}{\partial T} = 0. \tag{10.21}$$

The result is given by

$$V_{GS7} = V_T + 2\mu_n\left(\frac{\partial V_T/\partial T}{\partial \mu_n/\partial T}\right). \tag{10.22}$$

Substitute Equations 10.9 and 10.14 into Equation 10.22

$$V_{GS7} = V_T(T_o) - \alpha_{V_T}T_o + \alpha_{V_T}\left(1 + \frac{2}{m}\right)T. \tag{10.23}$$

The first two terms on the right-hand side of Equation 10.23 are the functions of the reference temperature, while only the last term varies with temperature. To ensure that V_{GS} is independent of temperature, $m = -2$ is required. It was shown in [12,13,14] that the value of m is indeed in the vicinity of -2. If we assume $m = -2$, Equation 10.23 becomes

$$V_{GS7}(T) = V_T(T_o) - \alpha_{V_T} T_o. \tag{10.24}$$

Equation 10.24 shows that V_{GS7} in this case is independent of temperature. The corresponding current is obtained from Equation 10.19

$$I_{DS7} = \frac{1}{2}\mu_n(T_o)C_{ox}S\alpha_{V_T}^2 T_o^2. \tag{10.25}$$

The operating point of the device in this case is independent of temperature, and is called the zero-temperature-coefficient (ZTC) bias point. Utilizing $m = -2$, Equation 10.18 is simplified to

$$I_{SD5} = \frac{1}{2}\mu_p(T_o)C_{ox}\left(\frac{W}{L}\right)_6 \left(\alpha_{V_T} T_o\right)^2. \tag{10.26}$$

It becomes apparent from Equations 10.25 and 10.26 that both the charging and discharging currents of the delay stage of delay line 2 can be made independent of temperature by operating the transistors at their ZTC bias points.

Let us now examine the temperature dependence of the delay of the delay line. To investigate the PTAT characteristics of the delay line, we examine the high-to-low and low-to-high propagation delays of the output voltage of a generic static inverter, denoted by τ_{PHL} and τ_{PLH}, respectively. They were given in Chapter 3 TDC fundamentals and are copied here for convenience

$$\tau_{PHL} = \frac{2C_iV_T}{k_n(V_{DD} - V_T)^2} + \frac{C_L}{k_n(V_{DD} - V_T)}\ln\left(\frac{3V_{DD} - 4V_T}{V_{DD}}\right), \tag{10.27}$$

and

$$\tau_{PLH} = \frac{2C_LV_T}{k_p(V_{DD} - V_T)^2} + \frac{C_L}{k_p(V_{DD} - V_T)}\ln\left(\frac{3V_{DD} - 4V_T}{V_{DD}}\right). \tag{10.28}$$

The temperature-dependent variables in Equations 10.27 and 10.28 are mobility and threshold voltage. τ_{PLH} and τ_{PHL} are both the functions of temperature, so is the average propagation delay $\tau = (\tau_{PHL} + \tau_{PLH})/2$.

The dependence of the mobility on temperature is given by Equation 10.9 whereas that of the threshold voltage is given by Equation 10.14. To quantify the effect of temperature on the threshold voltage, we calculate the variation of the threshold voltage with temperature varied from $-40°C$ to $+80°C$. At $T = -40°C$, using $\alpha_{V_T} = -2.4$ mV/K and $T_o = 27°C$ (300 K), we have

$$\alpha_{V_T}(T - T_o) = -24(-40-27) = 0.161 \text{ V}.$$

Similarly at $T = 80°C$, we have

$$\alpha_{V_T}(T - T_o) = -24(80-27) = -0.127 \text{ V}.$$

If the technology for realizing the delay line is a 130 nm 1.2 V CMOS with $V_{DD} = 1.2$ V and $V_T \approx 0.4$ V, when temperature is varied from $-40°C$ to $+80°C$, the threshold voltage varies from

$$V_T \big|_{-40°C} = 0.4 + \alpha_{V_T}(-40-27) = 04 + 0161 = 0.561 \text{ V}$$

to

$$V_T \big|_{80°C} = 0.4 + \alpha_{V_T}(80-27) = 04 - 0127 = 0.237 \text{ V}.$$

So the variation of the threshold voltage when temperature is varied from $-40°C$ to $80°C$ is given by: $\Delta V_T = V_T|_{-40°C} - V_T|_{80°C} = 0.561 - 0.237 = 0.324$ V. The variation of the threshold voltage is 81% !

To quantify the effect of temperature on mobility, we calculate the variation of mobility with temperature varied from $-40°C$ to $+80°C$ with mobility μ_o at the reference temperature 300 K. At $-40°C$, utilizing (10.9) and $m = -2$, we have

$$\mu \big|_{-40°C} = \mu_o \left(\frac{27}{40}\right)^2 = 0.4556\mu_o$$

and at $80°C$, we have

$$\mu \big|_{80°C} = \mu_o \left(\frac{27}{80}\right)^2 = 0.1139\mu_o.$$

The variation of mobility when temperature is varied from $-40°C$ to $80°C$ is obtained from: $\Delta\mu = \mu|_{-40°C} - \mu|_{80°C} = 0.4556\mu_o - 0.1139\mu_o = 0.3417\mu_o$. It is evident that mobility is a strong function of temperature.

It should be noted that the dependence of τ_p on mobility and threshold voltage is nonlinear. As a result, the delay line is not an exact PTAT delay line. As long as the dependence of τ_p on μ is monotonic, for each temperature, there will only be one corresponding propagation delay of the delay line. The set point temperature sensor will function properly.

A notable advantage of using dual delay lines rather than a single delay line to generate a pulse whose pulse width is PTAT is that the effect of PVT on the pulse width is minimized. This is because the pulse width used to sense temperature is the difference between the delay of the two delay lines. The dual delay line configuration therefore bears a strong resemblance to that of differential-pair amplifiers and therefore is capable of suppressing the effect of common-mode disturbances such as PVT effects on the delays. Once a temperature-dependent pulse width is generated, it can be digitized using a low-power TDC.

10.4   DIGITAL SET POINT TEMPERATURE SENSORS

As pointed out earlier, a voltage-mode integrated temperature sensor can be realized using a temperature-dependent circuit whose output voltage is a linear function of temperature, a temperature reference whose output voltage is independent of temperature, and a voltage comparator that compares the output voltage of the PTAT circuit and that of the temperature reference. The output voltage or the temperature set point of the temperature reference can be digitally adjusted such that for a given temperature, the output voltage of the temperature reference can be made the same as that of the temperature-dependent circuit. Once this occurs, the digital code used for adjusting the temperature set point of the temperature reference gives the digital representation of temperature. This approach resembles successive approximation analog-to-digital converters. The accuracy of voltage-mode set point temperature sensors is greatly affected by the available voltage headroom as the step size of voltage increment, which corresponds to the resolution of temperature, is given by $V_{FSR}/2^N$ where V_{FSR} is the full-scale range of the voltage headroom and N is the number of steps used to digitize V_{FSR}. Since the sensitivity of voltage comparators is finite, the smaller the value of V_{FSR}, the smaller is the value of N. The temperature resolution is given by $(T_{max} - T_{min})/2^N$. For a given $\Delta T = T_{max} - T_{min}$, the smaller the value of N, the worse the resolution of the temperature sensor.

Although scaling-induced voltage headroom loss greatly reduces voltage resolution, and subsequently the resolution of voltage-mode set-point temperature sensors, the principle of the preceding voltage-mode set-point temperature sensor is also applicable in time-mode; specifically, a time-mode integrated temperature sensor can be realized using a temperature-dependent delay line whose delay is a linear function of temperature, a temperature reference line whose delay is independent of temperature, and a time comparator that discriminates the delay of the two delay lines. For a given temperature, the delay of the temperature reference can be made identical to that of the temperature-dependent delay line by digitally adjusting the delay of the temperature reference, that is, the temperature set point of the temperature reference line. Once this occurs, the digital code for adjusting the temperature set point of the temperature reference line is the digital representation of the temperature [1].

Figure 10.5 shows the configuration of the set-point temperature sensor proposed in [1]. Its operation is briefly depicted here: The incoming signal START is fed to both the PTAT delay line whose per-stage delay is proportional to temperature and the reference delay line whose per-stage delay is independent of temperature. For a given temperature, the delay of the PTAT delay line is known. Depending upon the output of the time comparator realized using a simple D flip-flop, the output of the multiplexer connected to the reference delay line is adjusted until the output of the

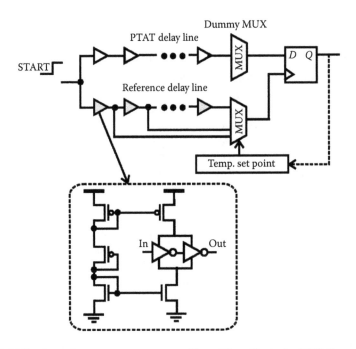

FIGURE 10.5 Set-point temperature sensor. (From Chen, P. et al., *IEEE Sens. J.*, 9(12), 1639, 2009.)

PTAT line and that of the reference line arrive at the D flip-flop at the same time. Once this occurs, the control word of the multiplexer associated with the reference delay line yields the digital representation of the temperature. This set-point temperature sensor is essentially a time-mode successive approximation temperature sensor and the function of the temperature set point block is similar to that of a successive approximation register (SAR) in successive approximation analog-to-digital converters.

The configuration of the PTAT delay line and that of the reference delay line are identical to those shown in Figure 10.4 [7,8]. Since a linear temperature-delay relation of the PTAT delay line is assumed, the accuracy of this temperature sensor is directly affected by the linearity of the temperature-delay relation of the PTAT delay line. If the PTAT delay line exhibits a nonlinear temperature-delay characteristic, the one-to-one mapping between temperature and delay might not exist. In [15], a linearization technique was proposed to compensate for the effect of the nonlinear relation between τ_p and temperature relation. Readers are referred to the cited reference for more information on this technique.

10.5 PERFORMANCE COMPARISON

Table 10.1 compares the performance of some recently reported time-mode temperature sensors. It is seen that the best temperature resolution achieved is approximately 0.1°C. The power consumption is largely sampling rate and topology dependent.

TABLE 10.1

Comparison of Time-Mode Temperature Sensors

Reference	Tech.	Range	Resolution (°C)	Power (µW)	Rate
[16]	0.35 µm	−40°C–60°C	0.1	1.5	5 samples/s
[17]	0.35 µm	−40°C–95°C	0.5	9	20 samples/s
[1]	0.35 µm	−40°C–80°C	0.5	9	20 samples/s
[15]	0.35 µm	0°C–90°C	0.1	36.7	2 samples/s
[18]	65 nm	0°C–60°C	0.14	150	10k samples/s
[19]	0.35 µm	0°C–100°C	0.2	1.5	10 samples/s

10.6 SUMMARY

Time-mode integrated temperature sensors were studied. The chapter started with the investigation of relaxation oscillator temperature sensors. The rationales of deploying relaxation oscillators were explored. We showed that relaxation oscillators are less sensitive to supply voltage fluctuation, making them particularly attractive for passive wireless microsystems where their operational power is harvested from external sources such as radio-frequency waves. It was followed with a study of ring oscillator–based temperature sensors. We showed although more sensitive to supply voltage fluctuation, ring oscillators are very attractive for low-power applications. Temperature sensors that utilize time-to-digital converters were studied. We showed that the absence of oscillators makes TDC-based temperature sensors strong candidates for applications where power consumption is of a critical concern. Digital set point temperature sensors whose operational principle resembles that of successive approximation analog-to-digital converters were also studied. Finally, the performance of some recently reported time-mode temperature sensors was compared.

REFERENCES

1. P. Chen, T. Chen, Y. Wang, and C. Chen. A time-domain sub-micro watt temperature sensor with digital set-point programming. *IEEE Sensors Journal*, 9(12):1639–1646, 2009.
2. S. Zhou and N. Wu. A novel ultra low power temperature sensor for UHF RFID tag chip. In *Proceedings of IEEE Asian Solid-State Circuits Conference*, Jejku, Korea, November 2007, pp. 464–467.
3. F. Yuan. *CMOS Circuits for Passive Wireless Microsystems*. Springer, New York, 2010.
4. C. Kim, B. Kong, C. Lee, and Y. Jun. CMOS temperature sensor with ring oscillator for mobile DRAM self-refresh control. In *Proceedings of IEEE International Symposium on Circuits and Systems*, Seattle, WA, May 2008, pp. 3094–3097.
5. S. Park, C. Min, and S. Cho. A 95 nW ring oscillator-based temperature sensor for RFID tags in 0.13 µm CMOS. In *Proceedings of IEEE International Symposium on Circuits and Systems*, Taipei, Taiwan, May 2009, pp. 1153–1156.
6. G. Yu and X. Zou. A novel current reference based on sub-threshold MOSFETs with high PSRR. *Microelectronics Journal*, 39:1874–1879, 2008.
7. P. Chen, C. Chen, C. Tsai, and W. Lu. A time-to-digital-converter-based CMOS smart temperature sensor. *IEEE Journal of Solid-State Circuits*, 40(8):1642–1648, August 2005.

8. P. Chen, C. Hwang, and W. Chang. A precise cyclic CMOS time-to-digital converter with low thermal sensitivity. *IEEE Transactions on Nuclear Science*, 52(4):834–838, August 2005.

9. Y. Tsividis. *Operation and Modeling of the MOS Transistors*, 2nd edn. McGraw-Hill, New York, 1999.

10. J. Rabaey, A. Chandrakasan, and B. Nikolic. *Digital Integrated Circuits—A Design Perspective*, 2nd edn. Pearson Education, Upper Saddle River, NJ, 2003.

11. P. Gray, P. Hust, S. Lewis, and R. Meyer. *Analysis and Design of Analog Integrated Circuits*, 4th edn. John Wiley & Sons, New York, 2001.

12. I. Filanovsky and A. Allam. Mutual compensation of mobility and threshold voltage temperature effects with applications in CMOS circuits. *IEEE Transactions on Circuits and Systems I*, 48(7):876–884, July 2001.

13. I. Filanovsky, A. Allam, and S. Lim. Temperature dependence of output voltage generated by interaction of threshold voltage and mobility of an NMOS transistor. *Analog Integrated Circuits and Signal Processing*, 27:229–238, 2001.

14. A. Bendali and Y. Audet. A 1-V CMOS current reference with temperature and process compensation. *IEEE Transactions on Circuits and Systems I*, 54(7):1424–1429, July 2007.

15. P. Chen, C. Chen, Y. Peng, K. Wang, and Y. Wang. A time-domain SAR smart temperature sensor with curvature compensation and a 3 σ inaccuracy of 0.4°C–0.6°C over a 0°C to 90°C range. *IEEE Journal of Solid-State Circuits*, 45(3):600–609, March 2010.

16. C. Chen, P. Chen, C. Hwang, and W. Chang. A precise cyclic CMOS time-to-digital converter with low thermal sensitivity. *IEEE Transactions on Nuclear Science*, 52(4):834–838, August 2005.

17. C. Chen, P. Chen, and Y. Shen. A low power CMOS time-to-digital converter based on duty cycle controllable pulse stretcher. In *Proceedings of IEEE European Solid-State Circuits Conference*, Montreux, Switzerland, 2006, pp. 316–319.

18. C. Chen and C. Yang. An auto-calibrated all-digital temperature sensor for on-chip thermal monitoring. *IEEE Transactions on Circuits and Systems II*, 58(2):105–109, February 2010.

19. C. Chen and H. Chen. A low-cost cmos smart temperature sensor using a thermal-sensing and pulse-shrinking delay line. *IEEE Sensors Journal*, 14(1):278–284, January 2014.

Index